Lecture Notes in Computer Science

Edited by G. Goos and J. Hartmanis

138

6th Conference on Automated Deduction

New York, USA, June 7–9, 1982

Edited by D.W. Loveland

Springer-Verlag
Berlin Heidelberg New York 1982

Editorial Board
W. Brauer P. Brinch Hansen D. Gries D. Luckham C. Moler
G. Seegmüller J. Stoer N. Wirth

Editor
D.W. Loveland
Duke University, Department of Computer Science
Durham, North Carolina 27706, USA

CR Subject Classifications (1979): E 2.3, F 4.1

ISBN 3-540-11558-7 Springer-Verlag Berlin Heidelberg New York
ISBN 0-387-11558-7 Springer-Verlag New York Heidelberg Berlin

This work is subject to copyright. All rights are reserved, whether the whole or part of the material is concerned, specifically those of translation, reprinting, re-use of illustrations, broadcasting, reproduction by photocopying machine or similar means, and storage in data banks. Under § 54 of the German Copyright Law where copies are made for other than private use, a fee is payable to "Verwertungsgesellschaft Wort", Munich.

© by Springer-Verlag Berlin Heidelberg 1982
Printed in Germany

Printing and binding: Beltz Offsetdruck, Hemsbach/Bergstr.
2145/3140-543210

FOREWORD

The SIXTH CONFERENCE ON AUTOMATED DEDUCTION, held on June 7-9, 1982, at the Courant Institute of Mathematical Sciences, New York University, included the presentation of 23 papers selected by the program committee, a keynote address by Lawrence Wos, and an invited talk by John McCarthy (the latter not included in this volume). The conference is the primary international forum for reporting research in all aspects of automated deduction. As such the conference addresses such issues as the design and implementation of (semi) automated theorem proving systems, experimentation with theorem provers, the capability and efficiency of various representations and control structures for automated deduction, and domains of application for automated deduction. Authors presenting papers at this conference come from Canada, France, Israel, Poland, the United Kingdom, the United States and West Germany.

Expert and non-expert alike are encouraged to read the keynote address, which reviews the application of a leading automated theorem proving system to open problems in mathematics and circuit design and confronts some "myths" regarding automated theorem proving.

The five previous conferences on automated deduction have been held at Argonne (Illinois), Oberwolfach (West Germany), Cambridge (Massachusetts), Austin (Texas), and Les Arcs (France). The proceedings of the 1980 conference at Les Arcs appear as Volume 87 of the Lecture Notes in Computer Science series.

PROGRAM COMMITTEE

P. Andrews (Carnegie-Mellon)
M. Ballantyne (Intel)
W. Bibel (TU Munchen); past co-chairman
F. Brown (U. Texas)
A. Bundy (U. Edinburgh)
R. Constable (Cornell)
J. Doyle (Carnegie-Mellon)
R. Kowalski (U. London); past co-chairman
D. Loveland (Duke); chairman
E. Lusk (Northern Illinois)
U. Montanari (U. Pisa)
D. Musser (GE)
R. Shostak (SRI)

LOCAL ARRANGEMENTS CHAIRMAN

M. Davis

CONTENTS

MONDAY MORNING

Solving open questions with an automated theorem-proving program
L. Wos (U.S.A.) (Keynote Address) 1

STP: a mechanized logic for specification and verification
R.E. Shostak, R. Schwartz, P.M. Melliar-Smith (U.S.A.) 32

A look at TPS
D.A. Miller, E. Cohen, P.B. Andrews (U.S.A.) 50

MONDAY AFTERNOON

Logic machine architecture: kernel functions
E.L. Lusk, W.W. McCune, R.A. Overbeek (U.S.A.) 70

Logic machine architecture: inference mechanisms
E.L. Lusk, W.W. McCune, R.A. Overbeek (U.S.A.) 85

Procedure implementation through demodulation and related tricks
S.K. Winker, L. Wos (U.S.A.) 109

The application of homogenization to simultaneous equations
B. Silver (U.K.) 132

Meta-level inference and program verification
L. Sterling, A. Bundy (U.K.) 144

An example of FOL using metatheory: formalizing reasoning systems and introducing derived inference rules
R. Weyhrauch (U.S.A.) 151

Comparison of natural deduction and locking
resolution implementations
S. Greenbaum, A. Nagasaka, P. O'Rorke, D. Plaisted (U.S.A.) . . . 159

TUESDAY MORNING
Derived preconditions and their use in program synthesis
D.R. Smith (U.S.A.) 172

Automatic construction of special purpose programs
C. Goad (U.S.A.) 194

Deciding combinations of theories
R.E. Shostak (U.S.A.) 209

Exponential improvement of efficient backtracking: a
strategy for plan-based deduction
T. Pietrzykowski, S. Matwin (Canada) 223

Exponential improvement of exhaustive backtracking:
data structure and implementation
S. Matwin, T. Pietrzykowski (Canada) 240

TUESDAY AFTERNOON
Non-monotonic reasoning
J. McCarthy (U.S.A.) (Invited lecture, not included in this volume)

Intuitionistic basis for non-monotonic logic
D.M. Gabbay (Israel) 260

Towards automatic deduction by analogy
Z. Zwinogrodzki (Poland)   . . . (Not included in this volume)

Knowledge retrieval as limited inference
A.M. Frisch, J.F. Allen (U.S.A.) 274

On indefinite databases and the closed world assumption
J. Minker (U.S.A.) 292

WEDNESDAY MORNING

Proof by matrix reduction as plan + validation
R. Caferra (France) 309

Improvements of a tautology-testing algorithm
K.M. Hörnig, W. Bibel (W. Germany) 326

Representing infinite sequences of resolvents in recursive
first-order Horn databases
L.J. Henschen, S.A. Naqvi (U.S.A.) 342

The power of the Church-Rosser property for string
rewriting systems
R.V. Book (U.S.A.) 360

Universal unification and a classification of
equational theories
J. Siekmann, P. Szabó (W. Germany) 369

SOLVING OPEN QUESTIONS WITH AN AUTOMATED THEOREM-PROVING PROGRAM[*]

by

L. Wos
Argonne National Laboratory
Argonne, Illinois 60439

Abstract

The primary objective of this paper is to demonstrate the feasibility of using an automated theorem-proving program as an automated reasoning assistant. Such usage is not merely a matter of conjecture. As evidence, we cite a number of open questions which were solved with the aid of a theorem-proving program.

Although all of the examples are taken from studies employing the single program, AURA [19] (which was developed jointly at Argonne National Laboratory and Northern Illinois University), the nature of the various investigations shows that much of the success also could have been achieved with a number of theorem-proving programs. In view of this fact, one can now correctly state that the field of automated theorem proving has reached an important goal. A theorem-proving program can now be used as an aid to ongoing research in a number of unrelated fields.

The open questions are taken from studies of ternary boolean algebra, finite semigroups, equivalential calculus, and the design of digital circuits. Despite the variety of successes, no doubt there still exist many who are very skeptical of the value of automating any form of deep reasoning.

It is the nature of this skepticism which brings us to the second objective of the paper. The secondary objective is that of dispelling, at least in part, some of the resistance to such automation. To do this, we discuss two myths which form the basis for the inaccurate evaluation of both the usefulness and the potential of automated theorem proving. Rejection of the two myths removes an important obstacle to assigning to an automated theorem-proving program its proper role--the role of colleague and assistant.

---

[*]This work was supported by the Applied Mathematical Sciences Research Program (KC-04-02) of the Office of Energy Research of the U. S. Department of Energy under contract W-31-109-ENG-38.

1. Introduction

There is one question which has been repeatedly asked of those in automated theorem proving. When will automated theorem-proving programs prove new theorems? The question is a natural one and exceedingly reasonable. The name of the field itself evokes the image of answering open questions. Fortunately, the field can now say that programs can and have answered various previously unanswered questions.

In this paper we shall briefly discuss some of the open questions which were solved with the assistance of an automated theorem-proving program. These questions are taken from ternary boolean algebra [21], finite semigroups [22], formal logic [26], and circuit design [20,24]. In solving them, the program was used in a variety of ways. Taken together, the uses suggest that the program behaves like a colleague. Colleagues sometimes provide the necessary information to complete a task, sometimes give but a clue to solving the problem, and sometimes fail completely. The program is equally valuable and equally frustrating. However, the material covered here shows that such a program can be used as an aid to reasoning--a reasoning assistant. It may well serve as the complement to the numerical calculator.

We shall be content with a brief account of the methodology employed in solving the various problems since the details can be found elsewhere. The discussion will, however, be sufficient to illustrate the interplay which may occur between the program and the researcher. This cooperation is quite similar to that which is required when working with a consultant or colleague. We shall concentrate on our work rather than presenting a survey of the field in general. In Section 2, we shall cover the achievements to date. In Section 3, we shall give the methods and techniques used to obtain the results. In Section 4, we shall explore certain myths that have interfered with the acceptance of the field as a whole. In Section 5, we shall discuss the current work and the future intentions.

Although all of the open questions cited here were solved by relying on a single theorem-proving program, much of the methodology which was employed to obtain the solutions is available to the field as a whole. The following vital fact should be kept in mind throughout the entire reading of this paper.

> No additional programming was required to
> solve the open questions cited here.

Thus, although each area of investigation required both the mapping of the problem into appropriate clauses and the formulation of an overall strategy of attack, all of the research was conducted with a standard theorem-proving program. The program, AURA [19], is that which has been developed jointly at Argonne National Laboratory and Northern Illinois University.

2. Open Questions Answered with the Aid of an Automated Theorem-proving Program

In this section we catalogue the various open questions which have been answered. In the next section we give a brief discussion of the theorem-proving methodology employed in obtaining the solutions.

2.1. Ternary Boolean Algebra

A ternary boolean algebra is a nonempty set satisfying the following 5 axioms:

1) $f(f(v,w,x),y,f(v,w,z))$
   $= f(v,w,f(x,y,z))$
2) $f(y,x,x) = x$
3) $f(x,y,g(y)) = x$
4) $f(x,x,y) = x$
5) $f(g(y),y,x) = x$

where the function, f, acts as a so-called "product" and the function, g, acts as a so-called "inverse".

Which (if any) of these 5 axioms is independent of the remaining set of 4? Axioms 4 and 5 had been known to be dependent, and proofs had been obtained by various theorem-proving programs such as that of Allen and Luckham [1] and that of G. Robinson and Wos [unpublished]. There remained the corresponding questions for each of 1, 2, and 3. A methodology was formulated [21] which established the independence of each of the 3.

While the dependencies could be established by the standard use of a theorem-proving program in its capacity for finding a proof, the independencies relied on a quite different usage. The program was made to find models--models of various sets of 4 axioms which failed to satisfy the fifth. These 3 models were very small, each consisting of but three elements. The models in effect took the form of a table in which one could check that the necessary "products" and "inverses" held or failed as dictated respectively by those axioms to be satisfied and violated. The significance was in the new usage, namely, that of having automated theorem-proving programs find models and counterexamples. This powerful methodology, formulated by S. Winker [23], was essential to answering the next question to be discussed.

Before turning to the problem covered next, we briefly answer one obvious question. The question is: Why not extract the desired models from a proof-finding run. The objection to this approach centers around the fact that many of the questions to be answered are of a second-order nature. Although one could use a standard trick to then map the second-order question to a first-order question, there still exists the serious problem of obtaining a completed run--a run which obtains the required proof. The nature of the model might require an extremely, and perhaps prohibitively, long run. Thus, we suggest an attack which directly seeks a model when that is the objective.

2.2. Finite Semigroups

I. Kaplansky brought the following (then) open question to our attention. Does there exist a finite semigroup which simultaneously admits of a nontrivial antiautomorphism while admitting of no nontrivial involutions? An antiautomorphism is a mapping, h, such that

1. h is one to one;
2. h is onto;
3. $h(xy) = h(y)h(x)$.

The mapping is nontrivial if it is not the identity mapping. A nontrivial involution is an antiautomorphism, different from the identity, whose square is the identity.

Here again we have the same dichotomy as occurred with the ternary boolean algebra questions. If the presence of such an antiautomorphism implies the presence of such an involution, a standard proof search is the method of attack. However, if this is not the case because of the existence of a semigroup of the desired type, then a model or counterexample must be found. The model generating capability was the method which we chose first to employ, and it was successfully applied.

In this application, it was imperative that the program not attempt to examine the various finite semigroups starting with the smallest. There are entirely too many to consider even of order 6. It was also important to avoid consideration of commutative semigroups, for only nontrivial answers were being sought. The approach, the details of which are given in Section 3.2, yielded a semigroup of order 83.

With this answer in hand, the next natural question to be asked was that of minimality. Again it was desirable and, more importantly, possible to avoid a saturation approach. The minimal semigroup of the type being sought has order 7.

Finally, there arose the question of the number of such minimal semigroups. The program found that there are exactly four, which of course required the ability to test for isomorphism.

2.3. Logical Design of Digital Circuits

2.3.1. Design

How can an automated theorem-proving program be made to design circuits? Can it then be asked to find ones which are more efficient than the existing ones?

The field of interest was 4-valued logic, and the technology which was of interest consisted of the employment of T-gates. The T-gate is a basic component for multi-valued logic as well as for 2-valued logic. In 2-valued logic it is the familiar multiplexor. There existed digital circuits employing these T-gates which were quite satisfactory and even thought to be minimal. W. Wojciechowski suggested that an automated theorem-proving program might be used in attempting to find more efficient digital circuits.

In most of the cases that were studied, suitable encoding of the problems [24] for the program resulted in obtaining more efficient digital circuits than were known. Two important points thus emerged. First, a new application for theorem-proving programs was found. The problems required an encoding of various input/output tables and axioms to describe the desired behavior of components. Second, a member not of the field of automated theorem-proving chose to use such a program in preference to writing a special-purpose program.

2.3.2. Hazards

One property to be avoided in the design of circuits is that of the presence of hazards. A hazard is a condition in which some signal arrives at some point sooner than expected and also sooner than desired. The presence of a hazard can, for example, cause all signals to be simultaneously 1, which may be specifically prohibited. The problem, suggested by A. Wojcik, was that of having the program detect hazards. S. Winker was able to implement [20] a sophisticated hazard-detection algebra [2]. He was able to then present this problem to the program and have it successfully detect hazards when presented with a circuit containing one.

There then arose the question of validation of the absence of such conditions. Again S. Winker was successful in devising a method [20] to have the program make such a validation.

These applications extend both the use of the standard input language for theorem-proving programs and the corresponding methodology. They also strongly demonstrate versatility. One obvious goal for such a system is that of practical application. The examples so far have been essentially only theoretical. However, just as has now been demonstrated in mathematics and logic, the circuit design studies may also pass from the theoretical to the applied. The difficulty of the size of circuits to be studied might at first appear to be insurmountable, but the examples cited next from formal logic may in part dispel this fear.

2.4. Formal Logic

We now come to the perhaps most interesting and certainly the hardest questions we have as yet solved with the program. The questions concern the possible existence of "new" or additional shortest single axioms for the equivalential calculus [6,16]. The elements of that calculus are the expressions which can be recursively well-formed from the variables, x, y, z, w, ..., and the 2-place function E. (The function, E, can be thought of as "equivalent".) The theorems of the calculus are just those formulas in which each variable therein occurs an even number of times [3]. The standard inference rules consist of a substitution rule and a rule of detachment. However, often studies are instead conducted by means of the inference rule called condensed detachment.

Definition. For the 2 formulas E(A,B) and A, detachment is that inference rule which yields the formula B.

Definition. For the 2 formulas E(A,B) and A' which are assumed to have no variables in common, condensed detachment is that inference rule which yields the formula B'', where B'' is that which is obtained from detaching E(A'',B'') and A'', and where E(A'',B'') and A'' are respectively obtained from E(A,B) and A' by the most general possible substitution whose application forces A and A' to become identical.

Thus condensed detachment, when it applies, takes two formulas and renames their variables so that they have no variables in common, then seeks the most general substitution that can be found whose application will permit a detachment, and finally applies detachment to the pair of formulas after the substitution has been made. For example, the condensed detachment of E(E(E(x,x),z),z) with E(E(x,y),E(y,x)) yields E(x,x) and also yields E(x,E(E(y,y),x)), depending on the order in which the premisses are considered.

In the equivalential calculus, there are 630 formulas in which each variable occurs exactly twice and in which five E's occur. Each of these 630 formulas, therefore, has length eleven, where length is the symbol count. There are formulas of length 11 which are strong enough to serve as a single axiom for the entire calculus, and thus all theorems are deducible from such a formula. When we began our study, there were seven of the 630 formulas which had not yet been classified with regard to the status of being a single axiom. The conjecture [4] existed that none of the seven was strong enough.

A typical method for showing that some given formula of a calculus is too weak to be a single axiom for that calculus is the following. Find a model which satisfies the formula in question but which also fails to satisfy some known axiom of the calculus. If one finds such a model, then the given formula is too weak to be a single axiom since one of the theorems of the calculus--one of its axioms--is provably not implied by the given formula. Employment of this standard approach might force one to examine many models which fail to have the desired properties. Because of the number of models which were found to be useless, and more importantly because we had begun to formulate a "new" powerful methodology, we chose to abandon consideration of models altogether.

We formulated a method, completely implementable within current theorem-proving technology, which allows the examination of the entire set of theorems deducible from certain of the formulas. For those formulas which turned out to be inadequate with regard to being a single axiom, the method for establishing such inadequacy is the following. For such a formula, find a characterization of all theorems deducible from it by condensed detachment. Then, by recourse to the characterization, show that some known axiom of the equivalential calculus cannot be so characterized and thus must be absent from the theorems yielded by the formula under study. Finally, conclude that the formula is therefore too weak to be a single axiom for the calculus. Perhaps more interesting than establishing the various inadequacies and certainly surprising, in view of the conjecture, is the fact that we found proofs which establish that two of the formulas are in fact (new) shortest single axioms.

Each of the two proofs in question is extremely long and complex if phrased in terms of condensed detachment. Yet we found a representation which permitted the program to generate each. Since the proofs were not in terms of a straightforward application of condensed detachment, we then formulated a method to convert them to such. The conversions were also done with the automated theorem-proving program. This action was taken to yield information in the form preferred by the logician.

To illustrate the difficulty and complexity of the study, note that one of the proofs consists of 162 condensed detachment steps. Among those steps, some are of length greater than or equal to 95--the number of symbols excluding commas and grouping symbols--while one step is of length 103. This proof would appear to be out of reach of a special-purpose program employing a standard approach.

3. Methodology

The organization of this section parallels that of the previous section. We concentrate here on the methodology from the viewpoint of automated theorem proving, beginning with the parallel of Section 2.1.

3.1. Ternary Boolean Algebra

The study of ternary boolean algebra led directly to the extension of the use of automated theorem-proving programs. Before this study, they had been almost solely used for the task of proof finding. With the completion of the study, the ability to directly find models and counterexamples had been added. A model or counterexample often takes the form of a set of clauses which, for example, can be translated into a table giving relations among the various elements. Examination of the table shows which "products" and which "inverses" hold or fail.

The method employed illustrates the cooperation that can occur between the researcher and the program. The first action which was taken was that of seeking a proof of, for example, the dependence of axiom 2 on the remaining set of four. As is now well-known, this failed because axiom 2 is independent of the remaining.

In order to show that axiom 2 is an independent axiom, we sought a model which simultaneously satisfies each of axioms 1, 3, 4, and 5, while violating 2. Nonunit clauses are input and used to ensure that axioms 1, 3, 4, and 5 are satisfied. A glance at the Appendix reveals the mechanism being employed and also shows that the run looks identical to one in which a proof rather than a model is being sought. These nonunit clauses were used as nuclei for a run employing hyper-resolution [12,18]. (Recall that a nucleus for hyper-resolution is a nonunit clause containing some negative literals, while the positive clauses required by the inference rule are called satellites.)

We began the model search with a guess at the number of elements that might suffice. We guessed three was enough. Since an early run had found that the lemma, $g(g(x)) = x$, holds for all x when axioms 1 through 3 are present, it was sufficient to force the lemma to fail for some element of the proposed model. This violation is sufficient to allow axiom 2 not to hold. Next, a partial model was conjectured. The program was given some guesses at the so-called products and so-called inverses.

The nuclei are used to force certain relations to hold. For example, with

-Q(x) -Q(y) EQUAL(f(x,x,y),x)

as the nucleus corresponding to axiom 4, $f(a,a,a) = a$ must hold, where a is one of the elements of the partial model. The predicate, Q, means "is an element of the proposed model". We therefore added to the input the units

Q(a)
Q(b)
Q(c)

so that tests could be applied to the proposed model. By using hyper-resolution, the given nucleus was used to test all appropriate triples against the fourth axiom. Similar nuclei were input corresponding to the remaining axioms to be satisfied.

Demodulators, which respectively express the "products" and "inverses" occurring in the partial model under development, were input and/or adjoined while the run was already under way. These demodulators, or rewrite rules, are automatically applied to the clauses yielded by hyper-resolution. If at any time a clause is generated which is not an instance of EQUAL(x,x), and if the nucleus occurring in its history is the correspondent of one of the axioms to be satisfied, we have two possibilities. First, we may have found another relation on the way to completing the model. Second, we may have violated the assumed distinctness of the given elements comprising the model. If the former occurs, we adjoin the new relation and proceed further with the goal of completing the model. If the latter occurs, we must replace some given relation(s) with the valuable knowledge that the particular choices made so far will not enable us to complete a model of the desired type. An example of such an occurrence is the derivation of EQUAL(b,a), which is unacceptable since we are assuming that b and a are distinct elements.

At that point, the program must back up and try a different value for one or more products and/or one or more inverses. By iterating with this procedure, the program did find a model of the desired type which showed that axiom 2 was an independent axiom.

Finally, to see which triple violated axiom 2, a nucleus of the form,

-Q(x) -Q(y) EQUAL(f(y,x,x),x),

was adjoined. If we are successful in finding a model of the desired type, which means that all products and all inverses have been determined, there must exist a triple which will not demodulate to an instance of reflexivity. That triple will be found when hyper-resolution uses this last nucleus and performs the various substitutions including the interesting one.

3.2. Finite Semigroups

Although a number of questions were eventually answered in the investigation, we began with the following question. Does there exist a finite semigroup which simultaneously admits of a nontrivial antiautomorphism while admitting of no nontrivial involutions? The first task was the selection of the branch to explore. We could have the program attempt to prove a theorem stating that no such semigroup exists, or we could attempt to refute such a theorem by finding a counterexample in the form of a model. We chose, fortunately, to concentrate on the second alternative.

As will be seen, the theorem-proving program was used as an assistant. The following is an outline of the approach which was taken.

1. Concentrate on reasonably large semigroups. Larger structures allow more possibilities for such a simultaneous existence.

2. Choose a representation of semigroups which permits consideration of fairly large ones.

3. Conduct some experiments with the program to find some conditions which force finiteness on the structure.

4. Since the presence of an antiautomorphism and/or an involution depends on a certain type of symmetry, impose much symmetry on the structure to allow for the former. Then interfere sufficiently with that symmetry to block the latter.

Adherence to the first point allows enough room to impose somewhat odd behavior on the semigroup. Obviously, if the semigroup were very small, its structure would be very limited. However, largeness can present a problem of representation.

One can in general choose between two methods of representation. One can either give a set of triples which encode the multiplication table, or one can give a set of generators and relations which determine the entire structure of the semigroup. The employment of generators and relations is well-known to the mathematician. With its use, one gets a compact notation which permits the easier study of large structures. (The models one might hope to extract from a proof-finding run would not be in terms of generators.) The elements are the "words" expressed in terms of the generators. Multiplication is simply concatenation. Demodulators are used to encode the relations and to perform the multiplications, see Appendix.

An argument, conducted by hand, shows that the demodulators perform their intended task. We decided to have four generators because of certain mathematical considerations. For example, since we wish to avoid commutativity of the structure, any nontrivial antiautomorphism must have even order. Next, it must not have order 2 since we are desirous of blocking involutions. Then, with four generators, we could consider an antiautomorphism which cycled the generators. Finally, the program must not be given the task of examining the awesome number of large semigroups.

Experiments with the program showed what was needed to guarantee finiteness. By adding a relation which forces all words of length greater than or equal to n, for some n, to be identical, finiteness is achieved. Early failures produced the semigroup of order 1 and a well-known one of order 8, both of which proved useless.

Experiments led to the imposition of the relation which forces all elements of length greater than or equal to four to be identical. The resulting semigroup, generated by the four generators and constrained by this one relation, was found by the program to have order 85. However, symmetry abounds. Not only do antiautomorphisms exist, but also do many involutions. The mapping, h, which maps b to c, c to d, d to e, and e to b, where b, c, d, and e are the generators, is the antiautomorphism on which we concentrated. The mapping, j, which interchanges b and c while leaving fixed both d and e is an example of an involution which we needed to block.

To attempt to block such involutions, we added a relation which would at least in part destroy some of the symmetry in this semigroup of order 85. When the program was given that relation, bc = de, it found it inadequate. Involutions still existed which said that the symmetry had not been sufficiently marred. However, the relation, bbc = dde, was sufficient.

The program was then required to explore the consequences of this addition. It found that the only other relation which was implied was dcc = bee. This was found by continual reference to the assumed presence of the mapping, h.

Of course, the program continues to check for associativity, for example. After all, we are seeking a semigroup. The program also continues to check for the viability of the antiautomorphism, h, and the avoidance of various involutions. The result of this researcher/program exchange was the discovery of a semigroup of order 83 of the desired type.

We then turned to the related questions of minimality. We found by heavy use of the program that there exists a semigroup of the desired type of order 7. There are in fact four such which are not isomorphic. The tests for isomorphism were also conducted by the automated theorem-proving program. At no time did we examine these questions by means of saturation. There are a myriad of semigroups even of order 6.

The important point to continually keep in mind is the role of the program. It is an aid to reasoning and can be assigned many and varied tasks of a logical nature.

3.3. Electronic Circuitry

3.3.1. Design

The object in this study was that of designing various digital circuits. We decided to emulate the approach taken by engineers. The engineer begins with a set of specifications for the digital circuit and a set of possible components to be used. The characteristics of the components are known. He often then replaces the original problem with a set of smaller and more manageable subproblems. The idea is that of giving specifications for the subcircuits, and then of using the components to combine the subcircuits into the circuit being sought.

The program was given the specifications in a straightforward manner, namely, in the form of an input/output table. It was then told how to decompose this table into useful subtables, each representing a subcircuit. It was told how to recognize very small circuits which were immediately constructible. Finally, using the basic components, it was told how subcircuits were to be combined into the larger circuit. The basic component of interest is the T-gate.

The program was then assigned the task of finding interesting and efficient digital circuits. The task took the form of finding a proof that some digital circuit of the required type exists. Since a number of classes of subcircuit were not of interest, demodulators were used to have these subcircuits, upon discovery, transformed to "junk". Such a transformation thus prevents the uninteresting subcircuits from interacting with the interesting ones.

Since the goal was that of examining a number of solutions, different circuits of the desired type, the program was instructed not to stop after a (single) proof had been found. Thus a single run yielded a number of circuits of interest.

The digital circuits were extracted from the "proofs" thus obtained. Most of the digital circuits found were better than those extant in the literature at the time.

We offer the following brief explanation for our choice of method for finding the circuit models. We were able to seek them with a proof-finding run rather than with a direct model search because of both the given information and the nature of the problem. We were, for example, able to rely on various decompositions into subcircuits.

3.3.2. Hazards

A hazard is a condition in which some signal arrives at some point in time sooner than expected and also sooner than desired. To detect the presence of such a condition, one needs to evaluate the consequences which may occur, over time, as the input changes. There is an algebra, called Fantauzzi hazard algebra [2], which studies the consequences at discrete points in time of changes in the input to the circuit.

The two questions of interest concern the presence and/or the absence of hazards for a given circuit. Since the number of situations to be considered is a small finite number, at least in the cases we examined, the program employed an exhaustive approach. The determination of the presence or absence was simply made by inspecting the output of the program upon completion of the run, see Appendix. The run terminates when no inferences are left to be made. A case analysis technique is employed to consider the various changes which are of interest--the changes of the input to the circuit.

Circuits possessing and circuits free of hazards were examined and correctly classified [20].

3.4. Formal Logic

The problem which created our interest in formal logic was that of finding a shorter proof for a known theorem. The theorem in question establishes that a particular formula, XGK [14], from the equivalential calculus is a single axiom for the calculus. As was explained in Section 2.4, the often used rule of inference is condensed detachment. An examination of that rule reveals that it is quite similar to UR-resolution [9] and also to hyper-resolution. (In UR-resolution, the nucleus is a nonunit clause, the satellites are unit clauses, and the required inference is a unit clause.) We found a proof [26] which is but half as long as the published one. It was found by having the program weight [10,26] certain formulas in such a way which caused some to be much preferred while others were totally avoided [26].

During this investigation, we found that demodulators could be written which would correctly classify formulas according to various syntactic properties. Other demodulators were found which counted the number of occurrences of the symbol, E, in each parent and in the resulting inference. Because of our success in finding the shorter proof, and because of the discovery of new and nonstandard applications of demodulation, we turned to certain open questions.

We first considered the formula, XBB, and attempted to examine those formulas which it yielded by application of condensed detachment. Although this formula was already known to be too weak to be a single axiom for the calculus, its study nevertheless proved valuable. The program found longer and ever longer clauses. This immediately led to the conjecture that this ever-increasing in length property held for all formulas obtainable by starting with XBB and repeatedly applying condensed detachment. If this were the case, we could characterize all the theorems implied by the formula under study. It proved to be the case. The proof was obtained by hand, but the conjecture resulted directly from the examination of one run with the automated theorem-proving program.

We then applied the conjecture to other formulas--ones which had not yet been classified as to being single axioms. The program yielded the fact that the conjecture of ever-increasing length did not hold for those formulas. The demodulators for counting symbol occurrences, mentioned above, were in part employed. The reading of the resulting clauses was quite cumbersome in view of their similarity and complexity. We thus introduced a demodulator to transform them into a more readable form. We also decided to first concentrate on the formula, XAK.

In the next run, all kept clauses were rewritten, where the rewrite rule was in the form of a demodulator--a demodulator designed to promote readability. What resulted was the startling fact that all kept clauses shared a single subexpression, up to alphabetic variance. We thus conjectured that all formulas implied by the now under study formula, XAK, had this property. If this were provable, then all theorems deducible by condensed detachment from XAK would contain this marvelous subexpression, and we would know that the formula was inadequate with respect to being a single axiom for the equivalential calculus. The inadequacy follows from the fact that there are known theorems in the calculus which do not contain this particular subexpression.

Since condensed detachment yields an infinite set of formulas to be studied, we had the program conduct its examination in terms of schemata rather than in terms of individual formulas. The program was then used to examine the possible schemata, forms for theorems, so deducible from XAK. We quickly concluded that an induction argument, relying on a case analysis study, might suffice to pin down the possible forms. The induction is on the length of the most general common instance of two unifiable expressions. Although the actual induction was not by means of the program itself, the needed cases were examined by it. Again the program proved invaluable and performed as a colleague.

We had a somewhat natural but, as it turned out, naive notion as to the nature of the case analysis. It turned out to be somewhat subtler than expected. The program runs showed that the original plan for the case analysis would continually yield new cases to be considered. An examination of the output in fact suggested the true nature of the case analysis.

Since the details are to be found elsewhere [26], we are content with the following summary. Use of the program led to various conjectures, refuted others, suggested notation, and pointed the way to the correct case analysis that was needed. With this much assistance, we could hardly fail. We were able to correctly classify each of the (at the time) remaining seven formulas. Five were found to be inadequate, while two were found to in fact be shortest single axioms for the calculus.

One more success with the program was to occur. The proofs of the existence of the two additional shortest single axioms were obtained in a nonstandard notation. The notation was oriented towards the schemata we had employed throughout. The logician might well be uninterested in this form of proof. We therefore had the program translate each of the two proofs into the standard form of proof, that which employs condensed detachment.

As has been mentioned in Section 2.4, the resulting proofs are extremely long and extremely complex. The longest proof consists of 162 steps of condensed detachment. The most complex step in that proof contains 103 symbols, exclusive of commas and grouping symbols. A straightforward use of an automated theorem-proving program would have probably failed very badly. But a straightforward attack by the logician would also probably have met with disaster. This entire study is a graphic example of our viewpoint that an automated theorem-proving program can be treated and used as a valuable automated reasoning assistant.

4. Myths and Responses

Two myths exist in regard to automated theorem proving. It is the first myth, the 0/1 myth, which is the more significant. Its essence is the belief that such programs must either solve all problems or be classed as useless, which is the explanation for the 0/1 designation. This myth demands that work on a given problem must, for example, yield a proof or result in the designation of failure. Adherence to the myth would mean that one merely looks at the last line of a run to see if the empty clause has been found. If not, then discard the run and term it useless. One would not, therefore, examine the clauses themselves to glean the possible gems contained therein.

The second myth says that special-purpose programs are in the main preferable to general-purpose. We deal first with the 0/1 myth.

This 0/1 view, which is often hidden, may take the following form. For example, if a (known) theorem is submitted to an automated theorem-proving program, and if the program then fails to find a proof, the conclusion is drawn that the program is essentially useless. Such a reaction is the parallel to the following: if a colleague is asked to help with some research, and if he has nothing to contribute after some time, then he would also be classed as useless. Of course, this is not what is done.

By way of expansion and also to see that the 0/1 myth does not apply to automated theorem proving, consider first what is expected of a colleague and then what may be expected of an automated theorem-proving program. A colleague might have the following properties:

1. He is in general willing to assist in the investigation of virtually any question. In many cases, he unfortunately requires an inordinate amount of tutorial discussion to give some needed and missing background. In those situations, he often contributes nothing significant.

2. In some few areas, he has demonstrated unusual research ability. He has, for example, solved a number of open questions. He has also provided information which has allowed others to solve some open questions.

Such an associate is clearly considered of value. An automated theorem-proving program also now has those same two properties, and thus should be similarly considered. The presence of the first property is not surprising in that obviously much work is often necessary to prepare a problem for the program. The presence of the second property, on the other hand, is that which is interesting. As has been seen in the previous two sections, the program has been used to solve a number of open questions and thus has been used as a research aid and not just as a logical calculator. In attempting to solve these questions, we saw the complete spectrum of results. In formal logic, for example, a straightforward use of condensed detachment was shown to be pointless. However, as with a consultant, we proceeded with an alternate approach. Even then we met with various failures but also with various successes. Had we adhered to the 0/1 myth, our first attempts would have caused us to abandon the problem entirely. But the myth is not applied to a colleague, and, as evidenced by the eventual results of the study of formal logic, it should not be applied to automated theorem-proving programs.

When consulting an associate, three obvious outcomes may occur.

1. He may solve the problem completely on his own.
2. He may instead supply some hints or some missing information which is intended to make the problem easier.
3. He may totally fail.

It is, of course, the third case which is of concern. Failure may consist of misunderstanding of the problem, of having no intuition on which to rely, or perhaps of even being obtuse. In any event, one does not write off the associate. One instead hopes that the next consultation will be more fruitful. In other words, the 0/1 myth is not applied. We treat the use of the theorem-proving program similarly. The crucial point here is the expectation which is present. In using the program as an aid or research assistant, one can expect results which range from complete success to total failure.

If a criticism of the field were made, it might, at least implicitly, rest on the 0/1 myth. Various obstacles might be cited as evidence of the difficulty of achieving any substantial results. Obstacles such as complexity of proof, difficulty of representation in the first-order predicate calculus, and the inability to apply intuition are but a few which are in fact often cited. The proof that these obstacles are insurmountable rests on the evidence of failure of the theorem-proving program to complete some assigned task. Failure to complete a proof, for example, is extremely common for theorem-proving programs, but such failure is also very common even for the expert. One can, therefore, take either of two views in judging usefulness.

1. The designation of usefulness is applied to a colleague or a theorem-proving program when either can contribute significantly to the solution of problems from a limited but interesting class.
2. The designation is applied to either only when there exists the ability to approach almost any problem with at least competence, if not actual brilliance.

The second of these views is reminiscent of that same 0/1 myth. It is of course the first view which we take and which we believe we have demonstrated as applicable to automated theorem-proving as a whole. The successful refutation of the conjecture which existed for the equivalential calculus is a case in point. In view of the nature and difficulty of the proofs which comprise the refutation of the conjecture, it seems clear that the conjecture was a reasonable one. One could not reasonably expect an individual to guess that a proof of the type found by the program existed, especially in view of its length and complexity.

The point which is becoming increasingly clear is that automated theorem-proving programs can be treated as colleagues. With such a program, one can, for example, make conjectures, test conjectures, and find holes in proposed methods of proof. Each of these activities is reminiscent of that which a colleague performs.

We now turn to the second myth--special-purpose programs are preferable to general-purpose. Of course the burden of refuting this myth lies with those advocating general-purpose programs. The arguments we give are in terms of ease of maintenance, variety of use, and convenience.

Our claim to the first of these three properties would fall short if it were not for the fact that new modules are not needed for attempts in new areas. All of the problems which we cited were solved without recourse to additional programming. We found that the standard procedures available to those in the field could be adapted to the required tasks. This does not imply that, for example, no new inference rules are needed. The formulation of such a new rule might require programming, and perhaps extensive programming. This possibility is but one of the reasons for the development of a new automated theorem-proving program, which is briefly discussed in Section 5.

As for the second property, variety, the problems which were solved with an automated theorem-proving program range from circuit design to formal logic. The tasks range from finding more interesting models than existed to generating very long and complex proofs. The examples given from circuit design were in the context of efficiency, detection of hazards, and validation of the absence of such conditions. From abstract mathematics, we discussed problems of finding semigroups which admit of certain mappings while lacking others. In that connection, we discussed solving the corresponding problem of minimality but without recourse to examination of the myriad of undesirable semigroups. Finally from formal logic, we examined questions of the existence of new shortest single axioms. This investigation required a flexible approach, case analysis, induction, and a representation which avoids the awesome consideration of millions of formulas.

As for convenience, we state categorically that much is still needed. Familiarity with the language required by the program in order to input a problem is but one of the difficulties. It is often at least cumbersome to learn to speak clearly in the language consisting of clauses. Then, since our system has innumerable options and possible parameter settings, its use is further complicated. However, the second problem is somewhat reduced by the existence of a translator [7]--a program written by E. Lusk--which allows reasonable direction of the theorem-proving program by fairly understandable conventions.

Examination of the work presented earlier focuses some doubt on the choice which favors special-purpose programs. The examples from formal logic at least suggest the difficulty of writing a special-purpose program for considering the questions that were answered. If a special-purpose program were written embodying the techniques preseented earlier, the effort might be formidable. A direct attack on the open questions would have been perhaps impossible. The reason for this suspected impossibility lies with the very large number of formulas which would be examined were condensed detachment employed.

5. **Now**

The natural question to ask is: where are we now? We are presently seeking more open questions. We wish to find those which require the finding of models and counterexamples, but also we wish to have those which require the finding of a proof. We are studying the feasibility of extending the studies of circuit design with the possible eventual application to industrial needs. In fact, the intention is that of eventually having a large group of users from various fields and with very divergent interests and purposes.

The latest application is concerned with the validation of the design of nuclear plants. The case analysis techniques are but one example of that which can be brought to bear. We have a small partially completed interactive theorem-proving program written in Pascal. A demonstration of its possible use in the examination of the specifications for a plant has been given.

In order to facilitate the use of theorem-proving programs by a wide range of individuals, we are presently concentrating on the development and implementation of a new automated theorem-proving program. The program, written in Pascal, will be portable and will allow great flexibility. For example, one need not concentrate on the clause form for representing information. Also, one can conduct almost immediately new experiments either in the nature of "pure" theorem proving or in the nature of new application. The user of the new system will, therefore, be able to shape the program and its properties and its application to his wishes.

The notion is that the new automated theorem-proving program will be accessible to many in research. Being written in Pascal will make its structure and properties available for inspection and for change. In short, it may well serve as a research tool for various fields of study.

6. **Conclusions**

The field has now demonstrated the ability to contribute to research in various areas. Open questions in both mathematics and formal logic have been answered with the use of an automated theorem-proving program. However, the applications are not limited to those of research. There is some real evidence that practical use is possible. The success with digital circuit design and the detection of hazards is an example.

Two extremely important points must be made. First, the results which have been cited here are in the main within reach of many automated theorem-proving programs. The techniques and methodology which led to these successes are accessible to the field as a whole. Second, all of the results presented here were obtained without any modification to the already existing automated theorem-proving program. In view of these two facts, it seems reasonable to conjecture that the field of automated theorem proving has entered a new phase. That phase might be described as the merging of contemplated eventual goals with the realization of some of those same goals.

There are of course still many of the same problems to solve. For example, too little is known about the interrelation of the representation of information, the choice of inference rules, and the use of strategy. However, automated theorem-proving programs can now often be treated as a colleague--an automated reasoning assistant. Colleagues are asked to help with, among others, the formulation, testing, proof, and refutation of conjectures. A theorem-proving program can be asked for the same kind of help. Although the program often totally fails, occasionally it obtains startling results. Even many of those runs which run out of time and/or memory contain very useful information. The user/program exchange method of investigation often proves profitable.

More open questions are needed. The attempts at solving such questions lead to the knowledge of the shortcomings of the present systems. The solving of any such question leads eventually to the submission of more open questions. This ongoing dialogue will, and already has in part, place the automated theorem-proving program in its proper role--the role of automated reasoning assistant.

References

[1] Allen, J. and Luckham, D., "An interactive theorem-proving program," Machine Intelligence, Vol. 5(1970), Meltzer and Michie (eds), American Elsevier, New York, pp. 321-336.

[2] Fantauzzi, G., "An algebraic model for the analysis of logic circuits," IEEE Transactions on Computers, Vol. C-23, No. 6, June 1974, pp. 576-581.

[3] Kalman, J., "A shortest single axiom for the classical equivalential calculus," Notre Dame Journal of Formal Logic, Vol. 19, No. 1, January 1978, pp. 141-144.

[4] Kalman, J., private communication.

[5] Lukasiewicz, J., "Der Aquivalenzenkalkul," Collectanea Logica, Vol. 1 (1939), pp. 145-169. English translation in [McCall], pp. 88-115 and in [Lukasiewicz Borkowski], pp. 250-277.

[6] Lukasiewicz, J., Jan Lukasiewicz: Selected Works, ed. by L. Borkowski, North-Holland Publishing Co., Amsterdam (1970).

[7] Lusk, E., "Input translator for the environmental theorem prover - user's guide," to be published as an Argonne National Laboratory technical report.

[8] McCall, S., Polish Logic, 1920-1939, Clarendon Press, Oxford (1967).

[9] McCharen, J., Overbeek, R. and Wos, L., "Problems and experiments for and with automated theorem proving programs," IEEE Transactions on Computers, Vol. C-25(1976), pp. 773-782.

[10] McCharen, J., Overbeek, R. and Wos, L., "Complexity and related enhancements for automated theorem-proving programs," Computers and Mathematics with Applications, Vol. 2(1976), pp. 1-16.

[11] Meridith, C., "Single axioms for the systems (C,N), (C,O) and (A,N) of the two-valued propositional calculus," The Journal of Computing Systems, i, No. 3 (July 1953), pp. 155-64.

[12] Overbeek, R., "An implementation of hyper-resolution," Computers and Mathematics with Applications, Vol. 1(1975), pp. 201-214.

[13] Peterson, J., "Shortest single axioms for the equivalential calculus," Notre Dame Journal of Formal Logic, Vol. 17(1976), pp. 267-271.

[14] Peterson, J., "The possible shortest single axioms for EC-tautologies," Auckland University Department of Mathematics Report Series No. 105, 1977.

[15] Peterson, J., "An automatic theorem prover for substitution and detachment systems," Notre Dame Journal of Formal Logic, Vol. XIX, Jan. 1978, pp 119-122.

[16] Peterson, J., "Single axioms for the classical equivalential calculus," Auckland University Department of Mathematics Report Series No. 78.

[17] Prior, A. N., Formal Logic, Second Edition, Oxford, 1962, Clarendon Press.

[18] Robinson, J., "Automatic deduction with hyper-resolution," International Journal of Computer Mathematics, Vol. 1(1965), pp. 227-234.

[19] Smith, B., "Reference manual for the environmental theorem prover," to be published as an Argonne National Laboratory technical report.

[20] Winker, S., Private Communication.

[21] Winker, S. and Wos, L., "Automated generation of models and counter examples and its application to open questions in ternary Boolean algebra," Proc. of the Eighth International Symposium on Multiple-valued Logic, Rosemont, Illinois, 1978, IEEE and ACM Publ., pp. 251-256.

[22] Winker, S., Wos, L. and Lusk, E., "Semigroups, antiautomorphisms, and involutions: a computer solution to an open problem, I," Mathematics of Computation, Vol. 37 (1981), pp. 533-545.

[23] Winker, S., "Generation and verification of finite models and counterexamples using an automated theorem prover answering two open questions," to appear in J. ACM.

[24] Wojciechowski, W. and Wojcik, A., "Multiple-valued logic design by theorem proving," Proc. of the Ninth International Symposium on Multiple-valued Logic, Bath, England, 1979.

[25] Wos, L., Robinson, G., Carson, D. and Shalla, L., "The concept of demodulation in theorem proving," J. ACM, Vol. 14(1967), pp. 698-704.

[26] Wos, L., Winker, S., Veroff, R., Smith, B. and Henschen, L., "Questions concerning possible shortest single axioms in equivalential calculus: an application of automated theorem proving to infinite domains," in preparation.

APPENDIX

We illustrate here the use of an automated theorem-proving program in three diverse areas of application. The three areas are those of ternary Boolean algebra [21,23], finite semigroups [22], and the design of digital logic circuits [20]. We begin by discussing certain aspects common to all three applications. First, in none of the three areas do we seek the proof by contradiction normally associated with the use of theorem-proving programs. Rather we seek various kinds of models: mathematical models and counterexamples, or (in the case of circuit design) situations in which a physical system may fail to perform as desired. Second, the clauses used in generating models are essentially indistinguishable in form from those used to seek proof by contradiction. We emphasize that no reprogramming is needed to generate the models presented here. We do not even set a switch specifying model generation as opposed to proof search. Rather the same inference rules and parameter options serve for both purposes. In particular, the runs are so similar that glancing at the parameter settings does not allow one to differentiate between the two objectives.

We turn now to the two mathematical studies, that for ternary Boolean algebra and that for finite semigroups. In each study an open question in mathematics was answered by discovery of an appropriate model [21,22]. Of course we did not know the models beforehand. Since we knew almost nothing about the desired models, we had to conduct an iterated search for them by making heavy use of the theorem prover. Each run exhibited here represents the culmination of such a search.

Two methods for the presentation of a model to the theorem prover will be illustrated. The first method involves the use of explicit tables which give the action of various functions upon the elements of the model. This method will be applied to the study of ternary Boolean algebra. The second method implicitly defines the action of the various functions by means of generators and relations. This second method will be applied to the study of finite semigroups.

Ternary Boolean Algebra

In the first study, that of ternary Boolean algebra, we present the results of two theorem-prover runs. The first run establishes the consistency of one particular model. The second run shows that another proposed model fails to satisfy one of the required axioms. The input for the first run is as follows.

* Each line beginning with the character * is a comment.
* We omit from this example those lines of input which give

```
* program parameter settings.
*
* The following input clauses list the elements of a proposed model.
*;
CL Q(a) ;
CL Q(g(a)) ;   * g(a) was abbreviated as b in section 3.1. ;
CL Q(g(g(a))) ;   * g(g(a)) was abbreviated as c in section 3.1. ;

* The following clauses state that these three elements are
* distinct.
*;
CL -EQUAL(g(x),x);
CL -EQUAL(g(g(a)),a);

* The following demodulator defines the action of the one-place
* function g upon the three elements.
* g of a is g(a) and g of g(a) is g(g(a)).
* only the action of g on g(g(a)) remains to be defined.
*;
CL EQUAL(g(g(g(a))),g(a)) ;

* The following demodulators define the action of the three-place
* function f upon the three elements.
*;
CL EQUAL(f(x,y,g(y)),x) ;
CL EQUAL(f(x,x,y),x) ;
CL EQUAL(f(g(x),x,y),y) ;
CL EQUAL(f(x,y,x),x) ;
CL EQUAL(f(x,g(g(y)),g(y)),x) ;
CL EQUAL(f(g(y),g(g(y)),x),x) ;
CL EQUAL(f(a,g(g(a)),g(g(a))),a) ;
CL EQUAL(f(g(g(a)),a,a),g(g(a))) ;
CL EQUAL(f(a,g(a),g(a)),g(a)) ;

* The following clause subsumes and deletes
* any trivial equalities which may be derived.
*;
CL EQUAL(x,x) ;

* The remaining clauses are all nonunits.
* Each serves as a nucleus for the inference rule of
```

```
* hyperresolution.
*;
* The following nucleus checks that the given set of elements
* is closed under the action of the function g.
*;
CL -Q(x) Q(g(x)) ;


* The following nucleus checks that the given set of elements
* is closed under the action of the function f.
*;
CL -Q(x) -Q(y) -Q(z) Q(f(x,y,z)) ;


* The following nucleus checks that axiom 1 for ternary Boolean
* algebra (see Section 2.1) is satisfied for all ordered quintuples
* which can be formed using the three elements given above.
*;
CL -Q(v) -Q(w) -Q(x) -Q(y) -Q(z)
    EQUAL(f(f(v,w,x),y,f(v,w,z)),f(v,w,f(x,y,z))) ;


* The following three nuclei check that axioms 3, 4, and 5 for
* ternary Boolean algebra are satisfied for all ordered pairs
* which can be formed using the three elements given above.
*;
CL -Q(x) -Q(y) EQUAL(f(x,y,g(y)),x) ;
CL -Q(x) -Q(y) EQUAL(f(x,x,y),x) ;
CL -Q(x) -Q(y) EQUAL(f(g(x),x,y),y) ;
```

When this entire set of clauses is submitted to the theorem prover, no inferred clauses are added to the clause space. Rather, every inferred clause is demodulated, subsumed, and deleted. The fact that no clauses in the predicate Q are added indicates that the input demodulators do in fact give complete function tables for the functions f and g. The fact that no new equality clauses are added indicates that axioms 1, 3, 4, and 5, which were checked by means of nuclei, are in fact satisfied in the model. Axiom 2 is falsified by the function table entry

```
CL EQUAL(f(a,g(g(a)),g(g(a))),a)
```

among others. Thus we have a model in which axioms 1, 3, 4, and 5 are satisfied but axiom 2 is not. Hence axiom 2 is independent of the remaining four.

Our second run in ternary Boolean algebra is a slight modification of the first. We replace the clause

```
    CL EQUAL(f(a,g(a),g(a)),g(a))
```

by the clause

```
    CL EQUAL(f(a,g(a),g(a)),g(g(a)))
```

in the input to the theorem prover. This alters the function table for f in just one position. The theorem prover now gives the following output.

```
   23   19 1 2 2 1 1   EQUAL(g(g(a)),a);
      DEMODULATORS:      15    10    14    9    10
   24   23 5   CONTRADICTION;
```

This indicates that the axioms checked are not satisfied in all instances. Rather, in a particular instance, one of the axioms (axiom 1) would force g(g(a))=a, contrary to the input condition that a and g(g(a)) are to be distinct. The reader may note that this second run could be regarded either as a model seeking run which failed, or as a run which successfully proves by contradiction that g(g(a))=a follows from the remaining conditions.

Finite Semigroups

We turn next to the second mathematical study, that of finite semigroups. We present the results of two theorem-prover runs. One develops, from the presentation of a semigroup in terms of generators and relations, a fuller set of relations. A particular antiautomorphism is built into this semigroup. The other run shows that a certain mapping of the generators fails to induce an involution of the entire semigroup.

The first run is the culmination of a series of experiments which sought a semigroup with the desired combination of properties. It develops from the given minimal set of relations a set of demodulators sufficient for making (if desired) an element list and a product table. Two inference rules are used. Paramodulation is used to examine interactions among the axioms and the relations. Hyperresolution is used, in conjunction with a particular nucleus clause, to impose the desired antiautomorphism upon the relations of the semigroup. The input to the theorem prover is as follows.

```
  * The following clause subsumes any trivial equalities
  * which may be generated.
  *;
```

```
CL EQUAL(x,x);
*
* The following clause states that the product
* operation in the semigroup is associative.
*;
CL EQUAL(f(f(x,y),z),f(x,f(y,z)));
*
* The following three clauses equate all products
* of four or more elements in the semigroup to the same element 1.
*;
CL EQUAL(f(x,f(y,f(z,w))),1);
CL EQUAL(f(1,x),1);
CL EQUAL(f(x,1),1);
*
* The following relation is introduced to reduce the
* amount of symmetry present in the semigroup.
* Elimination of symmetry can eliminate possible involutions.
*;
CL EQUAL(f(d,f(d,e)),f(b,f(b,c)));
*
* The following nucleus applies the antiautomorphism h to all
* relations.
*;
CL -EQUAL(f(x,y),z) EQUAL(f(h(y),h(x)),h(z));
*
* The following demodulators calculate the image of a term under
* the antiautomorphism h.
*;
CL EQUAL(h(f(x,y)),f(h(y),h(x)));
CL EQUAL(h(b),c);
CL EQUAL(h(c),d);
CL EQUAL(h(d),e);
CL EQUAL(h(e),b);
CL EQUAL(h(1),1);
```

The output of this semigroup-seeking run yields but a single additional equality, namely,

```
14  7 6  EQUAL(f(d,f(c,c)),f(b,f(e,e)));
  DEMODULATORS:      8   12   11   11   2   8   8   10   9
  DEMODULATORS:      9    2
```

In addition a message is printed indicating that no more inferences can be made using the selected inference rules. The presence of this message indicates that the set of demodulators has been fully developed as desired. With the aid of these demodulators a list of the distinct elements and the values of various products of the elements could now be obtained as detailed in [23].

The second run shows that a particular involution -- that which interchanges b and d and also interchanges c and e -- is not present in the semigroup. We provide the following input.

```
* Subsumption of trivial equalities
*;
CL EQUAL(x,x);
*
* Associativity
*;
CL EQUAL(f(f(x,y),z),f(x,f(y,z)));
*
* Demodulators defining the product operation in the semigroup.
* The last one of this set was obtained in the preceding run.
*;
CL EQUAL(f(x,f(y,f(z,w))),1);
CL EQUAL(f(1,x),1);
CL EQUAL(f(x,1),1);
CL EQUAL(f(d,f(d,e)),f(b,f(b,c)));
CL EQUAL(f(d,f(c,c)),f(b,f(e,e)));
*
* The following nucleus applies a possible involution j
* to all of the above equalities.
*;
CL -EQUAL(f(x,y),z) EQUAL(f(j(y),j(x)),j(z));
*
* The following demodulators define a particular involution j
* by giving a mapping of the generators and a rule for extending
* the mapping to the remaining elements.
*;
CL EQUAL(j(f(x,y)),f(j(y),j(x)));
CL EQUAL(j(b),d);
CL EQUAL(j(d),b);
CL EQUAL(j(c),e);
CL EQUAL(j(e),c);
CL EQUAL(j(1),1);
```

The output yields two additional equalities.

```
15  8 6   EQUAL(f(e,f(d,d)),f(c,f(b,b)));
   DEMODULATORS:      9   13   11   11    2    9    9   12   10
   DEMODULATORS:     10    2
16  8 7   EQUAL(f(e,f(e,b)),f(c,f(c,d)));
   DEMODULATORS:      9   12   12   11    2    9    9   13   13
   DEMODULATORS:     10    2
```

Their presence in the run proves that the given mapping J of the generators cannot be extended consistently to an involution of the entire semigroup, for such an extension would require that these additional derived equalities hold. But these equalities do not hold in the semigroup on which we are focusing. This fact follows from a simple argument based on its order. Thus we establish that the involution under investigation is not present in this semigroup. Comparable theorem-prover runs can be made to refute the existence of other possible involutions by defining each in terms of its action on the generators.

Digital Circuit Design

We turn finally to the study of digital circuit design. Our example [20] concerns the detection of possible hazards in a binary logic circuit. The term "hazard" refers to certain unsatisfactory behavior of a circuit during a change in the input to the circuit. We use nuclei and demodulators to generate various possible time-dependent inputs and to simulate the response of the circuit.

We begin with a classification of possible time-dependent signals given by Fantauzzi [2]. The classes are as follows:

both t and f denote steady, unchanging signals;
tf denotes a smooth transition from the t state to the f state;
t*t denotes a signal which is steady except for a transient change to f and back;
t*f denotes a transition from t to f which is not smooth, but instead wavers between f and t before settling to f;
ft, f*f, and f*t are defined analogously.

We assume that the only admissible input signals are t, f, tf, or ft. The admissible output signals, however, are permitted to include the other four signals, thus allowing for the possible presence of a hazard. We supply the following information to the theorem prover.

```
* We give the effects of combining signals
* through the use of standard logic components.
* We encode the tables given by Fantauzzi [2].
*
* and  |  f    ft   f*f  f*t  t    tf   t*t  t*f
* -----|-------------------------------------
* f     |  f    f    f    f    f    f    f    f
* ft    |  f    ft   f*f  f*t  ft   f*f  f*t  f*f
* f*f   |  f    f*f  f*f  f*f  f*f  f*f  f*f  f*f
* f*t   |  f    f*t  f*f  f*t  f*t  f*f  f*t  f*f
* t     |  f    ft   f*f  f*t  t    tf   t*t  t*f
* tf    |  f    f*f  f*f  f*f  tf   tf   t*f  t*f
* t*t   |  f    f*t  f*f  f*t  t*t  t*f  t*t  t*f
* t*f   |  f    f*f  f*f  f*f  t*f  t*f  t*f  t*f
*;
CL EQUAL(and(f,x),f);
CL EQUAL(and(x,f),f);
CL EQUAL(and(t,x),x);
CL EQUAL(and(x,t),x);
CL EQUAL(and(ft,ft),ft);
CL EQUAL(and(ft,f*f),f*f);
CL EQUAL(and(ft,f*t),f*t);
CL EQUAL(and(ft,tf),f*f);
*
* (and similar clauses, omitted here because of
*  space considerations)
*
*        |  f    ft   f*f  f*t  t    tf   t*t  t*f
* not   |-------------------------------------
*        |  t    tf   t*t  t*f  f    ft   f*f  f*t
*;
CL EQUAL(not(f),t);
CL EQUAL(not(ft),tf);
* (and so forth).
*
* Definition of "nand" logical function
* of one, two, and three arguments.
*;
CL EQUAL(nand1(x),not(x));
CL EQUAL(nand2(x,y),not(and(x,y)));
CL EQUAL(nand3(x,y,z),not(and(x,and(y,z))));
```

```
*
*   We input demodulators which will evaluate the output
*   of a particular circuit given a set of time-dependent inputs.
*   We assume in this example that there are three input wires
*   named a, b, and c.
*   We assume further that nand gates, represented by the
*   functions nandg1, nandg2, nandg3, are used.
*   The function "ipt" will be used to encode the signals input
*   to the three wires a, b, and c.
*   For example  ipt(t,f,ft)  represents inputs of  t  for  a,
*   f  for  b,  and  ft  (transition  f  to  t,  no hazard) for  c.
*
*   The function "eval" evaluates the output obtained from a given
*   circuit (first argument) with a given input (second argument).
*;
CL EQUAL(eval(nandg1(x),vipt),nand1(eval(x,vipt)));
CL EQUAL(eval(nandg2(x,y),vipt),nand2(eval(x,vipt),eval(y,vipt)));
CL EQUAL(eval(nandg3(x,y,z),vipt),
         nand3(eval(x,vipt),eval(y,vipt),eval(z,vipt)));
CL EQUAL(eval(a,ipt(xa,xb,xc)),xa);
CL EQUAL(eval(b,ipt(xa,xb,xc)),xb);
CL EQUAL(eval(c,ipt(xa,xb,xc)),xc);
*
*   We test all possible non-hazard inputs.
*
*   We exhibit first a table of the circuit for fixed inputs,
*   and then a table for one-step transitions.
*   The outputs under one-step transitions may be examined for
*   the presence of hazards.
*;
*   List of all possible steady inputs  ;
CL INPUT(f,fixed);
CL INPUT(t,fixed);
*   List of all possible non-hazard transition inputs ;
CL INPUT(ft,transition);
CL INPUT(tf,transition);

*   Nucleus to generate list of outputs for fixed inputs ;
CL -CIRCUIT(w) -INPUT(x,fixed) -INPUT(y,fixed) -INPUT(z,fixed)
     IO(input,a,x,b,y,c,z,output,eval(w,ipt(x,y,z)));
```

```
*   Nuclei to generate list of outputs for two fixed inputs and
*   one transition input ;
CL -CIRCUIT(w) -INPUT(x,transition) -INPUT(y,fixed) -INPUT(z,fixed)
    IO(input,a,x,b,y,c,z,output,eval(w,ipt(x,y,z)));
CL -CIRCUIT(w) -INPUT(x,fixed) -INPUT(y,transition) -INPUT(z,fixed)
    IO(input,a,x,b,y,c,z,output,eval(w,ipt(x,y,z)));
CL -CIRCUIT(w) -INPUT(x,fixed) -INPUT(y,fixed) -INPUT(z,transition)
    IO(input,a,x,b,y,c,z,output,eval(w,ipt(x,y,z)));
*
*   Circuit to test
*;
CL CIRCUIT(nandg2(nandg2(a,b),nandg2(nandg1(a),c)));
```

Among the clauses derived by the theorem prover is the following.

```
78   63 66 60 59 59   IO(input,a,ft,b,t,c,t,output,t*t);
```

This clause states that input signals of ft to the "a" wire and t to both the b and c wires produce the undesirable "hazard" behavior t*t in the output. By means of similar runs, one may test various alternative designs in searching for a design free of hazards. This concludes the Appendix.

STP: A Mechanized Logic for Specification and Verification

R. E. Shostak, Richard Schwartz, and P.M. Melliar-Smith

Computer Science Laboratory
SRI International
333 Ravenswood Avenue
Menlo Park, CA 94025 USA

1 Introduction

This paper describes a logic and proof theory that has been mechanized and successfully applied to prove nontrivial properties of a fully distributed fault-tolerant system. We believe the system is close to achieving the critical balance in man-machine interaction necessary for successful application by users other than the system developers.

STP is an implemented system supporting specification and verification of theories expressed in an extension of multisorted first-order logic. The logic includes type parameterization and type hierarchies. STP support includes syntactic checking and proof components as part of an interactive environment for developing and managing theories in the logic. In formulating each new theory, the user begins with a certain *core theory* that comprises a set of primitive types and function symbols, and extends this theory by introducing new types and symbols, together with axioms that capture the intended semantics of the new concepts. The mechanical proof component of the system is predicated on a fast, complete decision procedure for a certain syntactically characterizable subtheory. By providing aid to this component in the form of the selection of appropriate instances of axioms and lemmas, the user raises the level of competence of the prover to encompass the extended theory in its entirety. As a result of a successful proof attempt using STP, one obtains the sequence of intermediate lemmas, together with the axioms, auxiliary lemmas, and their necessary instantiations, which lead to the theorem. The system automatically keeps track of which formulas have been proved and which have not, so that the user is not forced to prove lemmas in advance of their application. The system also monitors the incremental introduction and modification of specifications to maintain soundness.

This paper is organized as follows: Section 2 discusses motivation for the form of man-machine interaction embodied by STP. Section 3 contains a formal description of the logic and the proof theory, and illustrates the description with an example. Section 4 discusses the use of STP in a large-scale effort to prove nontrivial properties of SIFT, a distributed Software-Implemented Fault-Tolerant operating system for aircraft flight control. Finally, Section 5 describes directions for further work.

2 Issues in Mechanized Verification Support

STP's design was guided to a considerable extent by our experience in attempting to formulate and reason about properties of SIFT. The following concerns were strongly influential.

This research was supported in part by NASA contract NAS1-15528, NSF grant MCS-7904081, and by AFOSR contract F49620-79-C-0099.

2.1 Property-Theoretic Specification

It is often desirable to specify program or system characteristics abstractly by stating properties possessed without defining the method of attainment. High-level system specifications represent, in effect, system *requirements*, rather than a prescription of implementation characteristics. That the specification method allow such partial specification of system properties is important; without this capability, one is forced to *overspecify* system descriptions. As a consequence, one both overconstrains possible implementation solutions and introduces spurious detail into the design specification.

2.2 Credible Specifications and Proofs

The intent of formal specification and verification is to increase one's confidence that a system will function correctly. As such, the specification and proof of system conformance must be *believable*. To produce credible specifications, the specification language must be sufficiently close to the user's conceptual understanding of the task the system is to perform. Specifications that are as long as or longer than the actual code for the system are likely to be harder to understand than the code itself.

Credibility of a verification effort requires that the end product be a proof that can be independently scrutinized and subjected to the social process. It must be possible to separate the process by which the proof was obtained from the proof itself.

2.3 Form of Verification Support

Mechanical theorem provers can be characterized by the style and level of user direction necessary to complete the proof. The spectrum of possibilities ranges from completely automatic "out to lunch" verification, where no user interaction is necessary to direct the proof to completion, to a proof checker (e.g., the FOL [Wey 80] system) where all steps are provided by the user. Between these extremes are interactive semi-automatic systems (such as LCF [Mil 79]) in which proofs are generated by a symbiosis of mechanically derived and user-provided steps. This is necessary because, in practice, theories that are sufficiently rich to be useful are usually either undecidable or have combinatorics that preclude practical decision procedures. Research in theorem proving has thus focused on methods by which the user can direct machine inference.

Mechanical deduction in most systems has taken the form of heuristic algorithms for searching large state spaces to determine the sequence of intermediate steps necessary to form a proof. Because of the difficulty in determining the ultimate success or failure of a heuristically chosen proof strategy without exhaustive search, the user is charged with the responsibility for monitoring the proof attempt and aborting an unpromising path. Where the user can determine that an inappropriate proof strategy has been chosen, he then introduces additional lemmas in an attempt to induce the system to follow a more fruitful path.

A drawback to heuristic theorem-proving attempts is that successful proof depends upon intimate knowledge of the heuristics employed. One must understand how very subtle changes in specification structure, even those that preserve semantic equivalence, can affect the direction and final outcome of the proof attempt. Lemma form becomes as important as content. In many cases the user may be aware of the proof steps necessary to justify the lemma within the supported theory, but he may be unable to suggest the lemmas in the form appropriate to guide the verification system down the proper path. This difficulty may be attributed to the inability to provide a succinct, yet complete, characterization of the heuristics employed by the theorem prover. Without this characterization, effective use of the system will depend not only on the understanding of the underlying theory, but also upon intimate knowledge of the theorem prover implementation.

Our experience has led us to believe that effective symbiosis between man and machine depends upon

a. The *predictability* of machine-supplied deduction
b. the user's ability *directly* to provide proof strategies and steps beyond the automatic deductive capability
c. machine interaction with the user in the style and level of conceptualization natural to him
d. the machine's ability to provide *responsive* deductive support to maintain continuity of user interaction.

Our proof experience indicates that the predictability of machine aid is far more important than the occasional burst of insight. Successful interaction with a theorem prover depends upon the user's having a clear picture of how the formula is deducible within the theory and when user assistance is necessary. It seems unlikely in systems supporting extensive, but incomplete, deduction that the user would succeed without this insight.

In a system involving extensive user interaction, one should not underestimate the importance of the user interface. It is crucial that all interaction be presented at the level of user input and in a natural and succinct notation. Management of information becomes a major problem during proof construction. That the user retain a clear intuitive understanding of the specifications is paramount. Techniques for aggregation of information, such as theory parameterizaton, as suggested by Goguen and Burstall [BuG 77], are extremely important. Data base management aids for organizing and retrieving theories are critical.

For the man-machine relationship to be symbiotic, machine response must keep pace with the user. The size of conceptual steps comprehensible to the user must be well matched to the computational efficiency of the theorem prover. Our experience indicates that a delay of more than on the order of one minute in machine response tends to cause loss of concentration.

2.4 Related Work

Much progress has been made in the last ten years on techniques for formal specification and verification. Early contributions to formal specification include Milner's work [Mil 67] on weak simulation and Parnas' [Par 72] on hierarchical methodology. The concepts introduced in this early work were further developed and incorporated in the HDM methodology [RoL 77, LRS 79]. The more recent research of Goguen and Burstall on Clear [BuG 77] and of Nakajima on Iota [Nak 77] introduced the notion of higher-order theories and theory parameterization.

At the same time, a great deal of research has focused on systems for mechanical verification. The earliest such systems depended strongly on heuristic-based, theorem proving strategies. The systems of King [Kin 69] and Levitt and Waldinger [LeW 75] are among these. The Boyer-Moore theorem prover [BoM 79] is one of the most striking examples of the power possible using heuristic techniques. The deductive component of STP, however, is more akin to the theorem provers of Bledsoe [Ble 74], Nelson and Oppen [OpN 78], Shostak [Sho 77], and Suzuki [Suz 75], all of which are founded on the use of decision procedures. The GYPSY system [Goo 79], the Jovial Verification System [Els 79], the Stanford Verifier [Luc 79], and the SDC system [Sch 80] are recent examples of program verification systems.

By and large, specification research has been pursued independently of work on verification. Only in the last few years has emphasis been placed on the interaction between the specification medium and the verification component. The Affirm system [Mu 80], for example, utilizes a term rewriting system, both as an algebraic specification medium and as a vehicle for mechanical proof. The system described in the current paper continues the emphasis on maintaining a close balance between the level of conceptualization supported in the specification and the level at which machine-aided deduction occurs.

3 The Logic of STP

Before presenting a formal description of the logic supported by the system, we present a simple example to give an intuitive feeling for the specification style.

We define a parameterized theory of Pairs of objects of two arbitrary type domains. We then use this theory to derive a theory of integer Intervals, represented by pairs of beginning and end points.

Figure 1 shows the specification of these theories in STP. The user declares the parameterized type PAIR.OF(T1 T2) in line 3, having previously declared type variables T1 and T2 in lines 1 and 2. The accessor operation FIRST is defined by the DS (Declare function Symbol) command in line 4 to take a value of type PAIR.OF(T1 T2) and return a value of type T1. The SECOND component accessor is analogously defined in line 5. A pair constructor MAKE.PAIR(T1 T2) is declared in line 6. Variables X and Y are declared to be of schematic types T1 and T2 (respectively) in lines 7 and 8. These declarations introduce new function and variable symbols, but attach no semantics. Lines 9 and 11 introduce two axiom schemes to define the properties of Pairs. Axiom A1 defines the accessor functions FIRST and SECOND to retrieve the first and second components (respectively) of a pair constructed by MAKE.PAIR. Axiom A2 extends the equality operation by defining two Pairs to be equal exactly when the corresponding components are equal. Equality is predefined over all domains.

INTERVAL is introduced as a subtype of PAIR.OF(INTEGER INTEGER) in line 13. Note that type variables T1 and T2 are thus both instantiated as ground type INTEGER. The subtype declaration declares Intervals to be an extension of the theory of Pairs of Integers. The type theory allows implicit type coercion from a subtype to a supertype (but not vice versa). Thus, all axioms defining Pairs of Integers are applicable to Intervals – in this case, instances of axiom schemes A1 and A2.

Lines 15 and 16 introduce derived Interval operations BEGINNING and END, defined as the selection of the first and second Pair values (respectively). The DD (Declare Definition) construct can be viewed as a means of conservative extension. Semantically, line 15 is equivalent to introducing the axiom (EQUAL (BEGINNING II) (FIRST II)). Operationally, the DD defining BEGINNING is automatically instantiated and applied as an axiom. Similarly, a MAKE.INTERVAL constructor is derived in line 20 in terms of the MAKE.PAIR operation of the supertype.

After introducing the signature for an Interval MEMBER operation in line 21, axiom A3 in line 22 begins to introduce Interval semantics. An Integer I is defined to be a Member of Interval II exactly when it lies between the beginning and ending points of the Interval. This completes our abbreviated definition of Integer Intervals.

1. (DTV T1)
2. (DTV T2)

(QUOTE "The following is a partial theory of Pairs")
3. (DT PAIR.OF (T1 T2))
4. (DS T1 FIRST ((PAIR.OF T1 T2)))
5. (DS T2 SECOND ((PAIR.OF T1 T2)))
6. (DS (PAIR.OF T1 T2) MAKE.PAIR(T1 T2))
7. (DSV T1 X)
8. (DSV T2 Y)
9. (DA A1 (AND (EQUAL X (FIRST (MAKE.PAIR X Y)))
             (EQUAL Y (SECOND (MAKE.PAIR X Y)))))
10. (DSV (PAIR.OF T1 T2) P)
11. (DSV (PAIR.OF T1 T2) P1)
12. (DA A2 (IFF (EQUAL P P1)
            (AND (EQUAL (FIRST P) (FIRST P1))
                 (EQUAL (SECOND P) (SECOND P1)))))

(QUOTE "The theory of Intervals is now derived as a subtheory of Pairs")
13. (DST INTERVAL (PAIR.OF INTEGER INTEGER))
14. (DSV INTERVAL II)
15. (DD INTEGER BEGINNING (II) (FIRST II))
16. (DD INTEGER END (II) (SECOND II))
17. (DSV INTEGER I)
18. (DSV INTEGER J)
19. (DSV INTEGER K)
20. (DD INTERVAL MAKE.INTERVAL (I J) (MAKE.PAIR I J))
21. (DS BOOL MEMBER (INTEGER INTERVAL))
22. (DA A3 (IFF (MEMBER I II)
            (AND (LESSEQP (BEGINNING II) I)
                 (LESSEQP I (END II)))))

Figure 1

3.1 Formal Description of the Language

The language of our logic is similar to that of conventional multisorted first-order logic, but provides for parameterized sorts and sort hierarchies. Before describing the structure of formulas in our logic, we need first to define the language of sort expressions (which, for reasons of conformance with the specification literature, we call *type* expressions).

3.1.1 Language of Type Expressions

The vocabulary of type expressions is very much like that of ordinary first-order terms. A theory in our logic has a countable set of *type variables*, and for each $n \geq 0$, a countable set of n-ary *type symbols*. Type symbols of degree 0 are said to be *elementary*, while those of nonzero degree are said to be *parameterized*. Every n-ary parameterized type symbol has associated with it a *parameterized type template* given by an n-tuple of (not necessarily distinct) type variables. The intended meaning of the templates will be clear shortly. A *legal type expression*, or simply *type expression* is a term recursively constructed from type variables and symbols in the following manner:

a. A type variable is a type expression.

b. An elementary type symbol is a type expression.

c. If t is an n-ary parameterized type symbol, $t_1, t_2, ..., t_n$ are type expressions such that $t_i = t_j$ whenever the ith and jth components of the type template of t are equal, then $t(t_1, t_2, ..., t_n)$ is a type expression.

Note that the template of a parameterized type symbol forces certain of the symbol arguments in a type expression to be identical. If, for example, INTEGER and REAL are elementary type symbols, U and V are type variables, and MIXEDTRIPLE is a trinary type symbol with template $<U,U,V>$, then MIXEDTRIPLE(INTEGER INTEGER REAL), MIXEDTRIPLE(REAL REAL REAL), and MIXED-TRIPLE(V V U) are all legal type expressions, but MIXEDTRIPLE(INTEGER REAL REAL) is not.

We refer to type expressions that contain type variables as *schematic types* and those that do not as *ground types*. By *type substitution*, we mean a substitution that replaces type variables by type expressions. By a *type instance* of a given type, we mean any type resulting from the application of a type substitution to the given type.

The *raison d'etre* for schematic types is to permit us to talk about many types of objects at once. For example, we may want to formulate and apply a certain property of SETs, in various contexts, to SETs of INTEGERs, SETs of FOOs, and so on. Rather than stating and proving the property separately for SET(INTEGER), SET(FOO), etc., we need only prove it about SET(U), where U is a type variable. As will be seen later, we will then be able to apply the property in the context of each specific instance of U.

In addition to a set of type variables and type symbols (and templates), each theory in our logic has associated with it a *subtype structure*, expressed as a binary relation over type expressions. The subtype relation is defined in the following way.

First, certain type symbols are designated as *subtype symbols*. Associated with each elementary subtype symbol is a ground type expression, said to be its *immediate supertype*. Associated with each parameterized symbol s is a schematic type expression t, said to be the *immediate supertype* of the type expression $s(t_1, t_2, ..., t_n)$, where $< t_1, t_2, ..., t_n >$ is the template of s. The type expression t is constrained to have exactly the same set of type variables as the set of type variables occurring in the template of s. As a further constraint, it must be possible to find a total ordering of all subtype symbols in such a way that each is junior in the ordering to every subtype symbol occurring in its associated immediate supertype. (This constraint is necessary to prevent circularity in the subtype structure, and is automatically satisfied in the mechanization by virtue of the chronological ordering of subtype declarations.) Now, the subtype relation is defined recursively as the coarsest binary relation over type expressions that:

i. contains $< s, t >$ for each elementary subtype symbol s with immediate supertype t.

ii. contains $< s(t_1, t_2, ..., t_n), t >$ for each parameterized subtype symbol s with immediate supertype t and template $< t_1, t_2, ..., t_n >$.

iii. is closed under reflexivity, transitivity, and type instantiation.

By "closed under type instantiation", we mean that if t is a subtype of t' (i.e., $< t, t' >$ is in the relation) and σ is a type substitution, then $\sigma(t)$ is a subtype of $\sigma(t')$.

Suppose, for example, that U and V are type variables, that SET and SETOFPAIRS are unary type symbols with template $<U>$, that HOMOGPAIR is a binary type symbol with template $<V,V>$, and that INTEGER, RATIONAL, and REAL are all elementary type symbols. Suppose also that INTEGER is a subtype symbol with immediate supertype RATIONAL, RATIONAL is a subtype symbol with immediate supertype REAL, and that SETOFPAIRS is a subtype symbol with immediate supertype SET(HOMOGPAIR(U U)). Then the following are true:

INTEGER is a subtype of RATIONAL and REAL

RATIONAL is a subtype of REAL

SET(INTEGER) is a subtype of SET(INTEGER)

SETOFPAIRS(V) is a subtype of SET(HOMOGPAIR(V V))

SETOFPAIRS(SET(INTEGER)) is a subtype of SET(HOMOGPAIR(SET(INTEGER),SET(INTEGER)))

SETOFPAIRS(SET(SET(U))) is a subtype of SET(HOMOGPAIR(SET(SET(U)),SET(SET(U))))

Note, however, that SET(INTEGER) is *not* a subtype of SET(REAL).

One can prove from the definitions that the subtype relation imposes a well-founded partial ordering on the type expressions of the theory. This partial ordering, moreover, is structured as a set of top-rooted trees (thinking of sons as subtypes of fathers). A type expression can have several sons, but no type expression can have two unrelated ancestors.

3.1.2 Primitive Types

Different theories in our logic can, of course, have quite different type vocabularies, supertype structures, or both. All, however, are considered to share certain primitive types. These include the elementary type BOOL and the elementary types INTEGER, RATIONAL, REAL, and NUMBER. INTEGER is a subtype symbol with immediated supertype RATIONAL, RATIONAL is a subtype symbol with immediate supertype REAL, and REAL is a subtype symbol with immediate supertype NUMBER. Neither BOOL nor NUMBER is a subtype symbol. As we will see later, these symbols are all interpreted, i.e., have *a priori* semantics in interpretations. We will see that the semantics are as one would expect, except that NUMBER, REAL, and RATIONAL are considered to have identical semantics. In addition, each theory is considered to have the type variable *T*. The inclusion of at least one type variable is necessary for defining certain primitive function symbols, such as EQUAL.

As a theoretical aside, it might be noted that BOOL is the only primitive type that is truly necessary to provide the bootstrapping power needed to define interesting theories. For once BOOL is provided, one has all the power of conventional first-order logic, and can axiomatize other concepts (such as INTEGERs). We have included the other primitive types as an important convenience. As the conventional semantics of other useful types (such as SETs, SEQUENCES, and so on) are mechanized, these types will also be considered as primitives.

3.1.3 The Language of Formulas

In conventional predicate calculus, formulas are constructed from atomic formulas and the familiar propositional and first-order connectives. The atomic formulas, in turn, are constructed from predicate letters and term expressions. All of the structure at or above the level of predicates in a first-order formula is of course Boolean, whereas all of the function symbols occurring beneath the predicate symbols are interpreted over an arbitrary nonempty set (said to be the *domain* of the interpretation).

Formulas in our logic are constructed similarly, except that the symbols occurring in terms can have arbitrary types, including type BOOL. There is therefore no reason to distinguish between "predicates" and "terms". In recognition of this point, it will be convenient simply to speak of *symbolic expressions*; formulas, in particular, will merely be symbolic expressions of type BOOL. As an abbreviation, we will sometimes simply say "expression" rather than "symbolic expression" when there is no possible confusion with type expressions. It is important to note, in this connection, that we want to draw a firm distinction between symbolic expressions and type expressions; in particular, "TYPE" is not itself a type, contrary to the viewpoint expressed in some programming languages.

These remarks having been made, we return to formal description. Beyond the vocabulary of type expressions described earlier, a theory in our logic has a countable set of *symbolic variables* (or just *variables*) $v_1, v_2, ...$, and for each integer $n \geq 0$, a countable set of *n-ary function symbols* $f_1^n, f_2^n, ...$

Associated with each variable and function symbol is a *signature*. The signature of a variable is an arbitrary type expression, said to give the *type of that variable*. The type signature of an *n*-ary function symbol is an $n + 1$-tuple of type expressions. The first component gives the *return type* of the function symbol, and the remaining *n* components the *formal argument types*. The only restriction placed on these type expressions is that for function symbols other than constants (i.e., of degree ≥ 1), *each type variable that occurs in the return type must occur in at least one of the argument types*. For example, if F is a unary function symbol with return type SET(U), then the formal argument type of F could be HOMOGPAIR(U U) but could not be HOMOGPAIR(INTEGER INTEGER). The intent here is that any ground binding of the type variables in the formal argument types should uniquely determine a ground instance of the return type. We will say that a variable or function symbol whose signature has at least one schematic type is *schematic*. The intuitive meaning is that a schematic symbol is a kind of abbreviation for an entire class of symbols, the signature of each member of which is a ground instance of the signature of the schematic symbol.

The *legal symbolic expressions*, or just *expressions* of a given theory are defined recursively as follows. We say that t is an *expression* if

i. t is a term, i.e., either a variable or of the form $f(t_1, t_2, ..., t_n)$, where f is an *n*-ary function symbol ($n \geq 0$) and each t_i is (recursively) a term, and

ii. t *typechecks*.

The only departure from predicate calculus is thus the type-checking restriction. Roughly speaking, the meaning of "typechecks" is what one would expect: that the arguments to a function symbol are of the appropriate type. Because of the presence of type variables and subtype structure, however, the exact meaning of "appropriate" needs some explanation.

Since the subtlety owes primarily to the type variables, let us first consider terms none of whose symbols is schematic. We will say that the *return type* of such a term is just the return type of the outermost function symbol (or, if the term is just a variable, the return type of the variable.) We will say that such a term typechecks if and only if the return type of each actual argument to a function symbol occurring in the term is a subtype of the corresponding formal argument type in the signature of the function symbol. For example, if QPLUS takes two RATIONALS and returns a RATIONAL, and if INTEGER is a subtype of RATIONAL is a subtype of REAL, then QPLUS(X X) typechecks if the type of X is INTEGER or RATIONAL, but not if the type of X is REAL.

The situation becomes more interesting if any of the symbols occurring in the term t is schematic. In this case, t typechecks if there is a way of instantiating the signature of each such symbol that causes t to typecheck in the sense just given for terms with no schematic symbols. For example, suppose that F is a schematic symbol with signature $<$SET(U),U,U$>$ (i.e, F takes two arguments of type U, U a type variable, and returns a SET(U)). Suppose also that I and X are variables with types INTEGER and RATIONAL, respectively. Then F(I X) typechecks, since applying the substitution {U / RATIONAL }to the signature of F (meaning that F now takes two RATIONALS and returns a SET(RATIONAL)) causes the type of each actual to be a subtype of the corresponding expected argument type.

In the case where a schematic symbol occurs more than once in a term, we permit a separate instantiation of its signature for each occurrence. For example, if G is a binary function symbol that takes a SET(BOOL) and a SET(REAL) as arguments, the term G(F(TRUE, TRUE), F(I, I)) typechecks, using the substitution {U / BOOL}for the left occurrence and {U / REAL}for the right occurrence of F.

Schematic variables and constants are treated specially. The substitutions associated with occurrences of variables and constants in an expression must agree on all type variables their signatures have in common. For example, if variables A and B are both of type SET(U), the term G(A,B) does not typecheck, since the substitutions associated with A and B must give U the same value.

Each assignment of substitutions to the symbol occurrences of a term that causes the typechecking requirement to be satisfied will be called a *syntactic interpretation* of the term. We can say, then, that a term is a legal expression if and only if it has at least one syntactic interpretation. Just as schematic symbols can be viewed as representing a class of symbols, terms involving schematic symbols can be considered as representing classes of terms, one for each syntactic interpretation.

It should be noted that the definition of "typechecks" just presented is completely nonconstructive; we have not said how to find the needed signature instances; indeed, we have not even explained how or whether it is possible to tell whether one type expression is a subtype of another. Clearly, without an effective way of testing for syntactic well-formedness, a logic is hardly of practical interest. Fortunately, both the subtype relation and the typecheck predicate are effectively (and easily) computable. We show in a forthcoming paper that the subtype relation can actually be computed using a finitely terminating canonical term rewrite system. A type expression t is a subtype of another type expression t' iff t (eventually) rewrites to t'; the canonical forms of the system are those types that are subtypes only of themselves.

We also give an algorithm for type checking that depends on the notion of *least general syntactic interpretation*. We say that one syntactic interpretation is *less general* than another if the type substituted for each type variable in the substitution corresponding to a given symbol occurrence in the first syntactic interpretation is a subtype of the corresponding substituted type in the second. (For example, the syntactic interpretation of F(I I) in which the substitution {U / INTEGER }is used for the occurrence of F is less general than that in which {U / RATIONAL }is used.) One can show that each expression has a unique least general syntactic interpretation modulo the assignment of types to the type variables in the signatures of variable and constant symbols.

3.1.4 Primitive Function Symbols

Each theory is considered to include the following function symbols as primitives. For each symbol, the return type is given first, then the function symbol followed by a list of argument types.

BOOL TRUE()

BOOL FALSE()

BOOL AND(BOOL BOOL)

BOOL OR(BOOL BOOL)

BOOL IMPLIES(BOOL BOOL)

BOOL NOT(BOOL)

BOOL IFF(BOOL BOOL)

BOOL FORALL(*T* BOOL)

BOOL EXISTS(*T* BOOL)

BOOL EQUAL(*T* *T*)

BOOL LESSEQP(NUMBER NUMBER)

BOOL LESSP(NUMBER NUMBER)

BOOL GREATEREQP(NUMBER NUMBER)

BOOL GREATERP(NUMBER NUMBER)

T IF(BOOL *T* *T*)

NUMBER PLUS(NUMBER NUMBER)

NUMBER DIFFERENCE(NUMBER NUMBER)

NUMBER MINUS(NUMBER)

NUMBER TIMES(NUMBER NUMBER)

The IFF construct shown above is interpreted as Boolean equivalence. The IF construct is the McCarthy three-placed *if* statement that is interpreted to return the value of the second or third argument, depending on whether the predicate in the first argument place is true or false, respectively.

In addition to those in the list above, there is a constant symbol for each integer and rational number. Note that arithmetic functions, such as PLUS, all take NUMBERs as arguments. Thus, if I is an INTEGER variable, and F is a monadic function symbol with argument type INTEGER, then PLUS(I I) is a legal expression, but F(PLUS(I I)) is *not*, even though the semantics of INTEGERs guarantee that the sum of two integers is always an integer. The desired effect can be obtained in practical applications by introducing a function symbol IPLUS with INTEGER arguments and return type, and endowing that symbol (using a definition) with the semantics of PLUS, restricted in integers.

The signatures of EQUAL and IF in the list above both use the primitive type variable *T*. It follows from the typecheck requirement that the types of the two actual arguments to EQUAL in a legal expresssion need not be the same; they must, however, have a common supertype. Similarly, the types of the two branch expressions of a three-placed IF construct must have a common supertype.

For the two quantifiers (FORALL and EXISTS), an additional syntactic restriction is imposed beyond the usual typecheck rule: the first argument to each of these in a legal expression must be a variable.

3.2 Semantics of Theories in the Logic

As one would hope, the meaning of formulas in the language reflects intuition. Once again, the only real departure from conventional quantification theory with equality stems from the presence of sorts.

Expressions have meaning only with respect to *interpretations*. An interpretation consists of the assignment of sets, called *domains*, to the type expressions of the theory, and of functions to the function symbols of the theory.

More specifically, an interpretation assigns to each ground type expression a nonempty set in such a way that for any two ground type expressions t and t'

i. The set associated with t is a subset of that associated with t' if and only if t is a subtype of t'.

ii. The set associated with t has a non-empty intersection with that associated with t' if and only if t and t' have a common supertype.

iii. The set of rational numbers is associated with types NUMBER, RATIONAL, and REAL, the set of integers is associated with type INTEGER, and the set of truth values (*true* and *false*) is associated with type BOOL.

Clause (i) says that the partial ordering of assigned domains under set inclusion must be isomorphic to the subtype partial ordering. Clause (iii) gives RATIONALS, INTEGERS, and BOOLS their standard interpretation, but identifies NUMBERS and REALS with rationals. REALs thus acquire the same interpretation they are given in most programming languages in recognition of the limitations of machine representation.

An intepretation also assigns to each n-ary nonschematic function symbol an n-ary function whose signature is obtained from that of the function symbol by replacing each type expression with its assigned domain. Constant symbols, in particular, are assigned elements from the domain associated with their return types. Primitive nonschematic function symbols are given their standard meanings.

Every schematic function symbol is assigned a multiplicity of functions – one for each ground instance of its signature. (Intuitively, one can think of schematic function symbols as abbreviations for a class of symbols.) The signature of the assigned function corresponding to a given signature instance is that obtained by replacing each type expression in the instance by its assigned domain. Once again, primitive symbols, including the quantifiers, are given their usual semantics. Note that each quantified variable ranges over a single domain.

Now, each closed expression (closed in the sense of having no unbound variables) takes a meaning, or *valuation*, with respect to a given interpretation I and a given assignment A of ground type expressions to the type variables occurring in the signatures of variable and constant symbols. For expressions involving only nonschematic symbols, the valuation is defined recursively in the way one would expect: the valuation of a constant is just the domain element assigned to it by I, and the valuation of a nonconstant is obtained by recursively applying the function assigned to its outermost symbol to the valuations of its arguments. For expressions involving schematic symbols, the valuation is defined as that of the least general syntactic interpretation (as defined earlier) modulo the assignment A. (Note that the assignment A is analogous to the assignment of domain values to free variables in interpretations of predicate calculus.)

The notions of validity and satisfiability are now defined in the usual way. A formula (i.e., an expression of type BOOL) is said to be *valid* if and only if its universal closure has valuation *true* in all interpretations; it is *satisfiable* if its existential closure is *true* in at least one interpretation.

3.3 The Proof Theory and its Mechanization

The connection between the model-theoretic semantics described in the previous section and the proof mechanism used in our system is made by a generalized form of the Skolem-Herbrand-Gödel theorem. Herbrandian semi-decision procedures, of course, have long been the most popular means of mechanizing quantified logics. Because of the extensions to ordinary quantification theory provided by our language, a generalized form of the theorem must be used and a mechanical proof procedure formulated on the basis of this generalization.

Before outlining the theorem and the derived procedure, it will be helpful to describe the deductive mechanism that lies at the heart of the prover. This mechanism consists of an efficient implementation of a decision procedure for unquantified formulas in an extension of Presburger arithmetic. The theory includes the usual propositional connectives: equality, rational and integer variables and constants, the arithmetic relations $<, \leq, >, \geq$, and addition and multiplication. Uninterpreted predicate and function symbols of type integer and rational are also permitted. The decision procedure is complete for the subtheory of this theory having no integer variables or function symbols, and containing no nonlinear use of multiplication. The procedure is sound, of course, for the entire theory, and is able to prove the vast majority of formulas involving integer constructs that are actually encountered in practice. The speed of proof, moreover, is fairly impressive: theorems occupying several pages in the SIFT effort were usually proved in well under a minute. (Indeed, as we noted elsewhere, such speed has proven to be essential to our methodology.) It should be noted, however, that this decision theory does not include *quantified* formulas, nor does it support function symbols of user-defined types.

The proof procedure derived from our generalization of the S.-H.-G. theorem can be viewed as a means of reducing the proof of formulas in the typed first-order theory that the user has formulated to the automatic proof of formulas in the underlying unquantified decision theory we have just described. Informally speaking, the S.-H.-G. theorem states that a formula of predicate calculus is valid if and only if some disjunction of instances of its validity Skolem form is tautological. The instances must replace variables in the Skolem form with terms in the Herbrand Universe of the formula. The generalization of the theorem to deal with formulas in our typed language is stated similarly, but requires that

i. Each variable be instantiated with an expression whose type is a subtype of that of the variable, and

ii. Each instance typechecks.

The generalization states that the given formula is valid according to the semantics defined in the last section if and only if some disjunction of instances satisfying (i) and (ii) above is *true, considered as a formula in the underlying decision theory*, where it is understood that symbols of a subtype of BOOL are considered to be of type BOOL, symbols of a subtype of INTEGER are considered to be of type INTEGER, and all other symbols are considered to be of type RATIONAL. In effect, the generalization holds that once the instances have been typechecked, all of the type information other than that distinguishing RATIONALs, INTEGERs, and BOOLs can be stripped away.

The theorem immediately gives rise to a proof procedure: with assistance from the user, appropriate instances of the Skolem form of the theorem to be proved (together with instances of any axioms or lemmas needed to prove it) are formulated, disjoined, and submitted to the underlying decision procedure. A detailed description of the decision procedure is given in [Sho 82].

Operationally , the user is saved from any details of this process other than selecting appropriate instances. In particular, the process of Skolemization is completely transparent to him, as are the disjunction of instances and the submission of this disjunction to the underlying decision mechanism. The user is thus free to reason exclusively at the level of his first-order typed theory.

As an illustration, let us return to the example from the theory of Intervals presented in Section 2.1.

We prove the simple theorem that every interval with an end point greater than or equal to the beginning point contains at least one number as a member. Figure 2 shows the expression and proof of this theorem. Line 23 encodes the theorem within the theory of Intervals previously defined. As expected, one proves a formula of the form $\exists x \; P(x)$ by demonstrating some value v such that $P(v)$. In this case, the user must observe that, from (GREATEREQP J I) in the antecedent, axiom A3 defining Interval membership, and the definition given in A1 of the Interval constructor MAKE.INTERVAL, one can determine that the beginning interval value I satisfies the formula (MEMBER I (MAKE.INTERVAL I J)). In line 24, the user invokes the PR command to construct this proof. The system then prompts the user for the needed substitutions.

---

23. (DF THEOREM (IMPLIES (GREATEREQP J I)
                       (EXISTS K (MEMBER K (MAKE.INTERVAL I J)))))

24. (PR THEOREM
    A1
    A3)

Want instance for THEOREM? Y
  K/ I
Want instance for A1? Y
  Y/ J
  X/ I
Want instance for A3? Y
  II/ (MAKE.INTERVAL I J)
  I/

———-Proving———-
160 conses
.2 seconds
Proved

Figure 2

---

4 The Proof of SIFT

SIFT (Software-Implemented Fault Tolerance)[Wen 78] is a reliable aircraft flight control computer system. SIFT uses five to eight Bendix BDX930 computers, each equipped with its own private storage. A broadcast communication mechanism allows each processor to transmit its results to a buffer area in each processor. The design is fully distributed, with no common buses, no common clocks, no common interrupt or synchronization registers, no common storage, and no physical means of isolating a faulty processor. The SIFT processors (physically) share only the ability to communicate with one another.

In SIFT, fault masking, detection, and reconfiguration are all managed by software. Safety-critical tasks are replicated on three or more processors, with all processors voting on the results of each redundant computation. A Global Executive task, which is itself replicated, is responsible for fault diagnosis on the basis of error reports from the voting software, and for selecting a new configuration excluding the processors deemed to be faulty. The result of this reconfiguration is that tasks are shifted from faulty processors to those still working. Every processor votes on the results of the Global Executive and adjusts its task schedule accordingly.

SIFT's processors run asynchronously; each contains its own private clock. The software must maintain a loose synchronization to within approximately 50 microseconds, and each processor runs a task periodically to resynchronize its clock with those of the other processors in the configuration. Care was taken in the design to ensure that, even under fault conditions, all working processors retain synchronization and remain consistent in their schedule and configuration.

4.1 The Design Hierarchy of SIFT

The problem of specification credibility in the proof of SIFT is addressed through the use of hierarchical design specification and verification. This approach allows incremental introduction and verification of design aspects – making a step-by-step connection between the high-level, abstract view of the system to the detailed control and data structures employed in the implementation. Figure 3 illustrates the SIFT design hierarchy. At present, the STP system does not provide specific mechanical support for the hierarchical specification structure; we discuss future work in this direction in Section 4.

The IO Specification, the most abstract functional description of the system, asserts that, *in a safe configuration*, the result of a task computation will be the effect of applying its mathematical function to the results of its designated set of input tasks, and that this result will be obtained within a real-time constraint. Each task of the system is defined to have been performed correctly, with no specification of how this is achieved. The model has no concept of processor (thus no representation of replication of tasks or voting on results), and of course no representation of asynchrony among processors. The specification of this model contains only 8 axioms.

The Replication Specification elaborates upon the IO Specification by introducing the concept of processor, and can therefore describe the replication of tasks and their allocation to processors, voting on the results of these replicated tasks, and reconfiguring to accommodate faulty processors. The specification defines the results of a task instance on a *working* processor based on voted inputs, without defining any schedule of execution or processor communication. This model is expressed in terms of a global system state and system time.

The Broadcast Specification develops the design into a fully distributed system in which each processor has access only to local information. Each processor has a local clock and a broadcast communication interface and buffers. The asynchrony among processors and its effect upon communication is modeled. The specification explicitly defines each processor's independent information about the configuration and the appropriate schedule of activities. The schedule of activities defines the sequence of task executions and votes necessary to generate task results within the required computation window. The Broadcast Specification is the lowest level description of the complete multiprocessor SIFT *system*.

The PrePost Specification consists of specifications for the operating system for a single processor. The specification, in terms of pre-condition/post-condition pairs, facilitates the use of sequential proof techniques to prove properties of the Pascal-based operating system as a sequential program. These specifications are very close to the Pascal programs, and essentially require the programs to "do what they do".

The Reliability Analysis is a conventional discrete semi-Markov analysis that calculates the probability that the system reaches an unsafe configuration from the rates of solid and transient faults and from the reconfiguration rates. Neither this Reliability Analysis nor the other Fault and Error Models will be described here.

A more detailed presentation of the SIFT specifications and their verification can be found in [MeS 82].

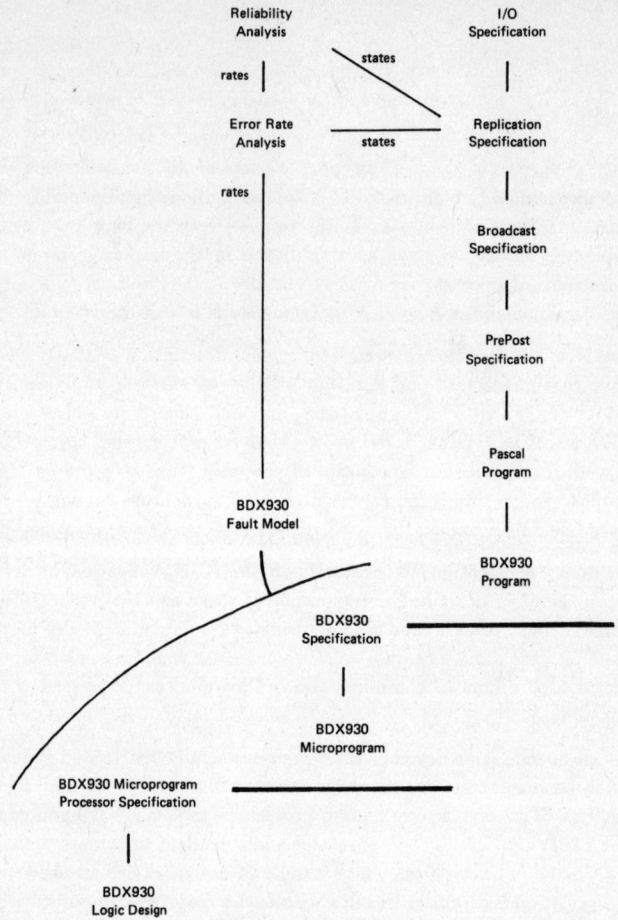

FIGURE 3

4.2 The Proof of the SIFT Design

Hierarchical proof of design involves axiomatically defining a mapping from functions and predicate symbols of a level of the design to terms of the level below. One then proves each axiom of the higher level as a theorem in the theory of the level below.

The most substantial portion of the verification of the IO Specification involved proof of the axioms that define the results of tasks in a safe configuration. To derive these from the Replication level involved a demonstration that task replication and majority voting suffice to mask errors in a safe configuration. To do so required approximately 22 proofs, with an average of 5 premises necessary per proof, and 106 instantiations of axioms and lemmas used overall.

The proof of the relationship between the Replication Specification and the Broadcast Specification was more challenging. This proof required demonstrating the consistency of information present in each working processor (the set of processors still in the configuration, in particular), of the correct schedule to be used, and of the results of past task iterations. Furthermore, the proof required showing that the various processors, operating independently and asynchronously with only local information, can communicate with each other without mutual interference, and can cooperate to produce a single global system defined by the Replication Specification. It was also necessary to show that the task schedules were such that the task results are always available in other processors when required. The derivation of the Replication axioms involved 56 proofs, with an average of 7 premises each, and 410 instantiations of axioms and lemmas overall.

The proof of the relationship between the Broadcast Specification and the PrePost Specification was easier. Most of the interest centered on the mapping between the properties of the changing values of variables in the Pascal system and the properties of the Broadcast model's more abstract state functions which are explicitly parameterized by time and processor. A futher complication concerned mapping the functional representation of data structures in the Broadcast model to the (finite) Pascal data structures. Derivation of the necessary Broadcast axioms involved 17 proofs, with an average of 9 premises each, and 148 instantiations overall. The proof of the Pascal programs from the PrePost Model specifications used conventional verification condition generation.

5 Future Work

As a result of experience with both design and code proof in the SIFT effort, several improvements and changes to STP are contemplated. We intend to implement decision procedures for an expanded set of primitive theories – all syntactically characterizable. We envision doing this at least for a large fragment of set theory (following the work of Ferro and Schwartz [FeS 80]), sequences, and tuples. Mechanical support for induction schemes is also needed. The user currently can introduce "induction" axioms, but with no implied semantics. We also anticipate providing direct mechanical support for hierarchical development.

Improvements to the user interface will take several forms. In the current system, the user is forced to specify theories in the abstract syntax of Lisp. For all but the heartiest of users, pure prefix syntax is clearly unsatisfactory. We are currently defining an external, enriched specification language and a corresponding reduction to the current type theory. The language will contain explicit support for hierarchical specification, theory encapsulation, scoping, and implicit state; it will support SPECIAL-like state-machine specifications [LRS 79] as a sublanguage.

A graphics interface is contemplated to assist in constructing and manipulating theories and libraries of theories. In addition, our experience with code proof indicates that forcing the user to reason at the level of formulas mechanically generated by a VCG is doomed to failure. Reasoning at this level is antithetical to our philosophy of man-machine discourse at the level of user conceptualization – in this case, the level of the program structure. We intend over the next year to experiment with user interaction at the level of path (and subpath) assertions within the program structure, hoping that this will achieve in the program domain what was gained in the present system by hiding the internal first-order/ground formula mapping from the user in present design proof.

Finally, we intend to supplant after-the-fact mechanical support for complete proof construction with incremental proof construction. Our goal is to replace, insofar as possible, explicit user instantiation of symbols with a formal language in which the user can incrementally construct and apply a high-level proof strategy. With this approach, we believe we have a more efficient proof strategy while requiring less effort on the user's part. In this way, we hope to preserve the man-machine balance as we increase the user's capability to formulate and reason about larger conceptual steps within the theory.

6 Acknowledgments

We want to acknowledge the contributions of Dwight Hare to the development of STP and the proof of SIFT, and of Leslie Lamport to the early formulations of the SIFT specifications.

References

[Ble 74] Bledsoe, W.W., "The Sup-inf method in Presburger arithmetic", Memo ATP-18, Mathematics Dept., Univ. of Texas, Austin, Texas, Dec. 1974.

[BoM 79] Boyer, R., J Moore, "A computational logic", Academic Press, 1979.

[BuG 77] Burstall, R., J. Goguen, "Putting theories together to make specifications", Proc. IJCAI, August 1977.

[Els 79] Elspas, B., et al. "A Jovial Verifier", SRI International, June, 1979.

[FeS 80] Ferro, A., E. Omodeo, J. Schwartz, "Decision procedures for some fragments of set theory", 5th Conf on Automated Deduction, July 1980.

[Goo 79] Good, D., R. M. Cohen, J. Keeton-Williams, "Principles of proving concurrent programs in Gypsy", *Proc 6th POPL*, 1979.

[Kin 69] King, J., "A program verifier", Ph.D. thesis, CMU, 1969.

[LeW 75] Levitt, K., R. Waldinger, "Reasoning about programs", *AI Journal* 5, 1974.

[LRS 79] Levitt, K., L. Robinson, B. Silverberg, "The HDM handbook", SRI International, 1979.

[Luc 79] Luckham, D., N. Suzuki, "Verification of array, record, and pointer operations in Pascal", *TOPLAS*, Oct, 1979.

[MeS 82] Melliar-Smith, P. M., R. L. Schwartz, "Formal specification and mechanical verification of SIFT: a fault-tolerant flight control system", *IEEE Transactions on Computers*, July 1982.

[Mil 71] Milner, R., "An algebraic definition of simulation between programs, CS 205, Stanford, 1971.

[Mil 79] Milner, R., "LCF: a way of doing proofs with a machine", *Proc 8th MFCS Symp*, 1979.

[Mus 80] Musser, D., "Abstract data type specification in the Affirm system", *IEEE TSE*, Jan. 1980.

[Nak 77]   Nakajima, R., M. Honda, H. Nakahara, "Describing and verifying programs with abstract data types", *Formal Description of Programming Concepts*, North Holland, 1977.

[OpN 78]   Oppen, D., G. Nelson, "A simplifier based on efficient decision algorithms", Proceedings of Fifth POPL, Tucson, Arizona, Jan. 1978.

[Par 72]   Parnas, D., "A technique for software module specification with examples", *CACM*, May 1972.

[RoL 77]   Robinson, L., K. Levitt, "Proof techniques for hierarchically structured programs", *CACM*, April 1977.

[Sch 80]   Schorre, V., J. Stein, "The interactive theorem prover (ITP) user manual", TM-6889/000-/01, SDC, 1980.

[Sho 77]   Shostak, R., "A practical decision procedure for arithmetic with function symbols", *JACM*, April 1979.

[Sho 82]   Shostak, R., "Deciding Combinations of Theories", *Proceedings of the Sixth Conference on Automated Deduction*, June 1982.

[Suz 75]   Suzuki, N., "Verifying programs by algebraic and logical reduction", Proc. Int. Conf. on Reliable Software, Los Angeles, 1975.

[Wen 78]   Wensley, J., et al., "SIFT: Design and Analysis of a Fault-tolerant Computer for Aircraft Control", *Proc IEEE*, Vol. 66, No. 10, Oct. 1978.

[Wey 80]   Weyhrauch, R., "Prolegamena to a theory of mechanized formal reasoning", *AI Journal*, 1980.

A Look at TPS

Dale A. Miller
Department of Mathematics
Carnegie-Mellon University
Pittsburgh, PA 15213

Eve Longini Cohen
Jet Propulsion Laboratory
California Institute of Technology
Pasadena, CA 91109

Peter B. Andrews
Department of Mathematics
Carnegie-Mellon University
Pittsburgh, PA 15213

Abstract

Certain aspects of the theorem proving system TPS are described. Type theory with λ-abstraction has been chosen as the logical language of TPS so that statements from many fields of mathematics and other disciplines can be expressed in terms accessible to the system.

Considerable effort has been devoted to making TPS a useful research tool with which interaction is efficient and convenient. Numerous special characters are available on the terminals and printer used by TPS, and many of the traditional notations of mathematics and logic can be used in interactions with the system. When constructing a proof interactively, the user needs to specify only essential information, and can often construct needed wffs from others already present with the aid of a flexible editor for wffs.

TPS constructs proofs in natural deduction style and as p-acceptable matings. Proofs in the latter style can be automatically converted to the former. TPS can be used in a mixture of interactive and automatic modes, so that human input need be given only at critical points in the proof.

The implementations of the algorithms which search for matings and perform higher order unification are described, along with some associated search heuristics. One heuristic used in the search for matings, of special value for dealing with wffs containing equality, considers in close sequence vertical paths on which mating processes are likely to interact. A very useful heuristic for pruning unification trees involves deleting nodes subsumed by other nodes. It is shown how the unification procedure deals with unification problems which keep growing as the search for a mating progresses.

It has been found that although unification of literals is more complicated for higher order logic than for first order logic, this is not a major source of difficulty.

An example is given to show how TPS constructs a proof.

This work is supported by NSF Grants MCS78-01462 and MCS81-02870.

§1 Introduction

In this paper we shall describe certain features of an automated theorem-proving system called *TPS* which is under development at Carnegie-Mellon University, and which implements the ideas in [1] and [2], with which we assume familiarity.

§2 Long term goals and design philosophy

There is abundant evidence of the need for new intellectual tools to aid mankind in dealing with the complexities of the modern world. Among these tools must surely be some designed to facilitate logical reasoning. Since the construction of proofs is one of the best understood aspects of logic, work on automated theorem proving provides a natural starting point for research on the more general problem of automating logical reasoning.

Naturally, one must expect that the early stages of work on automating reasoning will produce systems which can be successful only on fairly simple problems. Nevertheless, in designing such a system it is helpful to keep in mind the features and capabilities which one would like it to have in the long run, and to provide an adequate framework for the long term developments which are anticipated. One of the important decisions which must be made at an early stage concerns the language of logic to be used.

As one envisions the sophisticated question-answering systems of the future, it seems clear that scientific knowledge at various levels of abstraction will gradually be incorporated into them. Since mathematics is the language of science, artificial intelligence systems which use logic in a sophisticated way will surely come to be regarded as severely limited if they cannot handle mathematics on a conceptual as well as a computational level. Even if mathematics is not used explicitly, the sort of logical abilities needed for mathematics are vital for many sophisticated applications of logic. For this reason, one of the objectives in the development of *TPS* is to obtain a system in which mathematics can be expressed naturally and in which mathematical reasoning can be (partially) automated.

Of course, mathematics provides an ideal environment for research on logical reasoning. Mathematical statements are customarily expressed in ways which make their logical structure very clear, and mathematical arguments are customarily carried out with an attention to logical detail unparalleled in any other field. The theorems of mathematics exhibit an enormous variety of subject matter and complexity. Thus mathematics provides a wealth of ready-made examples which can be used in the study of ways to increase the efficiency of a computer's deductive apparatus.

In order to provide for a wide variety of possible future applications, *TPS* needs a universal language of logic in which virtually any mathematical statement can be expressed. For reasons which we shall discuss only briefly here, a formulation of type theory with λ-notation due to Church [3] has been chosen as the logical language used by *TPS*. Mathematical statements can be translated rather directly into this language, and many sets and functions have names in the language, so no axioms concerning set existence (except an Axiom of Infinity and the Axiom of Choice, where appropriate) are needed. Of course, to use type theory one must assign a type to each mathematical entity, but mathematicians naturally do make intuitive distinctions between different types of mathematical entities. Indeed, type symbols provide important syntactic clues which enable an automated theorem-prover automatically to eliminate from consideration expressions which would be permitted in languages such as axiomatic set theory, but not in type theory, and which a working mathematician would reject almost immediately as meaningless. This is particularly valuable for the unification algorithm, where many inappropriate substitutions are avoided via type considerations.

Actually, it has been found that unification algorithms for type theory ([4], [6]), which combine λ-conversion with substitution, are powerful tools which enable one to find proofs of certain intricate theorems (such as Cantor's Theorem for Sets) which are hard to prove automatically by other means. The availability of these algorithms constitutes a strong argument for the use of type theory.

The basic logical approach to theorem proving underlying the design of *TPS* is discussed in [1] and [2], where the formal discussion is limited to the problem of proving theorems of first order logic. While *TPS* is in principle logically complete in automatic mode only for proving theorems of first order logic at present, it has a number of facilities for handling wffs of higher order logic, such as the interactive construction of proofs in natural deduction style, and unification. While it may seem premature to be concerned with automating type theory when there is still so much to be done to develop systems which can handle first order logic satisfactorily, the use of type theory has proved to be very advantageous even at the current stage of research, since there are numerous theorems, such as Cantor's Theorem, which can be expressed in type theory much more naturally than in first order logic even though their proofs involve essentially only the techniques of first order logic combined with higher order unification.

TPS is both a system under development and a research tool. The program is written in

LISP and uses about 18,000 lines of code. In order to maximize its usefulness, we have tried to make it serve those who use it, rather than asking them to adapt to its requirements. Thus, we have devoted considerable effort to making the computer accept input convenient for people to provide, and provide output convenient for people to use. Much of the traditional notation of mathematics and logic can be used in communicating with *TPS*. In order to achieve this, *TPS* had to be given the ability to handle the many special symbols which are used in mathematics and logic. This required a sizable amount of hardware and software support.

§3 Support for special characters

TPS is written in CMU LISP, a descendant of UCI LISP. The system uses Concept 100 terminals and a Xerox Dover printer, both of which support multiple character sets. Two auxiliary sets of 128 characters were developed by us for the terminals. These sets contain Greek, script and boldface letters, plus many mathematical and logical symbols. Some characters are available as subscripts and superscripts. The Dover is capable of printing all these characters and many more. In order to use these characters, several improvements to the standard LISP input and output facilities have been made. For example, an input facility called RdC was added to LISP so that special characters could be entered and translated to a representation suitable for LISP.

The Concept terminals can also define windows, read and write the cursor position, and insert and delete both characters and lines. These capabilities permitted us to write two text editors for use in *TPS*. The *scratch pad* editor is a very simple editor relying on the terminal's local screen editing. A second, more powerful editor called VIDI is a screen-type editor, much like EMACS, which supports special characters, and permits such text to be stored and retrieved from disk storage.

TPS can also produce hard copy images of text containing special characters by preparing a manuscript file for the document compiler SCRIBE [7]. This typesetting language permits access to numerous character sets and many formating features. The result of compiling such files is a second file which can be printed on the Dover. Files can be made which permit the user's interaction with *TPS* to be redisplayed on the terminal screen.

§4 The TPS interface with the user

TPS has two representations for wffs. The internal representation is a linked list whose structure mirrors the structure of the wffs found in [3]. This representation is very convenient for computing, but it does not resemble conventional notation of mathematical logic because of its syntax and lack of special characters. The *TPS* user deals with an external representation of wffs which is much more convenient to read and type.

The external representation uses both brackets and Church's dot convention to specify scopes of connectives. Function symbols and predicates can be declared to be infix instead of simply prefix, which permits a more natural rendering of such symbols as $\subseteq$, $+$, and $\equiv$. Symbols are handled by first assigning each character a name and then giving each symbol a list of character names, called its *face*. For example, the quantifier $\exists_1$ has the internal name EXISTS1, and this has the face (EXISTS SUB1), for the two characters used to represent this symbol. *TPS* can both print this symbol with the correct characters and find its occurrences within text typed by the user. Text containing special characters can be input by both RdC and the VIDI editor. The internal name of a symbol is used if a device is being used on which special characters are not available.

The fact that wffs are represented within higher order logic proves quite useful, even when we restrict our attention to first order wffs. Abbreviations can be defined very conveniently by using λ-abstraction. For example, SUBSET is defined by the wff [$\lambda P_{o\alpha}$ $\lambda Q_{o\alpha}$ $\forall x_\alpha$. Px $\supset$ Qx]. Thus, instantiating SUBSET is done simply by substituting this term for the occurrences of $\subseteq$ in the wff, and then doing λ-contractions of the expression. Such definitions can be stored in libraries as polymorphic type expressions. Their actual type is determined from the context. Individual users of *TPS* can have their own libraries.

A structure editor is available which permits the user to edit the internal representation of a wff. In this editor the user can issue commands to move to any subexpression of the original wff and change it by placing it into various normal forms, changing bound variables, extracting subformulas, instantiating abbreviations, etc. Although the user is editing the internal representation, it is the external representation which is printed out to the user at each level of the editing. This editor also has a convenient way of using the two text editors on subformulas.

The user can specify a wff in many different ways. The user can refer to other wffs or their subformulas, or the result of processing other wffs. For example, all of the following ASSERT commands will place into the current proof outline a line numbered 4 containing the wff specified by the second argument:

```
>assert 4 "FORALL X(OA) . [POWERSET . X INTERSECT Y(OA)] SUBSET
           . POWERSET Y"
>assert 4 Pdc
RdC>∀X_{oα} . [P . X ∩ Y_{oα}] ⊆ . P Y
>assert 4  pad
>assert 4  thm1.
>assert 4  (neg (skolem thm1))
>assert 4  (loc 2)
```

In the first command, the wff is entered as a string without special characters. In the second command, the user requests the RdC prompt and then enters the wff. The third command places the cursor in the scratch pad editor and the user then either edits a wff already present in the pad or types another one. When the contents of the pad is the desired wff, pressing the appropriate key on the Concept will send the pad's contents to the wff parser. In the fourth and fifth commands a previously defined wff called THM1 is used. The last command calls the LOC function, which is much like the structure editor but is simply used to locate a subformula of a given wff; in this case, it returns a subexpression of the wff in line 2 of the proof.

TPS can print wffs in several different fashions. For example, they can be pretty-printed so that the arguments of infix operators are indented appropriately. Also, TPS can produce two-dimensional diagrams of wffs in nnf such as are found in [2]. When these are too big to fit on one sheet of paper, they are laid out so that several sheets can be pasted together to make one large diagram.

The user of TPS can make many choices about such matters as the way wffs are displayed, the amount of information that is printed out during the search for a proof, the search heuristics and parameters that are used, and the degree to which TPS functions in automatic or interactive mode. These choices are made by changing the values of certain variables called flags. A command called REVIEW provides help for the user who wishes to recall what flags are available, what they mean, and what their current values are.

TPS also has help facilities to remind the user of the correct format for various commands, the commands available in the editor, etc.

The user can cause a *status line* to be displayed on the terminal screen. This shows which lines of the proof are sponsored by each planned line, and is automatically updated as the proof is constructed.

§5 Organization of the proof process

Proofs in *TPS* can be constructed in natural deduction format or as *p-acceptable matings* (see [2]), and both styles of proof can be constructed automatically or interactively. When constructing natural deduction proofs, *TPS* builds and progressively improves *proof outlines* (see [1]), which are fragments of proofs in which certain lines called *planned lines* are not yet justified. The transformation rules described in [1] are implemented in *TPS* as commands which can be used interactively to fill out the proof. Since the context often suggests how these commands should be applied, by setting appropriate flags and letting *TPS* compute default arguments for commands, the user can let this process proceed automatically until decisions are required. When input is needed, wffs can be specified with minimal effort as described above, and *TPS* checks for errors and allows corrections.

Instead of proceeding interactively, one can send the wff to be proved in a planned line to the mating program. When a p-acceptable mating is found, it is converted into a *plan* and used to construct a natural deduction proof of the planned line as described in [1]. This process is illustrated with an example in §8. Even when a plan is being used, the user can intervene to make decisions concerning the order in which rules are to be applied and thus control the structure of the proof.

Thus, the user can construct some parts of the proof interactively or semi-automatically and others automatically. Plans as well as proofs can be constructed interactively.

The mating program can also be run as an independent entity.

All of the rules of inference of natural deduction mentioned in §2 of [1] are available as *deducing rules* (see [1]). In addition, there is an ASSERT command which can be used to insert into a proof a theorem which the interactive user obtains from a library of theorems or simply asserts. Deducing rules are used to construct proofs down from the top, while *planning rules* essentially provide for working backwards to build a proof up from the bottom. By combining these rules *TPS* or the interactive user can work both forward and backward. This facilitates the construction of the proof and the implementation of heuristics to control its style and structure. Of course, some of the deducing rules (such as Universal Generalization) are used only in interactive mode, since in automatic mode *TPS* inserts the same lines into the proof by the use of appropriate planning rules.

At least one transformation rule is available to deal with every possible form of an active or planned line which is not a literal. Thus, a number of special forms of Rule P (see [1]) are available, and are invoked when appropriate. Examples of these are to infer *A* and *B* from $[A \wedge B]$, to infer A from $\sim\sim A$, and to push in negations. At present, deduced lines of the form $[A \supset B]$ are replaced by $[\sim A \vee B]$, and disjuctions are broken into cases using the rule P-Cases of [1], though this sometimes yields inelegant proofs. When all quantifiers and definitions have been eliminated from a planned line and the active lines which it sponsors, TPS invokes the unrestricted Rule P to infer the planned line, thus providing the essential link between the top and bottom parts of the proof of the planned line. It is anticipated that more elegant proofs will be constructed in automatic mode as we learn to make more sophisticated use of the information in plans, especially the matings.

§6 Mating search

The mating program first processes the wff to be proved by removing all abbreviations, negating, miniscoping (when appropriate), skolemizing, and transforming to negation normal form. The final wff contains only universal quantifiers, conjunctions, disjunctions, and negations with atomic scopes. TPS then attempts to derive a contradiction from this wff by searching for a p-acceptable mating for each top-level disjunct in turn. As in [2], we call any set of pairs of literals a *potential mating*. A vertical path is *fixed* if it contains a pair of mated literals. A potential mating is *complete* if each vertical path is fixed, otherwise it is *partial*. A *p-acceptable mating* is a complete potential mating for which a substitution exists which makes mated literals complementary. Thus, a contradiction is derived, and by Herbrand's theorem the original wff is valid.

The first step in the mating search is to create a connection graph, which stores information useful in constructing the mating. As in a strictly first order connection graph, an arc is directed from literal A to literal B whenever $\sim A$ and B are unifiable, *i.e.*, whenever A and B are *c-unifiable*. Type theory adds additional complications to the construction of the graph:

- It is not *a priori* obvious which pairs of literals can be c-unified, or which is to be the "negative" literal. Because the substitution term may contain $\sim$, it may be possible to c-unify A and B when either, neither, or both of the literals is a negation, or to c-unify the literals in both orientations. Of course, a path can be fixed by mating the literals in either orientation, but different orientations produce different substitutions.

- Unification in type theory is in general undecidable, so the algorithm may not terminate. We deal with this problem by limiting the depth of unification search at the time the graph is created.

- Even when a unifier exists, there may be no most general unifier of two wffs in type theory. In this case there will be branching in the unification tree. However, during the construction of the connection graph we pursue only those parts of the unification problem requiring no branching.

Each arc in the connection graph is labeled with a partial c-unifier of its two literals.

The next part of the process is to choose a vertical path in the wff containing no mated pair, choose from the connection graph a pair of literals which fixes that path, add it to the mating under construction, partially unify (still allowing no branching in the unification tree) the set of disagreement pairs representing the partial mating, and then iterate the above process until a complete potential mating is constructed, backtracking as required by failures of mating or unification. The same partial mating may arise several times in the search process, but *TPS* considers it only once. When a completed mating is obtained, the full power of unification is allowed, occurring in parallel with the construction of a new mating. After all potential matings of a wff have been considered, *TPS* replicates certain quantifiers and their scopes, and seeks a mating for the enlarged wff. Thus it alternates between the construction of a single mating, and unification work on that partial mating and on any number of previously constructed completed matings, which may be from any number of replication levels. The user can control the relative effort devoted to the mating and unification procedures.

The search for a mating may cause certain quantifiers to be duplicated inappropriately, so after a p-acceptable mating has been found, *TPS* simplifies the replication scheme if possible, and makes appropriate adjustments of the substitutions and mating. Such simplification avoids certain ambiguities which might otherwise arise when Skolem functions are eliminated from substitution terms in the process of constructing a plan from the mating. It also eliminates redundancies which would otherwise occur in proofs constructed from the mating.

We have tried a number of different heuristics for the choice of an unfixed path and a pair to fix it. *TPS* tries to fix first the paths where the choices are most constrained. The default procedure for choosing paths is to focus on those with the smallest number of matable pairs with respect to the current partial mating.

Heuristics are also available to deal with certain special situations, such as those in which equality occurs. Equality is defined as $[\lambda x_\alpha \lambda y_\alpha \forall Q_{o\alpha} . Qx \supset Qy]$, and a positive equality statement occurs as $\forall Q[\sim QX \vee QY]$. When one mates one of these literals, one constrains the possible substitutions for Q and limits the possible mates for the other literal. This motivates the following heuristic.

Suppose that the wff for which one is seeking a mating contains a subformula [K ∨ L], where K and L are conjunctions of literals, that P is a path which passes through K, that A and B are literals on P, with A in K but B not in K, and that the mating process has tentatively decided to mate A with B. Of course, there is another path P' which is like P but which passes through L instead of K. P' must also be given a pair of mated literals, and we might as well assume that at least one of these is to be in L, since otherwise the mated pair which fixes P' also fixes P, and the mating of A with B is unnecessary. If each literal in L has a variable in common with A or with B, then the substitution required to mate A with B may constrain the possible ways of fixing P'. Indeed, if this constraint is so strong that there is no way to fix P', then one should not mate A with B. This leads to the following heuristic: having fixed P by mating A with B, consider next all variant paths P' which satisfy the given description, and see if they can each be fixed. This may require repeated calls on this heuristic. In this way, one may be able to reject bad partial matings sooner rather than later, which is the essence of good heuristic search.

§7 Unification search

The unification scheme used by TPS is that described by Gérard Huet in [4], with only minor modifications. Each mating is represented by a list of pairs of wffs which require unification. We call each such pair a *disagreement pair* (*dpair*), and the whole list a *unification node*. Each wff of type theory is of the form $[\lambda w^1,...,\lambda w^n . hE^1,...,E^m]$, where n, m $\geq$ 0. We call h the *head* of the wff, $E^1,...,E^m$ its *arguments*, and $\{w^1,...,w^n\}$ its *binding*. We call a wff *rigid* if its head is either a constant or a member of its binding. Non-rigid wffs are said to be *flexible*. The head of a rigid wff cannot be altered by substitution, so if two rigid wffs are to be unified, the heads must be identical (modulo alphabetic changes of bound variables). Huet's unification procedure employs two alternating subroutines, Simpl and Match. Simpl reduces a node by breaking up all compatible rigid-rigid dpairs into the subproblems represented by their arguments, or returns failure upon finding non-compatible rigid-rigid dpairs. Match suggests a set of substitutions for the flexible head of a flexible-rigid dpair. If that set contains more than one substitution, it necessitates branching in the unification tree to create one node for each possibility. The non-unit sets of substitutions are called *branching* sets. Since we are only interested in the existence of a unifying substitution, a node consisting only of flexible-flexible dpairs is considered to be a terminal success node.

SIMPLIFY is a procedure based on Huet's Simpl with the following additions: 1) we include only one dpair from each equivalence class with respect to alphabetic changes of bound variables; 2) we recognize the trivial unifiable fixed point problem $\langle x, E\rangle$ where x does not occur free in E, with the substitution of the most general unifier $x \rightarrow E$; 3) we recognize some non-unifiable fixed point problems, including those suggested by Huet in 3.7.3 of [5]. MATCH is precisely Huet's algorithm of the same name, offering a choice of either the η-rule or non-η-rule substitutions.

7.1. Search heuristics for full unification

Once we have a complete description of the unification problem, *i.e.*, the potential mating is complete, we search for a unifier with the full force of Huet's algorithm. For each variable which is the flexible head of a flexible-rigid dpair, one substitution set is computed by applying MATCH to the first flexible-rigid dpair in which that variable occurs as a flexible head. Of course, such substitutions can be stored between iterations of the algorithm to avoid recomputing. All substitution sets which reduce to singletons after SIMPLIFication are applied immediately. If the search must branch because all substitution sets have multiple members, we choose a set causing the least branching. Unapplied branching substitution sets are not discarded, but passed on to each descendant. Each node also has associated with it a list of compound substitutions for the variables in the original mated pairs.

Surprisingly often, one encounters a unification node **M** which subsumes another node **N**, *i.e.*, **M** is a subset of **N** (modulo alphabetic changes of bound and free variables): In this case any unifying substitution for **N**, with the appropriate alphabetic transformation, will also unify **M**; thus, since we are interested only in the existence of a unifier, it is sufficient to consider **M**. We detect this relationship between nodes quite efficiently using a hash table on dpairs. When **N** is a newly-created node there is no question as to the right course of action: we simply eliminate **N** from further consideration. Of course, if **N** is the sole descendant of **M** this terminates attempts to find a unifier for **M**. However, when **N**, a proper superset of **M**, already exists in the unification tree and possesses descendants, it may well be that the leaves of the tree under **N** represent more progress toward the ultimate unification than does node **M**. It is not clear that the best action is to delete **N** and all its descendants (except, possibly, **M**), though that is our current strategy.

7.2. Search heuristics for unification on partial matings

We also perform unification on nodes representing partial matings. In order not to have multiple nodes to which a dpair representing a newly mated pair must be added, we allow no branching during unification until the mating is complete. Since the mating is incomplete, so also is the unification problem, so a node containing only flexible-flexible dpairs is not a terminal success node. Also, since the partial mating might well be a subset of an already completed mating though it in no sense subsumes the complete mating, the subsumption procedure described in the previous paragraph is applicable only when the subsuming node represents a complete mating.

In building up a mating, one wants to be sure that the partial unifier for mating a new pair of literals is consistent with the substitutions associated with the partial mating to which the new pair is to be added, and one certainly wants to make use of the work towards unification that one performed in producing the connection graph. When a pair of literals is added to the mating, therefore, we want to merge the substitutions labeling the arc between them in the connection graph with the substitutions already associated with the mating. Fortunately, the substitutions produced by Huet's algorithm are very well behaved. Given two substitutions $(v \rightarrow T^1)$ and $(v \rightarrow T^2)$ for the same variable v, the substitutions are compatible iff the dpair $\langle T^1, T^2 \rangle$ does not fail under SIMPLIFication, since T^1 and T^2 contain only new and distinct variables.

7.3. Experience with higher-order unification

Because the higher order unification algorithm constructs a search tree with potentially infinite branches, it might be expected to place an unreasonable computational burden on a theorem prover. However, our experience suggests that this is not the case when the algorithm is implemented as described above. The difficulties we have encountered in trying to prove various theorems have generally been attributable more to the complexity of the search for a mating than to the complexity of the unification process.

§8 An example

We shall now demonstrate several features of *TPS* by having it prove THM87:

∃w ∀j ∃q .[j ∈ .[P a .h j] ∪ .P w .k j] ⊃ .j ∈ .P w q

Here ∈ is an abbreviation for [λX λP .P X] and ∪ is an abbreviation for [λP λQ λX .[P X] ∨ .Q X]. This theorem is neither interesting nor challenging, but permits us to briefly illustrate various features of *TPS* in the space available. Here h and k are function symbols and [h j] denotes the value of h on the argument j. For this example, we have suppressed the printing of type symbols. A dot denotes a left bracket whose mate is as far to the right as is consistent with the pairing of brackets already present. Except for the comments in italics, the remaining text of this section is essentially what the user sees when *TPS* processes this theorem.

The PLAN *command initiates the proof process by creating a proof outline whose only line is the theorem to be proved.*
```
*PLAN THM87
( 100 )          ⊢   ∃w ∀j ∃q .    [ j ∈ .[ P a .h j ] ∪ .P w .k j ]
                                ⊃ .j ∈ .P w q                              PLAN1
```
At this stage the justification PLAN1 *for line 100 is just an empty label.* TPS *chooses 100 to be the number of the last line in the proof. This choice is easily changed, and if a given choice does not leave enough room for the complete proof, all the lines can easily be renumbered.*

The mating search program now attacks the theorem in line 100. TPS *takes the wff in line 100, instantiates all definitions, places it in negation normal form, skolemizes it, and attaches names to the literals. The result is represented as a vertical path diagram:*

```
            ∀w q
                            LIT3                    LIT4
                 P a [h .jA w ][jA w] ∨ P w [k .jA w ].jA w
                                     LIT6
                                  .P w q .jA w
```

Here disjunctions are displayed horizonally and conjunctions are displayed vertically so that this diagram represents the wff:

∀w∀q .[[P a [h .jA w].jA w] ∨ .P w [k .jA w].jA w] ∧ ∼ .P w q .jA w

TPS now searches for an acceptable mating for this wff.

```
Path with no mates: (LIT3 LIT6)
    try this arc: (LIT3 LIT6)

Partial Mating 0:
```

(LIT3)-(LIT6)

Path with no mates: (LIT4 LIT6)
Path with no mates: (LIT3 LIT6)
No proof on this level

TPS has found that there is no acceptable mating for this particular amplification of the theorem. Hence, it must duplicate some quantifiers. It was proved in [2] that duplicating outer quantifiers is a complete although not sophisticated method to duplicate quantifiers. In the diagram below, variables and literals are given a suffix to show which copy of the orginal variable or literal they represent.
Duplicate outer quantifiers.

```
 ∀w↑1 q↑1
            LIT3↑1                              LIT4↑1
      P a [h .jA w↑1 ][jA w↑1] V P w↑1 [k .jA w↑1 ].jA w↑1

                        LIT6↑1
                   ¬ .P w↑1 q↑1 .jA w↑1

 ∀w↑2 q↑2
            LIT3↑2                              LIT4↑2
      P a [h .jA w↑2 ][jA w↑2] V P w↑2 [k .jA w↑2 ].jA w↑2

                        LIT6↑2
                   ¬ .P w↑2 q↑2 .jA w↑2
```

In this expanded wff, TPS quickly finds an acceptable mating.

Path with no mates: (LIT3↑1 LIT6↑1 LIT3↑2 LIT6↑2)
   try this arc: (LIT3↑2 LIT6↑1)

Partial Mating 0:
   (LIT3↑2)-(LIT6↑1)

Path with no mates: (LIT3↑1 LIT6↑1 LIT4↑2 LIT6↑2)
   try this arc: (LIT4↑2 LIT6↑2)

Partial Mating 0:
   (LIT3↑2)-(LIT6↑1), (LIT4↑2)-(LIT6↑2)

$q^2 \Leftarrow k$ .jA a $q^1 \Leftarrow h$ .jA a $w^2 \Leftarrow a$ $w^1 \Leftarrow a$

At this point, an acceptable mating and its associated substitution have been found.

The replication has been simplified.
Old: (w   2)   New: (q   2)
The substitution has also been changed.
Old: $q^2 \Leftarrow k$ .jA a $q^1 \Leftarrow h$ .jA a $w^2 \Leftarrow a$ $w^1 \Leftarrow a$

New: .w $\Leftarrow a$ $q^1 \Leftarrow h$ .jA a $q^2 \Leftarrow k$ .jA a

TPS now displays the new replicated wff and the result of instantiating its quantifiers using the new substitution.

```
∀w
  | ∀q↑1
  |           LIT7↑1                    LIT8↑1
  |     P a [h .jA w ][jA w] V P w [k .jA w ].jA w
  |
  |                      LIT9↑1
  |                 ¬ .P w q↑1 .jA w
  |
  | ∀q↑2
  |           LIT7↑2                    LIT8↑2
  |     P a [h .jA w ][jA w] V P w [k .jA w ].jA w
  |
  |                      LIT9↑2
  |                 ¬ .P w q↑2 .jA w
  |
  |           LIT7↑1                    LIT8↑1
  |     P a [h .jA a ][jA a] V P a [k .jA a ].jA a
  |
  |                      LIT9↑1
  |                 ¬ .P a [h .jA a ].jA a
  |
  |           LIT7↑2                    LIT8↑2
  |     P a [h .jA a ][jA a] V P a [k .jA a ].jA a
  |
  |                      LIT9↑2
  |                 ¬ .P a [k .jA a ].jA a
```

An acceptable mating is:
($LIT8^2$ $LIT9^2$)   ($LIT9^1$ $LIT7^2$)   ($LIT8^1$ $LIT9^2$)   ($LIT7^1$ $LIT9^1$)

Now PLAN1 *must be constructed by removing skolem functions from the substitution terms and computing ancestory information for variables and atoms as mentioned in [1].*

PLAN1 is:
$\exists w\ \forall j\ .\ [\exists q^1\ .[[P\ a\ [h\ j\ ]\ j\ ]\ v\ .P\ w\ [k\ j\ ]\ j\ ]\ \supset\ .P\ w\ q^1\ j\ ]$
$\quad\quad\quad\quad ATM3^1 \quad\quad\quad\quad\quad ATM4^1 \quad\quad\quad\quad ATM6^1$
$\quad v\ .\exists q^2\ .[[P\ a\ [h\ j\ ]\ j\ ]\ v\ .P\ w\ [k\ j\ ]\ j\ ]\ \supset\ .P\ w\ q^2\ j$
$\quad\quad\quad\quad ATM3^2 \quad\quad\quad\quad\quad ATM4^2 \quad\quad\quad\quad ATM6^2$

The substitution is: $w \Leftarrow a$ $q^1 \Leftarrow [h\ j]$ $q^2 \Leftarrow [k\ j]$

The mating is:
($ATM6^2$ $ATM4^2$)   ($ATM3^2$ $ATM6^1$)   ($ATM6^2$ $ATM4^1$)   ($ATM6^1$ $ATM3^1$)

The replication scheme is: (q   2)

TPS now proceeds to build the proof outline from this plan.
>P-EXISTS a 100
(99)   ⊢ ∀j ∃q .[j ∈ .[P a .h j] ∪ .P a .k j] ⊃ .j ∈ .P a q
                                                                    PLAN2

(100) ⊢ ∃w ∀j ∃q . [j ∈ .[P a .h j] ∪ .P w .k j]
                  ⊃ .j ∈ .P w q
                                                              ∃G: a 99

The Plan for line 99, PLAN2, is:
  ∀j . [∃q^1 .[[P a [h j] j] ∨ .P a [k j] j] ⊃ .P a q^1 j]
       ATM3^1 ATM7   ATM8
    ∨ .∃q^2 .[[P a [h j] j] ∨ .P a [k j] j] ⊃ .P a q^2 j
       ATM3^2 ATM9   ATM10

The substitution is: q^1 ⇐ [h j]   q^2 ⇐ [k j]

The mating is:
(ATM10 ATM9)   (ATM3^2 ATM8)   (ATM10 ATM7)   (ATM8 ATM3^1)

The replication scheme is: (q   2)

>P-ALL 99
(98)   ⊢ ∃q .[j ∈ .[P a .h j] ∪ .P a .k j] ⊃ .j ∈ .P a q
                                                              PLAN3
(99)   ⊢ ∀j ∃q .[j ∈ .[P a .h j] ∪ .P a .k j] ⊃ .j ∈ .P a q
                                                              ∀G: 98

PLAN3 *has been omitted since it differs only slightly from PLAN2. The reader may wish to construct it. TPS now recognizes that the present plan is existentially complex (see [1]), so the proof must now proceed in an indirect fashion. The symbol ⊥ is used to denote falsehood.*
>P-INDIRECT 98
(1)   1 ⊢ ~ .∃q .[j ∈ .[P a .h j] ∪ .P a .k j] ⊃ .j ∈ .P a q
                                                              Hyp
(97) 1 ⊢ ⊥   PLAN4
(98)   ⊢ ∃q .[j ∈ .[P a .h j] ∪ .P a .k j] ⊃ .j ∈ .P a q
                                                              Indirect: 97

The Plan for line 97, PLAN4, is:
  [~ . [∃q^1 .[[P a [h j] j] ∨ .P a [k j] j] ⊃ .P a q^1 j]
         ATM3^1 ATM7   ATM8
    ∨ .∃q^2 .[[P a [h j] j] ∨ .P a [k j] j] ⊃ .P a q^2 j]
       ATM3^2 ATM9   ATM10
  ⊃ ⊥

The substitution is: q^1 ⇐ [h j]   q^2 ⇐ [k j]

The mating is:
(ATM10 ATM9)   (ATM3^2 ATM8)   (ATM10 ATM7)   (ATM8 ATM3^1)

The replication scheme is: (q   2)

>D-NEG 1

(2)   1   ⊢ ∀q .[j ∈ .[P a .h j] ∪ .P a .k j] ∧ ~ .j ∈ .P a q
                                                                    Neg: 1
>D-ALL 2 [h j]
(3)   1   ⊢ [j ∈ .[P a .h j] ∪ .P a .k j] ∧ .~ .j ∈ .P a .h j
                                                                    ∀I: 2
>D-CONJ 3
(4)   1   ⊢ j ∈ .[P a .h j] ∪ .P a .k j   Conj: 3
(5)   1   ⊢ ~ .j ∈ .P a .h j   Conj: 3

>D-DEF1 5
(6)   1   ⊢ ~ .P a [h j] j   Def: 5

>D-ALL 2 [k j]
(7)   1   ⊢ [j ∈ .[P a .h j] ∪ .P a .k j] ∧ .~ .j ∈ .P a .k j
                                                                    ∀I: 2
>D-DEF1 4
(8)   1   ⊢ [[P a .h j] ∪ [P a .k j]] j   Def: 4

>D-DEF1 8
(9)   1   ⊢ [P a [h j] j] ∨ .P a [k j] j   Def: 8

>D-CONJ 7
(10)   1   ⊢ j ∈ .[P a .h j] ∪ .P a .k j   Conj: 7
(11)   1   ⊢ ~ .j ∈ .P a .k j   Conj: 7

>D-DEF1 11
(12)   1   ⊢ ~ .P a [k j] j   Def: 11

>D-DEF1 10
(13)   1   ⊢ [[P a .h j] ∪ [P a .k j]] j   Def: 10

>D-DEF1 13
(14)   1   ⊢ [P a [h j] j] ∨ .P a [k j] j   Def: 13

>RULEP 97 (14 9 6 12)
(97)   1   ⊢ ⊥   RuleP: 14 9 6 12

This completes the process of constructing the proof, which is displayed below:

(1)   1   ⊢ ~ .∃q .[j ∈ .[P a .h j] ∪ .P a .k j] ⊃ .j ∈ .P a q
                                                                    Hyp
(2)   1   ⊢ ∀q .[j ∈ .[P a .h j] ∪ .P a .k j] ∧ .~ .j ∈ .P a q
                                                                    Neg: 1
(3)   1   ⊢ [j ∈ .[P a .h j] ∪ .P a .k j] ∧ .~ .j ∈ .P a .h j
                                                                    ∀I: 2
(4)   1   ⊢ j ∈ .[P a .h j] ∪ .P a .k j   Conj: 3
(5)   1   ⊢ ~ .j ∈ .P a .h j   Conj: 3
(6)   1   ⊢ ~ .P a [h j] j   Def: 5

(7)   1   ⊢ [j ∈ .[P a .h j] ∪ .P a .k j] ∧ .~ .j ∈ .P a .k j
                                                              ∀I: 2
(8)   1   ⊢ [[P a .h j] ∪ [P a .k j]] j   Def: 4
(9)   1   ⊢ [P a [h j] j] ∨ .P a [k j] j   Def: 8
(10)   1   ⊢ j ∈ .[P a .h j] ∪ .P a .k j   Conj: 7
(11)   1   ⊢ ~ .j ∈ .P a .k j   Conj: 7
(12)   1   ⊢ ~ .P a [k j] j   Def: 11
(13)   1   ⊢ [[P a .h j] ∪ [P a .k j]] j   Def: 10
(14)   1   ⊢ [P a [h j] j] ∨ .P a [k j] j   Def: 13
(97)   1   ⊢ ⊥   RuleP: 14 9 6 12
(98)   ⊢ ∃q .[j ∈ .[P a .h j] ∪ .P a .k j] ⊃ .j ∈ .P a q
                                                              Indirect: 97
(99)   ⊢ ∀j ∃q .[j ∈ .[P a .h j] ∪ .P a .k j] ⊃ .j ∈ .P a q
                                                              ∀G: 98
(100)   ⊢ ∃w ∀j ∃q . [j ∈ .[P a .h j] ∪ .P w .k j]
                        ⊃ .j ∈ .P w q   ∃G: a 99

§9 Sample theorems

We here present some examples of theorems which have been proved automatically by *TPS*. Other examples were presented in Appendix B of [2]. First we give several definitions which are used in the theorems below.

[# $F_{\alpha\beta}$] $D_{o\beta}$ is the image of the set $D_{o\beta}$ under the function $F_{\alpha\beta}$. # is defined as:

$$\lambda F_{\alpha\beta} \, \lambda D_{o\beta} \, \lambda Y_\alpha \, \exists X_\beta \, .[D \, X] \wedge .Y = .F \, X$$

⊆ (set inclusion) is defined as:

$$\lambda P_{o\alpha} \, \lambda Q_{o\alpha} \, \forall X_\alpha \, .[P \, X] \supset .Q \, X$$

= is defined as:

$$\lambda X_\alpha \, \lambda Y_\alpha \, \forall Q_{o\alpha} \, . [Q \, X] \supset . Q \, Y$$

Theorems

THM30A

$[U_{o\beta} \subseteq V_{o\beta}] \supset .[\# F_{\alpha\beta} U] \subseteq .\# F V$

THM47A

$\forall X_\iota \forall Y_\iota .[X = Y] \supset .\forall R_{o\iota\iota} .[\forall Z_\iota .R Z Z] \supset .R X Y$

THM62

$[[\forall U_\iota .P_{o\iota\iota} A_\iota U] \vee .\forall V_\iota .P V B_\iota] \supset .\exists X_\iota .P X X$

THM76

$[\forall P_{o\iota} .[P Y_\iota] \supset .P X_\iota] \supset .\forall R_{o\iota} .[R X] \supset .R Y$

THM82

$[\quad [\forall x_\iota \exists y_\iota .F_{o\iota\iota} x y]$
$\wedge [\exists x \forall e_\iota \exists n_\iota \forall w_\iota .[S_{o\iota\iota} n w] \supset .D_{o\iota\iota\iota} w x e]$
$\wedge .\forall \varepsilon_\iota \exists \delta_\iota \forall x1_\iota \forall x2_\iota . \quad [D x1 x2 \delta]$
$\supset .\forall y1_\iota \forall y2_\iota . \quad [[F x1 y1] \wedge .F x2 y2]$
$\supset .D y1 y2 \varepsilon]$
$\supset .\exists y \forall e \exists m_\iota \forall w .[S m w] \supset .\forall z_\iota .[F w z] \supset .D z y e$

THM83

$[\forall X_\iota \exists Y_\iota .[P_{o\iota} X] \supset .\forall Z_\iota .[R_{o\iota\iota} X Y] \wedge .P Z]$
$\supset .\exists U_\iota \forall V_\iota .[P V] \supset .R W_\iota U$

§10 Acknowledgments

We would like to acknowledge the valuable assistance provided by Frank Pfenning, Edward Pervin, and Harry Porta.

References

1. Peter B. Andrews, "Transforming Matings into Natural Deduction Proofs," in *5th Conference on Automated Deduction, Les Arcs, France*, edited by W. Bibel and R. Kowalski, Lecture Notes in Computer Science, No. 87, Springer-Verlag, 1980, 281-292.

2. Peter B. Andrews, *Theorem Proving via General Matings*, Journal of the Association for Computing Machinery **28** (1981), 193-214.

3. Alonzo Church, *A Formulation of the Simple Theory of Types*, Journal of Symbolic Logic 5 (1940), 56-68.

4. Gérard P. Huet, *A Unification Algorithm for Typed λ-Calculus*, Theoretical Computer Science 1 (1975), 27-57.

5. Gérard Huet, *Resolution d'Équations dans les Languages d'Ordre 1,2,...,ω*, Thèse de Doctorat D'État, Université Paris VII, 1976.

6. D. C. Jensen and T. Pietrzykowski, *Mechanizing ω-Order Type Theory Through Unification*, Theoretical Computer Science 3 (1976), 123-171.

7. Brian K. Reid and Janet H. Walker, *SCRIBE Introductory User's Manual*, Third edition, UNILOGIC, Ltd., 160 N. Craig St., Pittsburgh, PA 15213, 1980.

Logic Machine Architecture:
Kernel Functions

Ewing L. Lusk

Northern Illinois University
Argonne National Laboratory

William W. McCune

Northwestern University

Ross A. Overbeek

Northern Illinois University

ABSTRACT

In this paper we present an attempt to abstract from the great diversity of approaches to automated deduction a core collection of operations which are common to all of them. Implementation of this kernel of functions provides a software platform upon which a variety of theorem-proving systems can be built. We outline the architecture for a layered family of software tools to support the development of theorem-proving systems and present in some detail the functions which comprise the two lowest layers. These are the layer implementing primitive abstract data types not supported by the host language and the layer providing primitives for the manipulation of logical formulas. This layer includes the implementation of efficient unification and substitution application algorithms, structure sharing within the formula database, and efficient access to formulas via arbitrary user-defined properties. The tools are provided in a highly portable form (implemented in Pascal) in order that a diverse community of users may build on them.

1. Introduction

There are currently a variety of theorem-proving research efforts, each oriented around a particular theorem-proving approach or application. Some of these approaches are clause-based resolution, natural deduction, equality-oriented systems, and Gentzen systems, as well as higher-order logics. Application systems include program verification, problem-solving, automated circuit design, and piping or wiring network analysis. Diverse as these approaches are, they all require solutions to certain common subproblems. Furthermore, implementation of solutions to these common subproblems constitutes a significant overhead in the implementation of systems designed to study new approaches and applications in theorem proving. The question we have attempted to answer in this paper is the following: Is it possible to define a kernel of functions dealing with the storage, retrieval, and manipulation of logical formulas which is common to a wide variety of theorem-proving environments?

In this paper we present such a collection of functions. They respond to the requirements outlined in [4] for any general-purpose system capable of dealing with very large numbers of formulas. They have all been implemented, and form the kernel of a new theorem-proving system being developed at Argonne National Laboratory and Northern Illinois University. We outline here the architecture of our new implementation and present in detail the two lowest layers of the system, the one which implements primitive abstract data types and the one which provides for the manipulation of logical formulas. The entire system we refer to as Logic Machine

Architecture (LMA), which is not a single theorem prover, but rather a carefully engineered set of layered functions from which many theorem provers can be constructed. We believe that this approach will be increasingly useful as more and more application-oriented theorem-proving systems are developed. We are currently using this package of tools to construct a variety of inference mechanisms — multiple resolution-based inference rules, equality-based inference mechanisms, the inequality resolution system of Bledsoe[1], and case analysis techniques. We further believe that the layers of LMA presented in this paper provide an integrated foundation for the implementation of several theorem-proving techniques (such as Gentzen and natural deduction systems) which we do not intend to implement immediately.

The LMA package of tools is designed for a wide range of users. The distribution package will include several distinct theorem provers, some of which are oriented towards specific applications. These are for users who wish to do no programming at all. It will allow such people to perform research in application areas without being forced to develop a substantial amount of software. In this regard it is worth noting that the open problems solved by Wos and Winker used only a standard general-purpose theorem prover. No extra programming was done to focus on peculiarities of any of the mathematical systems that were being studied.

Those who wish to utilize the many tools that exist in a large theorem-proving system (such as subsumption, demodulation, and the normal inference rules), but also wish to experiment with new strategies and inference mechanisms, can easily construct new theorem provers from the components included in the package. The layered architecture and the supporting tools included in the distribution package should dramatically reduce the effort required to subject new ideas to experimental verification.

Five distinct software layers have been defined within LMA:
- Primitive Abstract Data Types (Layer 0)
- Formula Manipulation Primitives (Layer 1)
- Inference Mechanisms (Layer 2)
- Theorem-Proving Processes (Layer 3)
- Interprocess Communication (Layer 4)

Layer 0 implements three primitive abstract data types not provided by the host language (which is Pascal). These are character strings of unbounded size, integer vectors of unbounded length, and long integers in a range fixed at compilation time but indefinitely extendible. This layer is discussed below.

Layer 1 implements the data type "object," which is used for representing and storing logical formulas, among other things. This layer provides a variety of operations which can be performed on objects, including unification, application of substitutions, and efficient access to objects via arbitrary boolean combinations of user-defined properties. This latter facility will become increasingly significant as theorem provers are applied to problems involving the very large sets of formulas typical of many applications. A discussion of the concepts implemented in Layer 1 provides the bulk of this paper.

Layer 2 provides mechanisms for generating new formulas from old and for detecting relationships among formulas. In particular, most resolution-based inference rules, paramodulation, demodulation, subsumption, and inequality inference rules are included in the initial set of functions provided at this level. These functions operate on objects which are clauses, but will not assume that all existing objects are clauses. Thus non-clausal inference mechanisms can eventually be integrated into the system in a harmonious fashion. Layer 2 will be described in detail in another paper.

Layer 3 defines a set of components from which a complete theorem prover can be configured. Each component will represent an independent theorem-proving process utilized in

Layer 4.

Layer 4 supports the creation of a configuration of communicating concurrent processes defined in Layer 3. It defines the primitives required to initiate processes and manage interprocess communication.

As this is written, Layers 0 and 1 are completely operational, implemented in Pascal within the UNIX environment. Layer 2 is undergoing final testing.

2. Layer 0: Primitive Data Types

Our last major theorem prover has been in existence for over seven years. It has been enhanced, modified, and utilized in ways that were never envisioned when several initial critical design decisions were made. In particular, certain limitations were imposed which simplified implementation and seemed at the time to be no restriction for users at all. For example, ten literals and sixteen distinct variables per clause seemed ample, and were satisfactory for much experimentation carried out during the life of the system. However, as more difficult problems were attacked and applications began to develop, both of these restrictions, as well as a variety of similar ones, had to be relaxed, often requiring substantial effort. For example, users of the system for automated circuit design indicated that they must be able to process clauses containing at least fifty distinct variables. Even an upper limit of two hundred was not considered "safe."

Painful experience over the last seven years indicates that a major implementation effort should avoid such limiting assumptions. Few of them can be expected to survive in a system with a long life expectancy, and only a long-lived system has a chance to evolve into one of real power. In LMA restrictions such as those discussed above do not exist. Subject only to the (virtual) memory size of the machine, the number of distinct objects, the number of literals or variables per clause, the depth of a term, and the length of the character representation of a formula are all unlimited. The removal of such restrictions required the implementation of three data types not supported by our host language.

Without the provision of special primitive abstract data types, the maximum length of a character string and the maximum size of an array would be fixed at compilation time, and the maximum value of an integer would be fixed by the compiler. We promoted each of these one step: the lengths of strings and sizes of arrays are completely dynamic, and the maximum value of an integer is fixed at compilation, but is indefinitely extendible.

These implementations are relatively straightforward. Both character strings and integer vectors are implemented as linked lists of fixed-size arrays, and long integers are implemented as fixed-length arrays of integers. In all cases we have provided the necessary manipulation routines which hide the details of the structure from the user of Layer 0.

The creation of these three data types is not of central interest in the creation of theorem provers, but we include it here to illustrate the approach of avoiding constraints. Appendix A lists the basic primitives and some functions composed from the primitives, which together make up Layer 0.

3. Layer 1: Formula Manipulation Primitives

3.1. Objects

The next layer of functions defines a family of operations on the most fundamental data type in our system — the *object*. Objects will be used to represent many higher level data types such as terms, literals, clauses, and lists. An object is general enough to represent formulas of the λ-calculus. While no immediate plans exist to implement operations on λ-calculus formulas, the basic functions in Layer 1 will not restrict the user to formulas of any less generality. There

are three types of objects:
- a) A *name* is a character string which can be used to represent predicate, function, or constant symbols.
- b) A *variable* is an integer which can be used to represent variable symbols.
- c) An *application* is an ordered set of objects.

The distinction between names and variables is reflected only in those operations which utilize substitutions or rename variables. A formula can be represented by a tree of application nodes, with name and variable nodes as the leaves. For example, the term $f(x_1,a)$ can be represented by the application object consisting of the ordered set $\{f,x_1,a\}$, where f and a are name objects and x_1 is a variable object.

To each object is assigned a long integer which is either 0 or is a unique identifier for the object.

In [4] and [7] we illustrated the effect of structure-sharing on fundamental theorem-proving algorithms. Keeping only one copy of each term and making all containing terms, literals, clauses, or lists accessible from it has contributed significantly to our previous program's ability to maintain performance levels in the presence of large numbers of clauses. In particular it allows the program to simultaneously generate sets of resolvents or paramodulants based on a single unification, and perform complete forward and backward subsumption checks.

The collection of objects which utilize structure-sharing will be called the *integrated* objects. Within the set of integrated objects, two occurrences of a formula are represented by a single object. The application objects containing the distinct occurrences each reference via links the object representing the formula, and are each locatable from it. An object is integrated when it is deemed appropriate to add it to the knowledge database. Integrating an object involves some overhead, but once this happens, all occurrences of the object in the database can be accessed very rapidly.

Finally, all objects have a standard external (character string) format, from which the internal form can be constructed. The motivation for such a format, as well as a proposed standard for use in communicating formulas among diverse theorem-proving systems, is discussed elsewhere[5].

In some ways Layer 1 can be thought of as the collection of functions which define and manage the object abstract data type. All higher layers will access objects through the Layer 1 functions, thus isolating the internal structure of an object from the theorem-proving algorithms in higher layers. Since higher-level layers contain algorithms to manipulate lists, clauses, literals, and terms which are encoded as objects, Layer 1 can also be thought of as a database management layer for theorem provers.

3.2. Attributes

The user of Layer 1 may attach to an object an integer-object pair called an *attribute*. The integer is thought of as a code identifying the type of attribute and the object as the value of the attribute. Layer 1 functions exist to attach, detach, and access the attributes of an object by code. This allows the user to associate with any object a set of more or less arbitrary data elements (encoded as objects). Uses for attributes include identification of qualifier literals[8], setting of lock values[6,2], designation of significant subterms (as in hyperparamodulation[3]), and a variety of others.

Enough algorithms require that flags be associated with objects to make useful a less general but more efficient form of attribute. Hence, with each object there is associated an array of integers called *user variables*, which can be accessed through Layer 1 functions. They are often used as short-lived flags to support particular theorem-proving algorithms, such as to mark fully demodulated subterms during demodulation.

3.3. Substitutions

The creation and application of substitutions is a basic component of any automated deduction system. In this section we describe our representation of substitutions. A restricted form of this representation, which has several interesting properties, has been in use for seven years in our existing theorem prover and has proven quite effective.

The representation of a substitution must possess three desirable but to some extent contradictory properties.

    a)   It must support efficient algorithms for unifying two objects and for applying a substitution to an object.

    b)   It must not consume inordinate amounts of memory. This will become more critical in a multiprocessing environment in which many substitutions may coexist.

    c)   It must avoid assumptions about the maximum number of distinct variables per object.

Displacements, Increments, and Multipliers

In order to describe our choice of representation for a substitution, it will be necessary to discuss a feature of our unification algorithm, which is designed only for first-order logic. In determining a unifying substitution for two objects O_1 and O_2, it is often necessary to keep the variables in the two objects distinct. The time required to generate variants can be saved by use of the following technique.

Suppose that r_1, r_2, r_3, and r_4 are integers such that for each variable v_j in O_1, $r_1 \leq j \leq r_2$ and for each variable v_k in O_2, $r_3 \leq k \leq r_4$. Choose a positive integer d (called the *displacement*), such that $r_1 + d > r_4$. Then the variant O_1' of O_1 created by replacing each v_j in O_1 by v_{j+d} has the property that its variables are distinct from those of O_2. The variant O_1' is not actually created; rather the displacement d is built into the algorithm which unifies O_1 and O_2 in the sense that the displacement is added to the number of any variable as it is being used. The variable number in the object itself is not altered.

This technique is generalized in the following way. In forming a hyperresolvent, it may be necessary to separate the variables in a nucleus clause N from those of satellites S_1 and S_2, as well as separating the variables of S_1 from those of S_2. If i (for *increment*) is larger than the number of variables in any of the three clauses, then we can choose $d_j = j \times i$ so that adding d_j to the variable numbers in S_j keeps the variables distinct. The number j is called a *multiplier*; it, in conjunction with the increment, is used to calculate appropriate displacements for the variables in an object.

Now we see how the property concerning limits of the number of distinct variables in an object comes in. One must not choose a global value for the increment i which is assumed to be large enough. Moreover, since a given object can be contained in any number of of other objects, it is impossible to calculate the maximum number of variables in a containing object (such as a clause) by examining a given object.

Our approach to this difficulty is to associate an increment value with each individual substitution table and to store entries in the table in a form independent of i. As will be seen below, this allows dynamic restructuring of substitution tables when variables with numbers greater than the current increment are encountered.

The Structure of a Substitution Table

A substitution table has the form of a vector of entries, one for each possible variable in the set of objects being considered. (Since a substitution table may need to be dynamically extended, it is actually implemented as a linked list of fixed-size arrays. Normally a substitution table will fit into one array.) If v_k is a variable in an object whose associated multiplier is j, and i is the increment associated with a particular substitution table, then the entry giving the substitution for v_k is the one with subscript $(j \times i) + k$.

The entry for a variable is a pair consisting of:

a) A pointer to the object being substituted for the variable (or nil, indicating that no substitution is being made for that variable), and

b) If the pointer is not nil, a multiplier to be used for the variables occurring in the object being substituted for the variable.

This is complicated enough to justify an example. Consider the substitution unifying

$$\text{multiplier} = 0 \quad f(f(v_1, v_2), f(v_2, g(v_2)))$$

with

$$\text{multiplier} = 1 \quad f(v_1, f(v_3, g(g(v_2)))).$$

A substitution table representing this substitution is:

increment = 5

variable	substituted object	multiplier
v_1	nil	0
v_2	$g(v_2)$	1
v_3	nil	0
v_4	nil	0
v_5	nil	0
v_6	$f(v_1, v_2)$	0
v_7	nil	0
v_8	v_2	0
v_9	nil	0
v_{10}	nil	0

Here the entry for v_2 indicates that v_2 is replaced by $g(v_n)$ where $n = (1 \times i) + 2 = 7$. The variable v_6 (which is v_1 from the second object) is replaced by $f(v_1, v_2)$, while v_8 (v_3 from the second object) goes to v_2, which in turn is instantiated to $g(v_7)$, so that the common instance is

$$f(f(v_1, g(v_7)), f(g(v_7), g(g(v_7)))).$$

At this point the reader may feel that a relatively simple concept has been implemented in a painfully obscure manner. We believe that the constraints imposed upon the solution at the beginning of this section are the source of the complexity. In any event the complexity can be isolated into the few routines which create, examine, and apply substitutions. The user of Layer 1 simply invokes unification functions to create substitutions and application functions to instantiate objects. Only the concept of multiplier is visible from above.

Unification

The unification algorithm implemented within Layer 1 is relatively straightforward, given the structure of a substitution. We do not give it here, but rather enumerate the principal design considerations for unification in a theorem-proving context.

The unification algorithm built into Layer 1 is currently useful only for objects representing formulas of first-order logic. The data structures of Layer 1 support greater generality (including, for example, the λ-calculus), but for the time being Layer 1 contains no manipulative procedures other than basic ones for such objects.

It is assumed that objects to be unified may have more than one unifier, and there are functions to sequentially process a set of unifiers of two objects. It is anticipated that this flexibility will be particularly useful in the case of commutative-associative unification and similar extensions.

A unification algorithm should be reasonably efficient. While some have claimed that the speed of the unification algorithm is the principal determinant of the overall speed of a theorem prover because of the frequency of its invocation, our experience indicates that this may be an oversimplification. By using techniques (presented in the next section) which efficiently select objects with a high probability of unifying with a given object, one can ensure that most invocations will produce a successful unification. In this environment a poor implementation could still absorb as much as fifty percent of overall theorem-prover execution time, so there is reason to implement unification efficiently, but there remains some leeway for trading off efficiency in return for flexibility.

The input for our unification algorithm consists of two objects, their multipliers, a substutution table, and a set of unification restrictions and options. The substitution table may already be partly filled in, in which case the substitution which unifies the objects is required to be an extension of the existing substitution. Unification returns a return code, the substitution table (possibly modified if unification was successful), and a list (called the *change list*), of the variables which were substituted for in that particular invocation of unification.

An example will illustrate these last two points. Suppose one is attempting to generate hyperresolvents from the nucleus

$$\neg L_1 \ \neg L_2 \ L_3$$

and that L_1 has been unified with, say, L_4. Then the substitution which unifies L_2 with another literal, say L_6, must be an extension of the substitution unifying L_1 and L_4. If such a substitution can be found, then it is applied to L_3 to create the resulting hyperresolvent. One then wants to undo the changes made to the substitution table made while unifying L_2 and L_6, and search for another literal which might unify with L_2 by means of another substitution extending the substitution unifying L_1 and L_4. This is not quite the algorithm for generating hyperresolvents implemented in Layer 2, but the need for the change list remains.

The format of a change list is a list of substitution table subscripts in the increment-independent (multiplier, variable-number) form. This allows the substitution table itself to be restructured (to utilize a larger increment) without requiring an update of the change list.

Finally, restrictions of unification are common. For example, determining whether one object is an instance of another may be viewed as attempted unification under the restriction that variables in only one of the objects may be instantiated.

4. Access to Objects via Properties

Layer 1 includes facilities for rapidly accessing sets of objects via user-defined properties of the objects. In particular it is possible to rapidly access those objects which can be unified with a given object. Objects accessed by property must have unique identifiers. Properties are not to be confused with attributes. A property is used to search for objects. There is no mechanism for determining from an object what properties are associated with it. Attributes, on the other hand, are user-accessible data items attached to an object.

Properties can be used in many algorithms to eliminate most sequential searching. There are beginning to appear applications characterized by thousands of clauses; these require rapid access to specific sets of objects. The unification-oriented properties maintained by our existing theorem prover have been a critical factor in its ability to cope with large numbers of clauses. The concept of property discussed here and implemented in LMA is a generalization of our existing implementation[7].

The user of Layer 1 is responsible for encoding into vectors of integers the properties in which he is interested, and for specifying that a given object has a particular property. (However, the properties used for rapid unification tests are assigned automatically by Layer 1 when an object is integrated.) The user can then process sequentially (with "get-first", "get-next" functions) the set of objects which have an arbitrary boolean combination of specified properties.

5. Updatable Pointers

Throughout both Layer 1 and Layer 2 there are algorithms which, when invoked, calculate a set of things — objects, substitutions, attributes, etc. LMA attempts to handle access to these sets in a consistent way. The user is provided with functions to

a) obtain the first result of some activity, which establishes position in the set of results, and

b) to obtain the next result, given a position.

Many of the algorithms are implemented in such a way that they do not ever create the entire set of results all at once, but rather maintain a "position" in their processing so that the computation of results is deferred until the individual result is requested by the user. This eliminates surges in memory usage which could impose design limitations.

A side effect of this approach is that the integrity of the above-mentioned "positions" must be maintained between calls to the appropriate "get-next" function, even in the face of changes to the database of objects. In this section we introduce a general-purpose mechanism for dealing with this problem. An appropriate solution to this problem significantly simplifies the set of services offered to layer 3, by making it possible to implement the "first-next" approach to the layer 2 services (such as the generation of new clauses by an inference rule). The failure to handle this problem properly in our current system has resulted in unnecessarily complex coding techniques in the higher layers to process one at a time clauses which are generated in groups by inference rules.

An example will perhaps best illustrate the need for this mechanism. It is drawn from the Layer 2 algorithm for hyperresolution. Layer 2 provides two service routines for the creation and processing of hyperresolvents:

Hyperf: return the first hyperresolvent generated from a given clause and clauses from a designated set of lists, and

Hypern: return the next hyperresolvent in the set whose formation was initiated by a call to hyperf.

Suppose that the given clause has the form:

$$C_1: \neg L_1 \quad \neg L_2 \quad L_3$$

The first hyperresolvent is found as follows:

1. Find the first positive literal L_4 which occurs in at least one positive clause in one of the approved lists, and which unifies with L_1. This is accomplished by invoking the function which returns the first object having these properties.
2. Find the first positive literal, say L_5, such that the unifier found in step 1 can be extended to unify L_5 with L_2.
3. Locate the first positive clause C_2 which contains L_4 (and record it as one of the three parent clauses).
4. Locate the first positive clause C_3 which contains the literal L_5 (and record it as the third parent).
5. Return the resulting hyperresolvent.

The positions in the sets of clauses which contain L_4 and L_5, and the positions in the sets of literals which unify with L_1 and L_2, must all be maintained in order to support a subsequent invocation of hypern. It is possible that the generated hyperresolvent subsumes C_2 or C_3. If so, and if deletion processing occurs before the next call to hypern, then the possibility of pointers referencing deleted objects is introduced. It is even possible that the only occurrence of L_4 or L_5 is in a deleted clause. These situations, which must be taken into account in most inference rules, necessitate the implementation of some mechanism similar to what we have called "updatable pointers."

An *updatable pointer* consists of a pair (flag,pointer), in which the pointer references an object, a position in the set of immediate subobjects of an object, or a position in the set of immediate superobjects of an object. The flag indicates whether or not the referenced object or position has been deleted or altered. It is maintained by the procedures which delete objects and which disconnect an object from an application object which contains it. In the event that the referenced entity is deleted, the pointer will be set to nil; a "deleted" position will be set to reference the next position in the set (of subobjects or superobjects).

The facility to manipulate updatable pointers allows the creation of service routines to sequence through the subobjects or superobjects of an integrated object, despite intervening deletions.

6. Summary

This paper presents an overview of the services provided in the lowest two layers of LMA. These services can be used to implement a wide variety of theorem-proving systems. It is our hope that this will provide a family of tools which will allow comparisons among various approaches to theorem proving within a single environment, as well as permitting the convenient implementation of new systems without the redesign and coding of routines which in some form are common to all theorem-proving programs.

The specific procedures which make up Layers 0 and 1 of LMA are given in the appendix. At this time most of Layer 2, the inference layer, has been implemented, and the services provided by our definition of Layer 1 appear to offer an appropriate set of basic tools. In a sequel to this paper, we will present the basic design of and services offered by Layer 2, including a highly integrated approach to inference rules.

Acknowledgement

This work was supported by the Applied Mathematical Sciences Research Program (KC-04-02) of the Office of Energy Research of the U.S. Department of Energy under Contract W-31-109-Eng-38, and also by National Science Foundation grants MCS79-03870 and MCS79-13252.

References

1. W. W. Bledsoe and Larry M. Hines, "Variable elimination and chaining in a resolution-based prover for inequalities," in *Proceedings of the Fifth Conference on Automated Deduction, Springer-Verlag Lecture Notes in Computer Science, v. 87*, ed. Robert Kowalski and Wolfgang Bibel, (July 1980).
2. R. S. Boyer, *Locking: a restriction of resolution*, Ph.D. Thesis, Univ. of Texas at Austin 1971.
3. Lawrence Henschen, R. Overbeek, and Lawrence Wos, "Hyperparamodulation: a refinement of paramodulation," in *Proceedings of the Fifth Conference on Automated Deduction, Springer-Verlag Lecture Notes in Computer Science, v. 87*, ed. Robert Kowalski and Wolfgang Bibel, (July 1980).
4. E. Lusk and R. Overbeek, "Data structures and control architecture for the implementation of theorem-proving programs," in *Proceedings of the Fifth Conference on Automated Deduction, Springer-Verlag Lecture Notes in Computer Science, v. 87*, ed. Robert Kowalski and Wolfgang Bibel, (1980).
5. E. Lusk and R. Overbeek, *Standardizing the external format of formulas*, (preprint). September 1980.
6. E. Lusk and R. Overbeek, "Experiments with resolution-based theorem-proving algorithms," *Computers and Mathematics with Applications*, (1981). (to appear)
7. R. Overbeek, "An implementation of hyper-resolution," *Computers and Mathematics with Applications* **1** pp. 201-214 (1975).
8. S. Winker, "An evaluation of an implementation of qualified hyperresolution," *IEEE Transactions on Computers* **C-25**(8) pp. 835-843 (August 1976).

Appendix A

1. Layer 0

1.1. Long Integers

The maximum value that can be kept in a long integer is determined at compilation. A long integer is kept as an array of integers. This allows object ids (among other things) to have a range that is independent of the compiler restrictions. The routines to manipulate long integers are as follows:

    lintstore - initialize a long integer with an integer

    lintincr - increment a long integer

    lintcpi - compare a long integer and an integer

    lintcomp - compare two long integers

    addlitovec - add a long integer to an integer vector

    getlifromvec - get a long integer out of a vector

1.2. Integer Vectors

Integer vectors can grow to any length. This allows arbitrarily long position vectors, property vectors, etc. The routines for manipulating integer vectors are as follows:

    ivecopen - initialize an integer vector

    ivecstore - put an integer into a vector

    ivecacc - access an integer in a vector

    ivecclose - deallocate an integer vector

    iveccomp - compare two integer vectors

    copyivec - copy an integer vector

    copyivpart - copy a section of a vector

    makevector - create a vector from a section of another vector

    intchar - convert an integer to a character string

    printivec - print an integer vector

    sizevec - return the size of a vector

    ivechash - hash an integer vector

1.3. Modules for Building Property Vectors

Property vectors are used in layer 1 to encode arbitrary Boolean properties and attach them to objects. They are simply integer vectors built up using the following basic operations:

    atomvec - make an ATOM vector for a given property

    notvec - make a NOT vector for a given property

    andvec - make an AND vector for a given property

    orvec - make an OR vector for a given property

1.4. Strings

Strings are simply a sequence of characters of arbitrary length. They are used as IO buffers and names, among other things. The basic operations for creating and manipulating strings are as follows:

    cstropen - initialize a string

    cstrgetc - get the next character from a string

    cstrputc - insert a character into a string

    cstrreset - reset current position in a string to the first character

    cstrwreos - write an eos (chr(0)) on the end of the string

    cstrclose - deallocate a string

    cstrposeq - are two positions in the same string equal?

    compname - compare a string with the contents of an array

    compstr - compare two strings

    copystr - copy a string

    catstr - add a copy of a string to the current position in another string

    charint - convert a character string to an integer

    hashcstr - hash a string

    isnull - check to see if a string is null

    isnumeric - does a string contain a valid number?

    readstr - read a string from the terminal

    writestr - write a string onto the terminal

    freadstr - read a cstring in from a file

    fwritestr - write a string to an open file

2. Layer 1

2.1. Construction of Objects

Layer 1 offers a number of routines that can be used to construct and relate objects. They are as follows:

    crappl - create an application node (with no subobjects)

    crname - create a name node

    crvar - create a variable node

    insfirst - insert object as first immediate subobject

    inslast - insert object as last immediate subobject

    insafter - insert subobject after a designated position

    insbefore - insert reference before the specified position

    copyobject - copy an object (without attributes)

    deliobj - delete an integrated object

    delnobj - delete a non-integrated object

    intobj - integrate an object

purgerelnd - disconnect an object from an immediate supobject

repsub - replace a subobject

repsup - replace a supobject

2.2. Positioning without Update Security

The routines in this set are used to traverse an object, but it is assumed that the user will not alter the object being traversed. If this assumption cannot be made, then the routines that utilize updatable pointers must be used. These routines tend to involve less overhead and slightly more ease of use.

firstsub - access the first immediate subobject of a given object

nextsub - access the next immediate subobject of an object

firstsup - access first immediate supobject of an object

nextsup - access the next immediate supobject of an object

lastsub - access the last immediate subobject of a given object

prevsub - access the previous immediate subobject of an object

lastsup - access the last iimmediate supobject of an object

prevsup - access the previous immediate supobject of an object

ithchild - access the ith immediate child of an application

numsub - get the number of immediate subobjects of a given object

numsup - get the number of immediate supobjects of a given object

whichsub - get the subscript of an immediate subobject

2.3. Traversing with Updatable Pointers

The following routines allow the user to access subobjects and supobjects without worrying about the effects of intervening insertions and deletions.

alupb - allocate an updatable pointer block

dealupb - deallocate an updatable pointer block

ufsub - updatable access to the first immediate subobject

unsub - updatable access to the next immediate subobject

ulsub - updatable access to the last immediate subobject

upsub - updatable access to the previous immediate subobject

cancusub - cancel position in updatable immediate subobject set

ufsup - updatable access to the first immediate supobject

unsup - updatable access to the next immediate supobject

ulsup - updatable access to the last immediate supobject

upsup - updatable access to the previous immediate supobject

cancusup - cancel position in updatable immediate supobject set

firstdownpos - access first subobject in the set of all contained objects

nextdownpos - access the next subobject in the set of all subobjects

cancdownpos - cancel position in the set of all contained objects firstuppos - get the first containing object in the set of all containing objects

nextuppos - advance to the next containing object in the set of all containing objects

cancuppos - cancel a position in the set of all containing objects

2.4. Processing Attributes of an Object

The following routine allows the user to assign, delete, and access attributes. The user variables are an array of "low-cost" integer attributes kept in the object node itself.

setattr - set an attribute on an object

delattr - delete an attribute from an object

getattr - get the attribute for a given code

getfattr - get the first attribute on a given object

getnattr - get the next attribute on a given object

setuvar - set a user variable for an object

accuvar - access the value of a user variable of a given object

2.5. Properties

The following routines allow the user to associate search properties with objects. These properties can be used to rapidly isolate pertinent objects from the database of objects (normally properties are associated with only integrated objects, but the only actual requirement is that any object must have an assigned id before properties are assigned to it.) When an object is integrated, the properties used for rapid unification checks are automatically assigned to it.

setprop - set a property for an object

delprop - delete a property of an object

locfp - locate the first of a set of objects by property

locnp - locate the next object which satisfies a condition

cancploc - cancel a property location position

locfunify - locate the first object that unifies and satisfies

locnunify - locate the next object that unifies and satisfies

cancpuloc - cancel position in search on properties and unification

2.6. Unification, Applying a Substitution, and Renaming

The routines to perform unification, apply a substitution, and rename the variables in an object are as follows:

unifyfobj - get the first unifier of two objects

unifynobj - return next unifier of two objects

cancunify - cancel position in a set of unifiers

apply - apply a substitution to an object

imgvar - get image of a variable

clearsub - clear entries in a substitution (only those in the change list)

delchlist - delete a change list

delsub - delete a substitution

normobject - rename the variables in an object

2.7. Miscellaneous Routines

These routines are required, but do not fall into any of the above categories:

    initcom1 - get an initialized common area for layer1
    assignid - assign an id to an object
    refid - access the id of the referenced object
    idref - access the object with the given id
    inputobj - convert <object>; to internal format
    outputobj - convert object from internal to portable format
    objis - returns what an object is (name, variable, or application)
    namendx - get the symbol table number corresponding to a name
    nameval - get the symbol corresponding to an object that is a name
    varnum - return the variable number of a variable

Logic Machine Architecture:
Inference Mechanisms

Ewing L. Lusk

Northern Illinois University
Argonne National Laboratory

William McCune

Northwestern University

Ross A. Overbeek

Northern Illinois University

ABSTRACT

Logic Machine Architecture (LMA) is a layered implementation of theorem-proving tools. The kernel of formula storage, retrieval, and manipulation primitives (layers 0 and 1) is described elsewhere[2]. The layer described here (layer 2) contains resolution- and equality-based inference rules, subsumption, and demodulation. It is designed to provide all of the tools required to create a theorem-prover with minimal effort. Although LMA is currently implemented in Pascal, an interface to LISP will be included in the original release. This paper includes the design principles and techniques used in layer 2, as well as two simple theorem provers which illustrate the services of layer 2 - one written in LISP and the other in Pascal.

1. Introduction.

Research in automated theorem proving has been hindered by a lack of powerful, widely available theorem-proving systems. The time and effort required to create useful programs has proven to be substantial. Many researchers simply do not have access to a system suitable for experimentation. Logic Machine Architecture (LMA) has been designed to provide a varied set of software tools suitable for use by everyone from researchers in theorem proving to developers of application systems based on theorem provers. The tools themselves are written in Pascal, but an interface to LISP will be included in the original distribution.

The purpose of LMA is to make it possible to easily and quickly prduce diverse theorem-proving programs, either to test new ideas in theorem-proving or to implement application systems based on theorem-proving principles. As an example of how easy it is to construct a theorem prover using the layer 2 primitives, we give a two-page Pascal theorem prover in Appendix B. The corresponding LISP version of the same theorem prover is in Appendix C.

Throughout the history of automated theorem proving researchers have expressed the desire for tools that would allow comparisons between varied approaches within a common environment. To achieve this level of experimentation will require a system that is well layered, easily modifiable, and well supported. It is hoped that LMA will evolve into such a system and will offer the environment in which comparisons between clause-based formalisms, natural deduction, and higher-order logics can meaningfully occur.

LMA has been conceived and implemented as a layered architecture. Each layer has an independent function and provides a set of well-defined services. This type of architecture allows a user to select the tools appropriate to his task. A user of theorem provers can implement a powerful theorem-prover in a few days (or simply use one of the standard programs included in the release). On the other hand, someone wishing to construct a system based on a new formalism will use only the tools provided in the lower layers to construct a set of new tools (which, if they follow the architectural guidelines of the higher layers, can be added to the services provided by the higher layers).

Layer 0 of LMA implements several abstract data types required in outer layers. Layer 1 contains the functions required to create and maintain a database of formulas. These two layers have been described in detail in [2].

Layer 2 is a package of routines that define the abstract data types *list*, *clause*, *literal*, and *term*. In addition it provides a package of inference rules, demodulation, and subsumption. These provide the basic services from which a theorem prover can be easily constructed.

Layer 3, when it is completed, will provide uniprocessing modules from which a multiprocessing theorem prover could be configured. Layer 4 will implement the logic required to configure and manage cooperating multiple processes.

This paper discusses the details of layer 2. It is broken into the following sections:

a) a discussion of the implementation of the basic data types - list, clause, literal and term - and the inference services provided for clauses,

b) an overview of how subsumption is implemented,

c) an overview of a generalized approach to implementing inference rules, and

d) a discussion of the interface to LISP.

The last section attains its significance from the fact that LMA is currently written in Pascal. It is included to accentuate our desire to provide a set of tools usable by AI researchers and accessible from any major language.

2. The Abstract Data Types of Layer 2

2.1. The Fundamental Role of Objects

Layer 1 of LMA was designed to be general enough to support almost any research requiring the manipulation of logical formulas. It provides the storage, retrieval, and manipulation services required to manage a large database of formulas. The "query" capability - that is, the ability to rapidly retrieve all formulas satisfying a specified set of properties - is particularly critical. The initial release of layer 2 of LMA will provide only the abstract data types required to support clause-based theorem-proving systems. However, it is designed with the specific goal that this set can be enriched, using the powerful services of layer 1, to include the features required to support other formalisms.

One explicit goal of LMA is flexibility based on a layered architecture. Throughout the implementation an attempt has been made to hide the underlying data structures through a well-defined set of services provided to the user. Hence, layer 2 will contain the services required to hide the exact structure of the data types list, clause, literal, and term.

While the internal structure of any abstract data type defined by layer 2 is hidden from the user of layer 2, one aspect of their implementation is not hidden - they must all be instances of the general data type *object*. Objects are defined in [2], and are the central data type supported by layer 1. Briefly, an object is a structural concept which allows a natural representation of logical formulas, among other things. Objects are of three types: names, variables, and applications. Applications are trees with names and variables for leaves. (For example, the term $f(x,a)$ can be represented by an application object with three subobjects: the name f, the variable x, and the name a.) The definition of lists, clauses, literals, and terms as objects has the following immediate implications, because of the object-oriented services provided in layer 1:

a) Each occurrence of these abstract data types can have *attributes* assigned to it. These attributes are themselves arbitrary objects. Similarly, each has a set of "low-cost" integer-valued attributes called *user variables* that can be set and accessed.

b) Occurrences of these data types can be *integrated* into the object database, in which structure-sharing occurs.

c) Occurrences can have search *properties* assigned to them. A user can then rapidly extract the objects that are pertinent to any given situation.

These concepts are presented in detail in [2].

2.2. Lists

A *list* is the data type used to create and maintain a set of objects. As such, it plays a central role in layer 2.

Traversing the elements in a list must work, even in the presence of intervening isertions and deletions. Consider the situation, for example, in which a user of layer 2

a) establishes position on a particular clause in a list of clauses,

b) uses it to generate a new clause,

c) deletes the clauses subsumed by the new clause (which may be every clause in some lists), and then

d) attempts to retrieve the next clause in the original list.

It is easy t see that the deletions may compromise the position in the list unless some mechanism is provided for maintaining the integrity of positions in lists in the face of alterations to the object database. The list traversal functions of layer 2 do provide this mechanism, using the "updatable pointer blocks" implemented in layer 1. The "access next list element" function is guaranteed to work even if deletions have occurred since position was established.

2.3. The External Format of Layer 2 Data Types

Layer 2 is designed to support an expanding set of abstract data types. Initially only the four mentioned above will be included. While the internal structure of each data type is effectively hidden, each data type other than lists has a well-defined external format. That is, each data type will have a specified string format.

The external format of a data item is not necessarily considered to be a format suitable for use as an interface for human users. Rather, it is an intermediate format into which user-input formulas are translated. Similarly, a formula in external format will have to be translated into a more readable format for output. It is not necessary or desirable that we standardize the user interface formats, since the formats appropriate to one application, such as automated circuit design, might well be unsuitable to other applications, such as group theory or symbolic logic. Instead, LMA routines will interface through the external format to a wide variety of users, both human users and other programs. The precise format of clauses, literals, and terms will be described separately elsewhere. It will be of interest to those implementing user interfaces or to those intending to coordinate data transfer between machines or software systems. It has become apparent to us that the applications for which LMA will eventually be used require communication between theorem provers and other large software packages. Specifically, we currently require the ability to transfer intermediate subproblems to packages such as MACSYMA or numerical analysis systems written in FORTRAN. We will achieve this by coding system interface routines that will (among other tasks) translate to and from the external format.

2.4. Services for a Layer 2 Data Type

For each of the layer 2 data types, other than lists, LMA will contain at least the following basic types of services:
1) Routines to convert back and forth between external and internal formats.
2) Routines to allow the construction of the data type.
3) Routines to traverse the data type and return its components, such as te predicate symbol of a literal or the subterms of a term.
4) Routines to extract or locate fields specific to the data type (such as the sign of a literal).

A short perusal of the routines included in Appendix A will clarify these comments.

Each data type will have the above basic services. In addition LMA includes inference services for some types. Currently inference services are provided only for clauses, but supporting services are provided which make these expandable to include other inference mechanisms such as natural deduction.

Layer 2 provides a number of services for deriving new clauses from existing clauses, as well as for carrying out complete subsumption checks. As rapidly as possible, we intend to add all of the usual resolution rules and inequality techniques [1]. Nonclause data types will be defined and inference mechanisms added as a user community which requires them develops.

Thus, it is understood from the beginning that layer 2 will continue to grow over the life of LMA. What must be established is the pattern for growth. There are essentially two ways in which layer 2 will change:

a) More fundamental abstract data types (besides clauses) will be included.
b) New inference rules will be added.

The first type of change will be rare and offers no apparent design difficulties, particularly since those data types which are implemented as objects will be able to utilize directly the services of layer 1. The second type will be common.

An inference rule will normally take the following two inputs:
1) A given object
2) A list of lists

The second parameter, called the *parent object*, specifies a set of lists from which other parents (besides the given one) may be chosen. Some rules, such as factoring or variable elimination[1], will not require a parent list. All inferred objects have the given object as a parent.

For each inference rule three routines are included. To illustrate let us consider the specific instance of hyperresolution:
1) *hyperf* returns the first hyperresolvent derivable from the given clause and clauses occurring in the parent list (i.e., in lists occurring in the parent list).
2) *hypern* returns the next hyperresolvent in the set whose construction was initiated by a call to hyperf.
3) *hypercanc* cancels position in the set of hyperresolvents.

The first routine computes the first in a set of inferred clauses and creates a *position* in the set. The position can be used by the second routine to resume the computation and produce the next inferred clause. If the user decides not to retrieve the entire set, the third routine is used to *cancel the position in the set*. This is required, since a position may contain a substantial amount of data - much of it in dynamically allocated memory.

This ability to retrieve inferred clauses one-at-a-time is convenient for the user of layer 2. However, it is also important that the user of layer 2 be able to process the clauses one-by-one. For example, in the theorem provers included later in this paper, after each resolvent is generated full subsumption tests are performed. This could result in numerous deletions from the object database. Thus, positions must remain valid through intervening insertions and deletions. This adds a burden to the implementor of new inference rules (in that it will often be necessary to use *updatable pointers However, it makes the resulting services particularly simple to use.*

Finally, it is worth noting that subsumption services are based on the same pattern of routines. Normally, only the first clause will be utilized in forward subsumption (the case where a given clause is subsumed). With back subsumption (the case in which the given clause subsumes existing clauses), the normal case consists of locating and deleting all of the clauses in the set.

2.5. Output from an LMA Theorem Prover

In most cases the main output of an LMA theorem prover will be a *log file* that records *events*. A post-processor will use this file to produce any desired user output. In an interactive environment the log file can simply be piped to an appropriate post-processor.

It is impossible to specify all of the events that will eventually be included in a log file. As new inference rules are added, new event types will be necessary. However, guidelines for the formats of log events can be established:

[1] Each event is represented as an object in external format. Hence, the log file is a sequence of objects which can, if desired, be directly processed by a tailored LMA theorem prover.

[2] The first subobject in an event is a name which represents the general category of the event.

The LMA theorem prover is responsible for writing out any events to the log file that are required to extract desired output. This requires that each inference rule return the information required to produce the corresponding log event. Each of the inference rules returns a vector containing such data. The returned vector can be converted into the corresponding log event (i.e., an object in external format) by invoking the layer 2 service *loginference*.

One specific post-processor will extract a proof from the log file. A proof will itself be an object in external format. The actual formatted output presented to a user will be created by a program that takes the proof as input.

There will probably be a number of desired final output formats, each appropriate to a class of applications. For example, in circuit design, a "theorem to prove" corresponds to a set of design specifications, and a "proof" corresponds to a circuit which meets those specifications. There is no reason to present the proof as a proof; instead, the pattern of inferences which produced it should be formatted to suit the circuit designer's taste. Other "proofs" represent answers to database queries, and should be presented as such. Our approach of defining the formats of events and proofs will hopefully allow a number of useful post-processors to evolve.

3. Implementation of Subsumption and Inference Rules

Since we intend to implement numerous inference rules, layer 2 includes a general exit-driven mechanism for defining new rules. The significance of this structure is that it makes the implementation of new inference mechanisms easy, tus encouraging both experimentation with new inference rules and the development of special-purpose rules for specific applications (such as the mechanism for extracting a proof from an event log). Currently, subsumption uses an exit-driven routine for locating relevant integrated clauses. The inference rules use this same routine, as well as an exit-driven routine that controls the invocation of inference-specific logic to create the derived clauses. In this section we cover the basic structure of both subsumption and the inference mechanisms.

3.1. Clash Positions

Many layer 2 algorithms (including subsumption and the currently implemented inference rules) search the object database to locate clauses that "clash" with a given clause. For example, in resolution the search is for clauses containing literals that clash with literals in a given clause. In paramodulation the search is for clauses that contain a positive equality literal with an argument that unifies with a term in the given clause. This type of search may be thought of as utilizing an algorithm with three nested loops. Within the innermost loop some operation (such as form a resolvent or create a paramodulant) is performed. The outer loop visits the appropriate terms (here term is used loosely to mean term or literal) of the given clause, the middle loop uses a search-by-property mechanism[2] to locate terms in the object database to "clash" with a specified term in the given clause, and the inner loop visits appropriate clauses that contain the current term of the middle loop. Such an algorithm generates a set of clauses (i.e., performs the operation within the loops an arbitrary number of times) before the algorithm exits. Following the "getfirst/getnext" LMA philosophy, we have broken such algorithms into two components, one which establishes a *clash-position*, and one which controls the operation that would otherwise be performed within the loop. A clash-position contains, among other things, three subpositions that correspond to the positions of the three nested loops. The format of a clash-position is:

(1) given-clause — a clause that must be a parent of any binary operation of the sequence;

(2) given-term — a term that occurs in given-clause;

(3) given-position — the position of given-term in the subterms of given-clause;

(4) found-term — an integrated term that clashes with given-term;

(5) query-position — the position in the sequence of terms that clash with given-term;

(6) substitution — a unifier for the clash of given-term and found-term;

(7) found-clause — a clause that contains found-term;

(8) found-position — the position of found-clause in the containing terms of found-term;

(9) exits — a vector of procedure addresses;

(10) clash-lists — a list of lists in which found clauses must occur;

(11) data-area — a data area that can be used for communication between the exits.

A single exit-driven procedure is used to advance clash-positions by advancing the appropriate subpositions. Exits 1, 2, and 3 correspond to the outer, middle, and inner loops, respectively, and direct the advancement of a clash-position. Each of the three exits receives as input a clash-position and returns a code indicating whether or not the subposition was successfully advanced.

- Exit 1 sets given-term to the first or next appropriate term of given-clause.
- Exit 2 sets found-term to the first or next term that clashes with given-term.
- Exit 3 sets found-clause to the first or next appropriate clause that contains found-term.

The algorithm for advancing a clash-position is straightforward. It is assumed that found-position and query-position are **nil** in a initial clash-position. The algorithm sets return-code to "advanced" (successfully located the desired position) or "done" (the end of the set of acceptable positions was reached).

```
procedure advance-clash-position(clash-pos,return-code)
    begin
    return-code ← "advanced";
    if found-position ≠ nil then
        call exit-3(clash-pos,sub-retcd);
    while sub-retcd = "done" and return-code = "advanced" do
        begin
        if query-position ≠ nil then
            call exit-2(clash-pos,sub-retcd);
        while sub-retcd = "done" and return-code = "advanced" do
            begin
            call exit-1(clash-pos,sub-retcd);
            if sub-retcd = "done" then
                return-code ← "done"
            else
                exit-2(clash-pos,sub-retcd)
            end
        if return-code = "advanced" then
            exit-3(clash-pos,sub-retcd)
        end
    end procedure.
```

3.2. Subsumption

Backward subsumption requires the location of all clauses in the object database that are subsumed by a given clause. Our implementation uses a clash-position as described in the previous subsection. The three exits for backward subsumption operate as follows:

- ex1-backsub visits only the first literal of given-clause;
- ex2-backsub visits all literals in the object database that are instances of given-term;
- ex3-backsub visits all of the desired clauses in the object database that contain found-term.

The three top level algorithms for backward subsumption follow:

```
procedure backsub-first(given-clause,clash-lists,return-clause,clash-pos)
    begin
    initialize clash-pos with given-clause;
    set the first 3 exits of clash-pos to the back subsumption exits;
    call backsub-next(clash-pos,return-clause)
    end procedure.
```

```
procedure backsub-next(clash-pos,return-clause)
    begin
    return-clause ← nil;
    return-code ← "advanced";
    while return-clause = nil and return-code = "advanced" do
        begin
        call advance-clash-position(clash-pos,return-code);
        if return-code = "advanced" then
            begin
            let i be the index of the literal given-term in given-clause;
            call sub-test(found-clause,given-clause,{i},substitution,rc);
            if rc = "subsumed" then
                return-clause ← found-clause
            end
        end
    end procedure.

procedure backsub-cancel(clash-pos)
    begin
    free clash-pos (i.e., clash-pos is no longer an "active" position)
    end procedure.
```

The procedure sub-test is a recursive routine that attempts to map the remaining literals of given-clause into found-clause by extending the substitution. The algorithm is given in [3].

Forward subsumption requires finding clauses in the object database that subsume a given clause. The implementation is similar to that for backward subsumption with the following exceptions:

(1) ex1-forsub visits all literals of given-clause;

(2) ex2-forsub visits literals that are more general than given-term;

(3) in the call to sub-test, the arguments found-clause and given-clause are reversed and the index of found-term is sent.

3.3. Inference Rules

A single exit-driven controlling mechanism is used to implement most binary-based inference rules of layer 2. By a binary-based inference rule, we mean one that can be achieved by performing a sequence of binary operations (e.g., creating a hyperresolvent as a sequence of P1-resolvents). This inference mechanism requires that a clash-position contain eight exits - the three position exits required for advance-clash-position and five more to control the generation and disposition of inferred clauses. Rather than operate on a single clash-position, the algorithms create and maintain a stack of clash-positions. Each clash-position in the stack represents an intermediate point in the computation of the completely formed result. The exact functions of the five added exits (exits 4-8) are as follows:

- Exit 4 is called if advance-clash-position cannot advance a clash-position. Exit 4 decides whether or not given-clause of the clash-position is returned to the user.
- Exit 5 makes a binary inference from a clash-position.
- Exit 6 receives a newly inferred clause and the stack. Any number of new clash-positions are initialized with the inferred clause as given-clause, and the new clash-positions are pushed onto the stack.

- Exit 7 decides whether or not a newly inferred clause is returned to the user.
- Exit 8 receives the stack and a clash-position. The clash-position is either freed or pushed onto the stack.

The algorithms for the three top level routines of the inference mechanism are as follows:

procedure infer-first(given-clause,exits,stack,return-clause)
    begin
    Create an initial clash-position with given-clause and exits;
    Push the clash-position onto the empty stack;
    call infer-next(stack,return-clause)
    end procedure.

procedure infer-next(stack,return-clause)
    begin
    return-clause ← **nil**;
    while return-clause = **nil and** the stack is not empty **do**
        begin
        clash-pos ← pop(stack);
        advance-clash-position(clash-pos,return-code);
        if return-code = "done" **then**
            begin
            call exit-4 (clash-pos,return-code);
            if return-code = "return" **then**
                return-clause ← given-clause
            end
        else
            begin
            call exit-5 (clash-pos,inferred-clause);
            call exit-6 (inferred-clause,clash-pos,stack);
            call exit-7 (inferred-clause,clash-pos,return-code);
            if return-code = "return" **then**
                return-clause ← inferred-clause;
            call exit-8 (clash-pos,stack)
            end
        end
    end procedure.

procedure infer-cancel (stack)
    begin
    free the stack of clash-positions
    end procedure.

Exit 6 deserves special attention. After a new clause is inferred by exit 5 it is given to exit 6 which creates any number of new clash-positions with the new clause as given-clause. This gives the ability to spawn a new type of "reasoning" within the inference mechanism. For example, if a new clause is generated by resolution and exit 6 decides that the new clause satisfies some syntactic criterion or was derived in a particular way, then a new task may be created with the new clause as given-clause and with exits that say "start paramodulating into given-clause."

The above mechanism has been used to implement hyperresolution, UR-resolution [4], demodulation, and several versions of paramodulation.

Hyperresolution

The following exits direct the inference mechanism to return hyperresolvents when the given clause is a nucleus (non-positive clause):

- ex1-hyper-nuc visits only the first negative literal of given-clause;
- ex2-hyper visits positive literals in the object database that clash with given-term and occur in at least one positive clause;
- ex3-hyper-nuc visits all appropriate positive clauses that contain found-term;
- ex4-hyper says "no", do not return given-clause;
- ex5-hyper performs a binary (P1) resolution;
- if the inferred clause is not positive, then ex6-hyper creates one new clash-position with this set of exits and the inferred clause as given-clause;
- ex7-hyper says return an inferred clause iff it is positive;
- ex8-hyper always stacks the clash-position.

The resulting algorithm infers hyperresolvents via iterated P1-resolution — intermediate clauses are generated then freed. This is in contrast to the hyperresolution algorithm described in [6] which does not create intermediate clauses. For a more complete discussion (including a comparison of the algorithms) see [5].

Example 1

We now present a detailed example to illustrate both the inference mechanism and the way in which the above exits direct the formation of hyperresolvents. For simplicity we present a propositional example. This means that exit 2 visits at most one literal since we need exact matches rather than unifiable literals.

Let the given clause be the nucleus
$$NUC: \neg P \ \neg Q \ R$$
and let the only integrated positive clauses be
$$SAT1: \ P \ S1$$
$$SAT2: \ P \ S2$$
$$SAT3: \ Q \ T.$$
Note that there are two hyperresolvents
$$H1: \ R \ S1 \ T$$
$$H2: \ R \ S2 \ T.$$
The routine infer-first is called with NUC and the above exits. The stack is initialized to
given-clause: $\neg P \ \neg Q \ R$, no positions yet.

The routine infer-next is then called. The clash-position is popped and advance-clash-position is called. Exit 1 finds $\neg P$, exit 2 finds P, and exit 3 finds $SAT1$. A resolution in performed. Because the resolvent is not positive, a new clash-position is created and stacked. The advanced clash-position is stacked. After the first iteration of the infer-next loop, the stack is
given-clause: $\neg P \ \neg Q \ R$, pos1=$\neg P$, pos2=P, pos3=$SAT1$
given-clause: $\neg Q \ R \ S1$, no positions yet.

After the second iteration of the loop the stack is
given-clause: $\neg P \ \neg Q \ R$, pos1=$\neg P$, pos2=P, pos3=$SAT2$
given-clause: $\neg Q \ R \ S2$, no positions yet
given-clause: $\neg Q \ R \ S1$, no positions yet.

During the third iteration the popped clash-position cannot be advanced so it is discarded. After the third iteration the stack is

given-clause: $\neg Q\ R\ S2$, no positions yet
given-clause: $\neg Q\ R\ S1$, no positions yet.

During the fourth iteration the positive clause $H1: R\ S2\ T$ is generated and returned to the user as a hyperresolvent. The stack that is returned for subsequent infer-next calls is

given-clause: $\neg Q\ R\ S2$, pos1=$\neg Q$, pos2=Q, pos3=$SAT3$
given-clause: $\neg Q\ R\ S1$, no positions yet.

The first (user) call to infer-next makes two loops and returns the hyperresolvent $H2: R\ S1\ T$ and the stack

given-clause: $\neg Q\ R\ S1$, pos1=$\neg Q$, pos2=Q, pos3=$SAT3$

The second (user) call to infer-next pops the stack, advance-clash-position fails, the stack is empty and no hyperresolvent is returned signaling that there are no more.

4. The LISP Interface

4.1. Choice of Language

LMA is currently implemented in Pascal. A number of colleagues have questioned the choice. Certainly we were forced to laboriously build up service functions in lower layers that are standardly available in LISP. Furthermore, LISP is clearly the predominant language in AI research.

The languages that were considered were LISP, Pascal, C, and ADA. The decision to use Pascal was based on the following observations:

a) It was difficult to determine the relative efficiencies of implementations in LISP and implementations in the other languages. However, those LISP users that we consulted indicated that if efficiency were a critical issue, LISP might not be an appropriate choice.

b) We felt a strong need to formulate the tools required to construct theorem provers based on coordinated multiple processes. ADA seemed by far the most promising language for such a project.

c) An automated Pascal to ADA translation appears quite possible.

It was the desire to build a theorem prover for a multiprocessing environment that made ADA a likely candidate. Since ADA is not yet widely available, we decided to use the language that would most likely be translatable into ADA.

Having decided on implementing LMA in Pascal, we decided to create an interface to LISP. From many points of view this seemed to us to represent the most viable approach. Ideally this would allow us to more directly confront efficiency issues, while allowing users the convenience of LISP.

4.2. Construction of the Interface

When interfacing routines written in two languages, memory management almost inevitably becomes a central issue. If dynamically acquired data elements move across the boundary, a host of compiler dependencies are introduced. Rather, we have adopted the approach of moving only copies across the interface. These copies are moved into data elements owned, acquired, and controlled by the receiving modules.

Besides moving data across the interface, it becomes necessary for modules on one side to simply retain pointers to data elements owned by modules on the other side. For example, the LMA routines (because Pascal lacks static variables) reference a common area, which is a parameter to most routines. The pointer to this common area must be retained by LISP and passed as a parameter to Pascal service routines. Since the pointer does not reference a valid

LISP value, it must be disguised (e.g., as an integer). This is necessary in order to avoid problems during LISP garbage collection.

Clearly an implementation of the above concepts is somewhat compiler dependent. However, assuming that there is a common system allocation/deallocation mechanism used by both LISP and Pascal (so dynamically acquiring memory is possible from both sides of the interface), the details are reasonably straightforward. There is one unfortunate consequence of the above strategy - the Pascal services must operate as subroutines, rather than functions (to avoid passing a Pascal-allocated node back to a LISP function). Since this is quite counter to the outlook imposed by LISP, it is fair to say that a significant cost in terms of programming style exists for using the interface. In particular the main LISP function, which must retain Pascal pointers, will have to use the PROG feature. The LISP theorem prover in Appendix C illustrates the problem, and in many ways might irritate a LISP purist.

4.3. Example Theorem Provers

To illustrate the use of LMA services, we have included in Appendix B a simple theorem prover that includes both hyperresolution and subsumption. Such a program can clearly be programmed and debugged in a few days (assuming that the services are properly documented and debugged). Although we do not intend for the reader to understand the details of each variable and service invocation, the basic level of complexity and power should be apparent. Note that it utilizes an integrated (structure-shared) object database, access of formulas via properties, and the other features that we have discussed. They are, for the most part, effectively hidden from the user.

In Appendix C we include a directly translated (manually) LISP version of the program in Appendix B. By using a package of macros, we have hidden most details of the interface. This short example does not illustrate the movement of data across the interface (such as examing a predicate symbol from LISP). Such features are included in the complete interface.

5. Summary

The goal of LMA is to provide convenient access to tools for the creation and use of theorem-proving systems. It is currently implemented under UNIX. An interface for programs written in LISP (Franz Lisp) has been prepared. We hope that, as a user community forms, a large body of tools can be made widely available. The next steps we anticipate in this direction are:

    a)   to develop a good input translator and a post-processor;

    b)   to develop interfaces to MACSYMA and numerical analysis software running under UNIX;

    c)   to add arithmetic simplification to the LMA inference rules; That is, LMA inference rules should automatically simplify a wide class of arithmetic expressions.

    d)   to add more inference rules to layer 2;

    e)   to install functions which provide at least a limited set of heuristics for evaluating the "worth" of clauses, literals, and terms. Such a mechanism has been essential to the success of our present system.

    f)   to proceed then with the development of layers 3 and 4 of LMA.

Acknowledgement

This work was supported by the Applied Mathematical Sciences Research Program (KC-04-02) of the Office of Energy Research of the U.S. Department of Energy under Contract W-31-109-Eng-38, and also by National Science Foundation grants MCS79-03870 and MCS79-13252.

References

1. W. W. Bledsoe and Larry M. Hines, "Variable elimination and chaining in a resolution-based prover for inequalities," in *Proceedings of the Fifth Conference on Automated Deduction, Springer-Verlag Lecture Notes in Computer Science, v. 87*, ed. Robert Kowalski and Wolfgang Bibel, (July 1980).

2. E. Lusk, William McCune, and R. Overbeek, *Logic machine architecture: kernel functions*, preprint.

3. E. Lusk and R. Overbeek, "Data structures and control architecture for the implementation of theorem-proving programs," in *Proceedings of the Fifth Conference on Automated Deduction, Springer-Verlag Lecture Notes in Computer Science, v. 87*, ed. Robert Kowalski and Wolfgang Bibel, (1980).

4. J. McCharen, R. Overbeek, and L. Wos, "Complexity and related enhancements for automated theorem-proving programs," *Computers and Mathematics with Applications* 2 pp. 1-16 (1976).

5. William W. McCune, *An inference mechanism for resolution-style theorem provers*, Master's Thesis, Northwestern University 1981.

6 R. Overbeek, "An implementation of hyper-resolution," *Computers and Mathematics with Applications* 1 pp. 201-214 (1975).

Appendix A

1. Layer 2

1.1. Lists

The following routines define the abstract data type *list*.

    lstcreate - create an empty list
    lstinsfirst - insert an object at the head of a list
    lstinslast - insert an object at the end of a list
    lstinsbefore - insert an object ahead of a designated position in a list
    lstinsafter - insert an object after a designated position in a list
    lstaccfirst - access the first element in a list
    lstacclast - access the last element in a list
    lstaccnext - access the next element in a list
    lstaccprev - access the previous element in a list
    lstcancpos - cancel a position in a list
    lstnumel - find the number of elements in a list
    lstdisconnect - remove an element from a list
    lstaltpos - has the object referenced by a position been disconnected?
    lstloc - locate an object in a list, if it is there
    lstdelete - delete an empty list
    lstcopy - copy a list

1.2. Clauses

The following routines describe the abstract data type *clause*.

    clinput - convert a clause from external to object format
    clread - read in a clause from a file
    cltread - read in a clause from the terminal
    cllstread - read in a list of clauses from a file
    cllsttread - read in a list of clauses from the terminal
    cloutput - convert a clause from object to external format
    clwrite - write a clause to a file
    cltwrite - write a clause to the terminal
    cllstwrite - write a list of clauses to a file
    cllsttwrite - write a list of clauses to the terminal
    clcreate - create a null clause
    clinslit - insert a literal (at a specified subscript)
    clacclit - access a literal via a subscript
    clnumlit - find the number of literals in a clause

cldisconnect - remove a literal from a clause
clintegrate - integrate a clause
cldelint - delete an integrated clause
cldelnon - delete a nonintegrated clause
clcopy - copy a clause

1.3. Literals

The following routines implement the abstract data type *literal*.

litinput - convert a literal from external to object format
litoutput - convert a literal from object to external format
litcreate - create an "empty" literal
litinsarg - insert an argument (by subscript)
litdisconnect - remove argument (by subscript) from literal
litsign - check the sign of a literal
litpred - access the predicate symbol for a literal
litaccarg - access argument via subscript
litnumarg - access the number of arguments in a literal
litintegrate - integrate a literal
litdelint - delete an integrated literal
litdelnon - delete a nonintegrated literal
litaccatom - access the atom of a literal
litcopy - copy a literal

1.4. Terms

The following services define the abstract data type *term*.

trminput - convert a term from external to object format
trmoutput - convert a term from object to external format
trmcrvar - create a variable
trmcrcon - create a constant
trmaccvar - access variable number for a variable
trmacccon - access constant symbol
trmcrcomplex - create an "empty" complex term
trminsarg - insert an argument (by subscript) into a complex term
trmfunc - access the function symbol in a complex term
trmaccarg - access an argument by subscript
trmnumarg - get the number of arguments in a complex term
trmdisconnect - disconnect an argument from a complex term (by subscript)
trmintegrate - integrate a term
trmdelint - delete an integrated term

trmdelnon - delete a nonintegrated term

trmcopy - copy a term

1.5. Demodulation Services

The following routines allow a user to make a clause a demodulator, end its role as a demodulator, demodulate it (forward demodulation), or use it to demodulate existing clauses (back demodulation). Note that the parent is not deleted (i.e., the user must explicitly delete the parent, if that is what is desired).

clsetdemod - make an integrated clause a demodulator

clenddemod - end the use of a clause as a demodulator

bdemodf - back demodulation, first

bdemodn - back demodulation, next demodulant

bdemodcanc - cancel position in a set of "back" demodulants

fdemodf - forward demodulation, first

fdemodn - forward demodulation, next demodulant

fdemodcanc - cancel position in a set of "forward" demodulants

1.6. Subsumption Services

The following routines allow the user to locate clauses subsumed by the given clause (back subsumption) or to find clauses that subsume the given clause (forward subsumption).

bsubfirst - get first clause subsumed by given clause

bsubnext - get next clause subsumed by the given clause

cancbsub - cancel position in set of clauses subsumed by given clause

fsubfirst - get first clause that subsumes given clause

fsubnext - get next clause that subsumes the given clause

cancfsub - cancel position in set of clauses that subsume given clause

1.7. Inference Rules

The following routines implement inference rules. Hopefully, this set of functions will increase substantially over time.

loginference - convert a derivation vector to a log element

hyperf - generate hyperresolvents (first)

hypern - generate the next hyperresolvent

hypercanc - cancel position in a set of hyperresolvents

urf - generate URresolvents (first)

urn - generate the next URresolvent

urcanc - cancel position in a set of URresolvents

paraff - get the first paramodulant from the given clause

paranf - get next paramodulant from the given clause

parafcanc - cancel position in a set of "from" paramodulants

parafi - get the first paramodulant into the given clause

        parani - get next paramodulant into the given clause
        paraicanc - cancel position in a set of "into" paramodulants
        funitconflict - test for unit conflict (first)
        nunitconflict - generate the next null clause (using unit conflict)
        cunitconflict - cancel position in a set of null clauses
        ffactor - generate the first factor of the given clause
        nfactor - generate the next factor of a clause
        cfactor - cancel position in a set of factors
        fbinary - generate the first of a set of binary resolvents
        nbinary - generate the next binary resolvent
        cbinary - cancel position in a set of binary resolvents
        funit - generate the first of a set of unit resolvents
        nunit - generate the next unit resolvent
        cunit - cancel position in a set of unit resolvents
        fp1 - generate the first of a set of p1 resolvents
        np1 - generate the next p1 resolvent
        cp1 - cancel position in a set of p1 resolvents

1.8. Qualification and Locking

    The following routines control the use of qualification and locking. Locks are of two kinds. A literal can be locked (since structure sharing is used, this applies to all occurrences), or a single occurrence of the literal in a clause can be locked. A literal lock is an integer value. An occurrence lock is either on or off.

        setqual - add a qualification template
        setqwopt - set qualification warning msg options
        qualcl - mark qualifiers on a clause
        setcllock - lock an occurrence of a literal in a clause
        delcllock - unlock an occurrence of a literal in a clause
        getcllock - get the lock character for the occurrence of a literal
        setlitlock - set a lock value on a literal (gets all occurrences)
        getlitlock - access the literal lock value
        setiglock - set the flag that determines whether locks are ignored

Appendix B

```
const
#include 'l2constants.i';
type
#include 'l2types.h';

#include 'l2externals.h';

procedure tp;

var axlist,soslist,hbglist,clashlists,allclauses,
      givencl,resolvent,subsumer,subsumed,clause: objectptr;
    com2: common2ptr;
    histvec: ivector;
    retcd,listretcd,numlits: integer;
    hyperpos,subsumerpos,subsumedpos: stackentptr;
    sospos,listpos: upbptr;

    done: boolean;

begin{tp};

{first acquire the common area for layer 2 services}
initcom2(com2);

{read in the list of axioms}
cllsttread(axlist,retcd,com2);
if (retcd = 0) then
    begin
    writeln('axioms are as follows:');
    cllsttwrite(axlist,retcd,com2);
    end
else
    writeln('input of axioms list failed');

{now integrate the axioms - that is add them to the formulae database}
lstaccfirst(axlist,clause,listpos,listretcd,com2);
while (listretcd = 0) do
    begin
    clintegrate(clause,retcd,com2);
    lstaccnext(axlist,clause,listpos,listretcd,com2);
    end;

{now read in the set-of-support list}
cllsttread(soslist,retcd,com2);
if (retcd = 0) then
    begin
    writeln('set of support clauses are as follows:');
    cllsttwrite(soslist,retcd,com2);
    end
else
    writeln('input of set of support list failed');

{integrate the set-of-support clauses}
lstaccfirst(soslist,clause,listpos,listretcd,com2);
while (listretcd = 0) do
    begin
    clintegrate(clause,retcd,com2);
    lstaccnext(soslist,clause,listpos,listretcd,com2);
    end;

{make clashlists a list containing axlist and hbglist.
 make allclauses a list containing axlist, soslist, and hbglist}
lstcreate(hbglist,retcd,com2);
lstcreate(clashlists,retcd,com2);
```

```
lstcreate(allclauses,retcd,com2);

lstinslast(axlist,clashlists,retcd,com2);
lstinslast(hbglist,clashlists,retcd,com2);

lstinslast(axlist,allclauses,retcd,com2);
lstinslast(soslist,allclauses,retcd,com2);
lstinslast(hbglist,allclauses,retcd,com2);

{This is the main loop.  Select a clause from the set-of-support,
 generate all hyperresolvents between it, axioms, and clauses on the
 hbglist.  Put the generated hyperresolvents that are not subsumed onto
 the soslist.  When that is all done, move the given clause from the
 soslist to the hbglist (and start over - until no more clauses in the
 soslist or the null clause is generated)
}
done := false;
while not done do
    begin
    {select a "given clause}
    lstaccfirst(soslist,givencl,sospos,retcd,com2);
    if (retcd <> 0) then
        begin
        done := true;
        writeln('no more clauses in set of support');
        end
    else
        begin
        write('given clause is: ');
        cltwrite(givencl,retcd,com2);
        {generate the first hyperresolvent}
        hyperf(givencl,clashlists,resolvent,histvec,
               hyperpos,retcd,com2);
        {This loop processes generated hyperresolvents}
        while (retcd = 0) and (not done) do
            begin
            write('resolvent: ');
            cltwrite(resolvent,retcd,com2);
            {just pitch the derivation information}
            ivecclose(histvec,com2^.ivecavl);
            {now check for the null clause}
            clnumlit(resolvent,numlits,com2);
            if (numlits = 0) then
                begin
                writeln('null clause found');
                done := true;
                end
            else
                begin
                writeln('checking forward subsumption');
                fsubfirst(resolvent,allclauses,subsumer,
                          subsumerpos,retcd,com2);
                if (retcd = 0) then
                    begin
                    writeln('resolvent subsumed');
                    {cancel position in the set of clauses that subsume
                     the generated hyperresolvent}
                    cancfsub(subsumerpos,com2);
                    {delete the nonintegrated hyperresolvent}
                    cldelnon(resolvent,retcd,com2);
                    end
                else
                    begin
                    writeln('checking back subsumption');
```

```
                        bsubfirst(resolvent,allclauses,subsumed,
                                    subsumedpos,retcd,com2);
                        {This loop deletes clauses subsumed by the new
                          hyperresolvent}
                        while (retcd = 0) do
                            begin
                                writeln('resolvent subsumes existing clause:');
                                cltwrite(subsumed,retcd,com2);
                                cldelint(subsumed,retcd,com2);
                                bsubnext(subsumedpos,subsumed,retcd,
                                          com2);
                            end;
                        {add the hyperresolvent to the integrated formulae
                          database}
                        clintegrate(resolvent,retcd,com2);
                        lstinslast(resolvent,soslist,retcd,com2);
                    end;
                hypern(hyperpos,resolvent,histvec,retcd,com2);
                end{while};
            end;
        {If the given clause was not deleted (due to subsumption), move it
          to the hbglist}
        if not done then
            begin
                lstaltpos(sospos,retcd);
                if retcd = 0 then
                    begin
                        lstdisconnect(sospos,com2);
                        lstinslast(givencl,hbglist,retcd,com2);
                    end;
            end;
        end;{while}
end;{tp}
```

Appendix C

```
(defun lisptp ()
 (prog (axlist soslist hbglist clashlists allclauses
     givencl resolvent subsumer subsumed clause
     com2
     histvec
     retcd listretcd numlits
     hyperpos subsumerpos subsumedpos
     sospos listpos
     )

   (initvar axlist)
   (initvar soslist)
   (initvar hbglist)
   (initvar clashlists)
   (initvar allclauses)
   (initvar givencl)
   (initvar resolvent)
   (initvar subsumer)
   (initvar subsumed)
   (initvar clause)
   (initvar com2)
   (initvar histvec)
   (initvar retcd)
   (initvar listretcd)
   (initvar numlits)
   (initvar hyperpos)
   (initvar subsumerpos)
   (initvar subsumedpos)
   (initvar sospos)
   (initvar listpos)

   (initcom2 com2)

   (cllsttread axlist retcd com2)
   (cond
      ((zerop (valueof retcd))
       (print "axioms are as follows:")
       (terpr)
       (cllsttwrite axlist retcd com2)
      )
      (t
       (print "input of axioms list failed")
       (terpr)
      )
   )

   (lstaccfirst axlist clause listpos listretcd com2)
   (do ()
```

```
      ((not (zerop (valueof listretcd))) nil)
      (clintegrate clause retcd com2)
      (lstaccnext axlist clause listpos listretcd com2)
)

(cllsttread soslist retcd com2)
(cond
   ((zerop (valueof retcd))
    (print "set of support clauses are as follows:")
    (terpr)
    (cllsttwrite soslist retcd com2)
   )
   (t
    (print "input of set of support list failed")
    (terpr)
   )
)

(lstaccfirst soslist clause listpos listretcd com2)
(do ()
    ((not (zerop (valueof listretcd))) nil)
    (clintegrate clause retcd com2)
    (lstaccnext soslist clause listpos listretcd com2)
)

(lstcreate hbglist retcd com2)
(lstcreate clashlists retcd com2)
(lstcreate allclauses retcd com2)

(lstinslast axlist clashlists retcd com2)
(lstinslast soslist clashlists retcd com2)

(lstinslast axlist allclauses retcd com2)
(lstinslast soslist allclauses retcd com2)
(lstinslast hbglist allclauses retcd com2)

(setq done nil)
(do ()
   (done nil)
   (lstaccfirst soslist givencl sospos retcd com2)
   (cond
      ((not (zerop (valueof retcd)))
       (setq done t)
       (print "no more clauses in set of support")
       (terpr)
      )
      (t
       (print "given clause is: ")
       (terpr)
       (cltwrite givencl retcd com2)
```

```
(hyperf givencl clashlists resolvent histvec
     hyperpos retcd com2)

(do ()
   ((or (not (zerop (valueof retcd))) done) nil)
   (print "resolvent: ")
   (terpr)
   (cltwrite resolvent retcd com2)
   (ivecclose histvec com2)
   (clnumlit resolvent numlits com2)
   (cond
      ((zerop (valueof numlits))
       (print "null clause found")
       (terpr)
       (setq done t)
      )
      (t
      (print "checking forward subsumption")
      (terpr)
      (fsubfirst resolvent allclauses subsumer
            subsumerpos retcd com2)
      (cond
         ((zerop (valueof retcd))
          (print "resolvent subsumed")
          (terpr)
          (cancfsub subsumerpos com2)
          (cldelnon resolvent retcd com2)
         )
         (t
         (print "checking back subsumption")
         (terpr)
         (bsubfirst resolvent allclauses subsumed
               subsumedpos retcd com2)
         (do ()
            ((not (zerop (valueof retcd))) nil)
            (print "resolvent subsumes existing clause:")
            (terpr)
            (cltwrite subsumed retcd com2)
            (cldelint subsumed retcd com2)
            (bsubnext subsumedpos subsumed retcd
                  com2)
         )

         (clintegrate resolvent retcd com2)
         (istinslast resolvent soslist retcd com2)
         )
      )
      )
   )
   (hypern hyperpos resolvent histvec retcd com2)
```

```
        )
        (cond
          ((not done)
            (lstaltpos sospos retcd)
            (cond
              ((zerop (valueof retcd))
                (lstdisconnect sospos com2)
                (lstinslast givencl hbglist retcd com2)
              )
              (t nil)
            )
          )
          (t nil)
        )
      )
    )
   )
  )
 )
```

PROCEDURE IMPLEMENTATION THROUGH
DEMODULATION AND RELATED TRICKS[*]

by

S. K. Winker and L. Wos
Argonne National Laboratory
Argonne, Illinois 60439

Abstract

In writing a computer program, the programmer often uses procedures or subroutines to carry out frequently-occurring subsidiary tasks. These tasks range from simple bookkeeping and updating to the seeking of the generalization of two given formulas. Problems that one wishes to solve with the assistance of an automated theorem-proving program likewise often involve subsidiary tasks and the corresponding need for procedures (or their equivalent) to accomplish them. It is reasonable to conjecture that employment of procedures either requires the use of external programs or requires substantial modification of the existing theorem-proving program itself. In this paper we give examples that show that neither is the case. Demodulation is the key to implementing the procedures that accomplish such frequently-occurring tasks.

We give here sets of demodulators for accomplishing a wide variety of tasks. The most complex given demodulator set enables the program to find the least general subsumer of two given unit clauses. We also consider questions of counting and classifying, bookkeeping and updating, and cleanup after case analysis. Finally, we give a set of demodulators for coping with set-theory in an efficient and natural way.

The material presented here is but a sample of the additional possibilities for the use of demodulation. Since it was not known how to accomplish some of the cited tasks purely within the context of automated theorem proving, the examples given here may lead to an expanded use of automated theorem-proving programs in general. Thus, although the use of demodulation for both simplification and canonicalization demonstrates its value, it turns out that demodulation is even more powerful than was realized.

---

[*]This work was supported in part by the Applied Mathematical Sciences Research Program (KC-04-02) of the Office of Energy Research of the U.S. Department of Energy under Contract W-31-109-Eng-38 (Argonne National Laboratory, Argonne, IL 60439) and in part by NSF grant MCS79-03870 (Northern Illinois University, DeKalb, IL 60115).

1. Introduction

Demodulation [24] plays an important role in automated theorem proving. There are many examples of problems in the literature [8,17,18,19,21,26] whose proofs would not have been obtained without the heavy use of demodulation. Demodulation is used most often to simplify clauses, but it also is valuable in its use for canonicalization [16]. In this paper, we present a number of applications that are quite distinct from these two familiar uses.

The applications cited here arise from the desire to solve a problem that frequently occurs with the use of automated theorem-proving programs--but to solve it strictly within the context of automated theorem proving. The problem is that of getting the theorem-proving program to accomplish various subsidiary tasks that, in standard programming, would be handled by means of some procedure or subroutine. These tasks, which arise while attempting to solve some larger problem, vary from the counting of the occurrences of some symbol within a formula to finding the generalization of two given unit statements. It is reasonable to conjecture that employment of procedures to accomplish subsidiary tasks either requires the use of external programs or requires substantial modification of the existing theorem-proving program itself. The examples given here illustrate that neither is the case. We show how the procedures needed to accomplish such tasks can be implemented by means of demodulation. The corresponding sets of demodulators may at first appear somewhat unnatural. The sets do, however, accomplish the tasks in question easily and without disrupting the main search being conducted by the program.

A natural question that can be asked is: Why employ demodulation to accomplish the various tasks to be discussed? After all, there are two immediately obvious alternatives, that of additional programming and that of reliance on inference mechanisms. As for the first alternative, we wish to avoid the necessity of additional programming. The adjunction of code to carry out some task burdens the user with debugging, for example. In addition, such code often has the disadvantage of only being useful for just the task at hand. Demodulation, on the other hand, is a common feature of many existing automated theorem-proving programs. Employment of it thus avoids the need for additional programming. In addition, as will be seen from the examples to be given, one merely replaces one set of demodulators by another set when the required task changes. As for the second alternative, reliance on an inference mechanism does not in general provide the desired control over the scheduling of subsidiary tasks. If the scheduling is not properly handled, the program can be so diverted from the main objective that it is unable to complete the required work. By relegating the lesser tasks to demodulation, one prevents such interruptions from occurring. In fact the tasks are accomplished almost as asides.

We give sets of demodulators for accomplishing a number of unrelated tasks. The tasks include:

1. finding the least general subsumer of two given unit expressions;
2. classifying expressions according to some classification scheme;
3. counting the occurrences of some given symbol within expressions;
4. bookkeeping of various types;
5. cleaning up after case analysis; and
6. treating set-theory in an efficient and natural fashion.

Employment of sets of demodulators to handle tasks of the type just cited allows an automated theorem-proving program to conduct the main inquiry without hazardous diversion. Such sets of demodulators act as subroutines and provide the corresponding control over scheduling.

With the exception of certain built-in functions for arithmetic, the approach underlying this work relies on standard theorem-proving techniques. It is the combination of certain notation, uses of such functions as the "list" function, and the nonstandard employment of demodulation that causes us to view some of what we present as "tricks". Although the tricks vary widely in complexity, taken as a whole they suggest the possibilities for having an automated theorem-proving program rather easily implement diverse procedures. We suspect that these examples will readily suggest the solutions to comparable problems with the use of an automated theorem-proving program--solutions that employ demodulation rather than relying directly on some inference rule.

2. Demodulation: Treatment and Discussion

Before turning to the heart of the paper, we give both a brief account of the treatment accorded to demodulation within our program and the answer to the following obvious but necessary question. In the examination of the notation employed in the various examples included in this paper, it is natural to ask: How does the information get mapped into the required form? The choice of method for producing the required notation depends on the purpose at hand. One can merely input the information in the required form, or one can arrange for the program to transform the information as needed. The former is sufficient when, for example, the sole purpose of the run is to obtain some counts of some symbol occurrence as in 5.2. The latter is required, on the other hand, when the task is but one of many to be performed during a long and complex run. Such might be the case if the task were that of finding the least general subsumer of various pairs of unit clauses. In this case, the input is expanded to include nuclei for transforming the information from one form to another and back. With this small but important point out of the way, we can turn to the general treatment of demodulation within our program.

As is typical of virtually any feature of our system [5,6,8,9,10,15,25], there exist various parameters that permit many choices for the use of demodulation. For example, one can prevent any demodulation whatsoever even though many equality units are present. But one can instead cause each input equality unit and each generated equality unit to be adjoined and treated as a potential demodulator.

Next, one has the option of having or blocking the back demodulation of clauses. Back demodulation is simply the demodulation of already retained clauses by "newly" adjoined demodulators. One can also select some subset of clauses and place them on a protected list--protected from back demodulation.

Next, one can choose between two rather different algorithms for the application of demodulators. The algorithm that is used in the examples cited here consists of applying demodulators in the order in which they occur in the clause space. One can choose, instead, to have the demodulators applied by a "best fit" criterion [9]. With this latter algorithm, the program chooses to apply that demodulator that most nearly matches the term being demodulated.

The various features just cited can be used in combination and can be turned on and off repeatedly throughout the run. Demodulation is applied from the innermost term toward the outer. It is applied otherwise in a left to right fashion. Only one path of demodulation is considered, which is in contrast to the Allen and Luckham approach [1] of generating the full set of demodulants. In choosing the form in which to store a generated equality and in deciding whether or not to apply certain demodulators, certain lexical conventions are observed for tie-breaking and ordering. Upon generation, each clause is immediately treated with the available demodulators before any other process is applied. If a term is successfully demodulated, all of its arguments are revisited for demodulation and the term itself is again considered for further demodulation. If demodulation or back demodulation is successfully applied to a clause, the demodulated or back demodulated version is retained while the undemodulated version is purged. Certain built-in constraints prevent looping from occurring.

At this point, one might well ask for some justification for this set of built-in choices. Among others, the following two comments are especially relevant to the material contained in this paper. First, back demodulation must be avoidable in order to prevent interference with certain given equalities which occur in certain examples given later. If this were not the case, then less control of the implementation of various procedures would exist. On the other hand, back demodulation must be available both for canonicalization of kept clauses and for the deletion of certain clauses that are only relevant to particular subcases of a case analysis. Second, the one path treatment is vital to such packages as that for counting which is given below. For example, the second demodulator in the set for counting is applied only if the first does not. If multiple path demodulation were allowed with this set of demodulators for counting, then a number of demodulants would result. They would yield counts ranging from 0 to the correct count, which of course defeats the purpose of the set. In addition, we feel the combinatorics of multi-path demodulation is in general to be avoided.

The expanded use of demodulation has led us to extend the first-order predicate calculus to include the integers. We chose to do this in preference to either the cumbersome use of the successor function or to the use of various axioms for arithmetic. This in turn led us to add certain built-in functions to the program. Such built-in functions are always preceded by "$". An example of such is $SUM. When, for example, $SUM(1,5) occurs, it demodulates to 6. (The only example occurring in this paper of such special functions is with the counting demodulators.)

3. Bookkeeping, Updating, and Scheduling

3.1. The problem Set

In various investigations, one often is forced to do a fair amount of bookkeeping and updating. In solving puzzles, for example, a common occurrence is that of making some list of possibilities and crossing off elements of the list with the appropriate designation of "good" or "bad".

As an example, consider the following incomplete description of a puzzle. There are four people who hold eight jobs between them. Each holds exactly two jobs. The object of the puzzle is to determine who holds which jobs. Information is given that enables one to pair off certain people with certain jobs. More importantly, information is given that allows certain pairings to be eliminated thereby eventually leading to the appropriate coupling.

This puzzle solving by the process of elimination is a common element in much of reasoning. With pencil and paper, the corresponding bookkeeping procedure is obvious. With a program written with the specific purpose of solving the puzzle, a procedure is written to continually update the database. However, with a theorem-proving program, it may not be obvious how to encode a corresponding procedure, especially if efficiency is required.

Of course, such updating can be accomplished through the usual means of inference. This approach, however, often interferes with the main train of thought, for it often inadequately handles the scheduling of the various subtasks. It can have the side-effect of causing the program to get lost, at least for a while. Thus it is desirable to have the program not leave the main line of inquiry but rather handle such bookkeeping chores as asides.

To achieve this end, a set of demodulators can be written, to be used in conjunction with certain "list" clauses, that allows the program to concentrate on the main search essentially uninterrupted. This use of demodulation permits the assigning of a high priority to the updating of information and thus, among others, solve each of the subproblems presented by the four people in a convenient fashion. In addition, since a problem of this nature does not naturally contain a "contradiction" to be sought, the approach to be given has the advantage that the program will stop as soon as all four subproblems have been solved.

3.2. Bookkeeping Demodulators

There are three kinds of lists used in working this puzzle: a list of the possible jobs which a given person might hold; a list of the people who might hold a given job; and a list of the people whose jobs have not yet been determined. An example of the first type of list is the following.

POSSJOBS(L(PJ(R,chef),L(PJ(R,guard),L(PJ(R,nurse),L(PJ(R,pol),
L(PJ(R,pwr),L(PJ(R,tchr),L(PJ(R,tel),PJ(R,cwtr))))))))).

The interpretations of the various symbols are as follows. The predicate POSSJOBS identifies this list as one of the first type: possible jobs for a given person. The person (R for Roberta) and the various jobs appear within the single argument of this predicate. The two-place function L may be referred to as a "listing function". Its sole purpose is to collect a number of subterms together into one large term. The subterms are items on the list. The large term is the whole list. Each subterm PJ(person,job) is one item on the list, one possible job for Roberta. The function PJ takes as its two arguments a person and a possible job which that person might hold. The jobs, in some cases abbreviated, are: chef, guard, nurse, police officer, professional wrestler, teacher, telephone operator, and waiter. (Names beginning with V-Z are avoided since our program reserves such names for names of variables.) There will be one list of possible jobs kept for each of the four people.

A natural question to now ask is: Why repeat Roberta throughout such a list? The answer is that such repetition permits the removal of the appropriate single piece of information from Roberta's list. If each item named only a job, then the demodulators that follow would not select only the entry desired but would instead attack all the lists indiscriminately.

The procedure for "crossing off" items begins with

EQUAL(L(crossed,x),x)
EQUAL(L(x,crossed),x)

which apply when the way has been prepared. To prepare the way, we do the following.

Whenever a possibility is to be crossed off the list, we force the program to derive a demodulator of the following form

EQUAL(PJ(R,pwr),crossed)

which, if applied, crosses off the possibility that Roberta is the wrestler. But, in order that this demodulator be considered for application, the user must have the program parameters set to ask for back demodulation. The feature of "back demodulation", when selected by the user, enables the theorem prover to detect that the new demodulator

    EQUAL(PJ(R,pwr),crossed)

could in fact be applied to the already existing clause

    POSSJOBS(L(PJ(R,chef),L(PJ(R,guard),L(PJ(R,nurse),L(PJ(R,pol),
    L(PJ(R,pwr),L(PJ(R,tchr),L(PJ(R,tel),PJ(R,cwtr)))))))).

The theorem prover then applies the new demodulator, obtaining

    POSSJOBS(L(PJ(R,chef),L(PJ(R,guard),L(PJ(R,nurse),L(PJ(R,pol),
    L(crossed,L(PJ(R,tchr),L(PJ(R,tel),PJ(R,cwtr))))))))

in which the possibility PJ(R,pwr) ("Roberta might be a professional wrestler") has been crossed off. But the theorem prover does not stop here in its demodulation effort. Rather it demodulates the clause as far as possible before considering it for retention. The input demodulator

    EQUAL(L(crossed,x),x)

is now applied, yielding

    POSSJOBS(L(PJ(R,chef),L(PJ(R,guard),L(PJ(R,nurse),L(PJ(R,pol),
    L(PJ(R,tchr),L(PJ(R,tel),PJ(R,cwtr))))))).

The effect of this latter demodulator is to "clean up" the list, making it clearer that only seven possibilities remain. Eventually only two jobs remain on the list. The program easily detects this fact and uses it to make other needed inferences. The use of these demodulators enables the program to purge the outdated list in favor of the updated information.

The crossing-off process just described has the following advantages over a standard process relying on some inference mechanism.

(1) The crossing-off takes place immediately rather than awaiting a turn in a long queue of possible inferences.

(2) The crossing-off method is similar to what a person might do by hand.

(3) The proof is shortened as compared with a standard first-order predicate calculus formulation in which the crossing-off would require several inference steps involving Skolem functions.

(4) Readability of the proof is correspondingly improved.

(5) The scheduling is easier to control as compared with the complex task of assigning high priority to the several steps of a complex inference sequence.

For the second type of "list", we employ the same functions as in the first list--that which holds the person fixed and varies the possible jobs. But, since we now intend to list the people who might hold a particular job, we use the predicate, POSSPPL, and vary the people who might hold a job while holding the job fixed. Eight such lists are maintained, one for each of the eight jobs. For example:

POSSPPL(L(PJ(R,pwr),L(PJ(S,pwr),L(PJ(T,pwr),PJ(CV,pwr))))),

which lists the four people who might hold the job of professional wrestler. Observe that the same demodulator discussed previously,

EQUAL(PJ(R,pwr),crossed),

serves also to cross Roberta off the professional wrestler list.

We turn finally to the third type of list--the list of people whose jobs remain to be determined. This list initially has the form

STILLTODO(L(JOBSOF(R),L(JOBSOF(S),L(JOBSOF(T),JOBSOF(CV))))).

When the two jobs of Roberta (R) have been determined, we derive the clause

EQUAL(JOBSOF(R),crossed).

Back demodulation then yields, in the manner discussed previously, the clause

STILLTODO(L(crossed,L(JOBSOF(S),L(JOBSOF(T),JOBSOF(CV)))))

which immediately demodulates as before to

STILLTODO(L(JOBSOF(S),L(JOBSOF(T),JOBSOF(CV)))).

When the jobs for Steve, Thelma, and Vince have also been determined, this list attains the form

STILLTODO(crossed)

meaning that all four subproblems have been crossed off. The following clause is input to stop the run by contradiction:

-STILLTODO(crossed).

The contradiction is detected immediately and ends the run. This is of course preferable to having the program continue in tangential directions for awhile before realizing that it's already done.

4. Classification of Formulas

4.1. Problem Description

The area on which we concentrate is the equivalential calculus [3,4,7], a field of formal logic. This field is concerned with the abstract study of the notion of equivalence. Its elements are the wffs that can be composed from a 2-place function, E, and the variables, x, y, z, ... There are single formulas in the calculus which are so strong that they can serve as single axioms [2,3,11]. For example, all of the theorems of the calculus are deducible from the single formula,

XGK = E(x,E(E(y,E(z,x)),E(z,y))).

A commonly used rule of inference in the calculus is that of condensed detachment [12]. Condensed detachment behaves rather like hyper-resolution [10,14] or like a restricted form of UR-resolution [8,9].

The first problem to be solved in our study of equivalential calculus [26] was that of finding a shorter proof than that which existed [2]--a proof that a particular formula is a single axiom for the calculus. By examining the given proof as well as others of the type existing in the literature, a classification of formulas was arrived at. The notion was that employment of the classification would strongly aid the theorem-proving program when it was assigned the problem of finding a shorter proof. Briefly, the idea was that of discarding clauses of certain classes while emphasizing the use of clauses of certain other classes.

The following is a brief description of the classification. Designate those formulas that are of the form, E(x,t), to be of class 1, where x is some variable and t is some expression. Designate the formulas of the form, E(s,y), which are not of class 1 to be of class 2, where s is an expression and y is a variable. (We wish the classes to be disjoint, so we force E(x,x), for example, to be of class 1 and not of class 2.) Class 3 formulas have the form, E(E(x,y),t). Class 4 formulas have the form, E(s,E(x,y)). The definitions of classes 5, 6, and so on proceed similarly. Of course, no assumption is made in classes 3 and 4 of distinctness of variables. We continue to impose the requirement that the classes be disjoint.

The task is that of having the theorem-proving program implement a procedure to correctly classify each formula as it is generated. The generated clauses that are of an undesirable class would then be immediately discarded. The task is accomplishable through demodulation.

4.2. Classification Demodulators

Keep in mind that the only expressions that can occur during the investigation into the equivalential calculus are of the form E(s,t) for terms, s and t. Further, such terms can only be some variable or some expression of the form, E(q,r), and so on. (In a typical theorem-proving run, we would ordinarily enclose each such formula in the unary predicate, P, and thus the symbol, E, is treated as a function symbol.)

For each term and subterm of each generated clause, we must be able to determine its status--that of being a variable or not a variable--in order to correctly classify the clause. We also must be able to assign the class number according to the scheme presented above which, among others, assigns class 1 to those expressions whose leftmost argument of the major function symbol is a variable. In order to easily deal with the two subtasks, we have the program generate the formulas in the following fashion. Rather than having a formula simply reside in some unary predicate, we instead use a binary predicate that contains the particular formula in question twice. Thus, for example, if E(x,x) is the formula found next in the search, the program is made to generate the clause

   Q(E(x,x),CLASS(E(x,x))).

The second occurrence of the formula occurs as the argument of the function, CLASS. Thus, upon generation, the second argument is immediately demodulated to the appropriate class number of the contained formula.

For the first subtask, that of separating variables from nonvariables, we have but two demodulators. These demodulators, which are applicable only after the way has been prepared (see below), are:

   EQUAL(IFVAR(E(x,y)),false),

which applies when the term is not a variable, and

   EQUAL(IFVAR(x),true),

which applies when the term is a variable. The function, IFVAR, means "if its argument is a variable". It may seem that, at least intuitively, the two demodulators are in the wrong order. However, it is vital that these be applied in the order given so that the correct status will be found. After all, the second applies to all formulas. So if the order were reversed, terms that are not variables would be classed as variables. If we permitted multiple path demodulation, the fact that a term was not a variable would be assigned both "true" and "false". If the first demodulator does not apply, then the formula (or subformula) being demodulated must not have the form, E(s,t). Hence it must be a variable. Because of the limited forms the formulas can take, one of the two demodulators must always apply.

The following are used for the other subtask, that of assigning the correct class number. Here we shall assume that those formulas that are not in the union of classes 1 to 4 are to be assigned class number 5.

We begin with

   EQUAL(CLASS(E(x,y)),EVAL(IFVAR(x),IFVAR(y),E(x,y)))

where the 3-place function, EVAL, will with help return the class number. But EVAL also holds the formula in question. The two previous demodulators will now apply to test the two arguments of the (sub)formula under study. To proceed further with the task of assigning the correct class number, we need demodulators to use the information yielded by the two demodulators for the function, IFVAR. We have

   EQUAL(EVAL(true,v,w),1),

which applies when the wff should be assigned the class number of 1;

    EQUAL(EVAL(false,true,w),2),

which applies when the class number is 2;

    EQUAL(EVAL(false,false,E(E(x1,x2),E(y1,y2))),
    EVAL2(AND(IFVAR(x1),IFVAR(x2)),AND(IFVAR(y1),IFVAR(y2)))),

which applies when the class number is greater than 2.

For those wffs that are not of class 1 or 2, we need to process them further. We therefore have demodulators to use the information yielded by the last demodulator, and we have

    EQUAL(AND(true,x),x)
    EQUAL(AND(false,x),false)
    EQUAL(EVAL2(true,v),3),

which assigns the class number of 3 for the wffs whose first argument is an E of two variables, (and of course not of class 1 or 2);

    EQUAL(EVAL2(false,true),4),

which assigns class 4 for the wffs whose second argument is an E of two variables; and

    EQUAL(EVAL2(false,false),5),

which assigns class 5 to the remaining wffs. Notice that the requirement of demodulating inside out is important, for the EVAL demodulators are applied only after those for IFVAR.

5. Counting Symbol Occurrences

5.1. A Counting Problem

The next area of investigation [26] concerned the status, with respect to being a single axiom, of various formulas of the calculus. With respect to each of the corresponding (at the time) open questions, we sought to prove or disprove that the formula was too weak. During the investigation, it was conjectured that certain relations held between the parents of an inference and the inference itself. The relations were in terms of the number of occurrences of the function, E, occurring therein. If the conjecture could be shown to hold, then each of the (at the time) unclassified formulas would be proved too weak to be a single axiom for the equivalential calculus.

Since we wished to avoid writing a special-purpose program to generate formulas and count symbol occurrences, we were left with two choices. On the one hand, we could of course take the output of a standard theorem-proving run and make the various counts by hand. But this is tedious and rather prone to error. On the other hand, we could attempt to devise a way for the theorem-proving program itself to make the appropriate counts during the actual run.

The task was that of finding a set of demodulators, if such exists, which encodes a procedure for both counting the appropriate symbol occurrences within a formula and then planting the count in the formula.

5.2. Counting Demodulators

Were we just studying the set of inferences yielded from some starting formula, we would have the program merely generate each such inference and use the one-place predicate, P, to hold the corresponding formula. However, to prepare the way for the application of the counting demodulators, we instead have the program generate

$$Q(f,COUNT(f))$$

for each new formula, f. (This clause is quite similar to that occurring in the classification of formulas, but the function CLASS is replaced with COUNT because of the task under discussion.) Then, by means of the following demodulators, COUNT of an expression will demodulate to the count of the number of E's occurring therein.

$$EQUAL(COUNT(E(x,y)),\$SUM(\$SUM(COUNT(x),COUNT(y)),1)),$$

which applies when the expression is not a variable; and

$$EQUAL(COUNT(x),0),$$

which applies when the expression is a variable.

Recall that the function, $SUM, is a built-in function for arithmetic. When its two arguments are both integers, it demodulates to their sum. The second of the two given demodulators is applied only when the first fails to apply. This occurs only when the term being demodulated is a variable. If they were interchanged, then terms that were not variables would be given counts of 0.

Notice that the first demodulator is recursive in the function, COUNT. Access to recursion of this type is often needed in procedures or subroutines in standard programming. By this use of recursion, we need only prefix the entire formula under study by the one occurrence of COUNT, thus avoiding the necessity of placing COUNT in front of each of the subterms. In a certain sense, the demodulation then works from outside in, for the recursion propagates the COUNT function appropriately to the various subterms.

6. Least Common Generalization

6.1. Problem Description

A common problem that occurs in mathematics is that of generalizing a result or set of results. A related problem in automated theorem-proving is that of finding a clause, if one exists, that is the least common generalization of two given clauses. (Note that this is precisely the opposite of the heart of automated theorem proving, namely, unification [13].) Given two unit clauses, for example, one might wish to know what clause, if any, is the least general subsumer of the two given clauses.

Given P(a,a) and P(x,a), the desired clause is P(x,a). Given P(a,y) and P(x,b), P(x,y) is the desired clause. However, with P(a,a) and P(b,b), one wishes P(x,x) and not P(x,y).

In view of the frequent use of subsumption [13], the value of such a generalization clause is clear. Its presence might materially reduce the size of the clause space retained during a proof search. The reason for wanting the "least" general subsumer is simply that the least generalization might be provable while a more general result might not. Employment of a procedure for generalization could be of substantial value as an editing strategy.

A more grandiose objective is the following. By means of this procedure for generalization, one may be able to find the general form of some sought after algorithm. Having seen the form for n = 1 and n = 2 and n = 3, the theorem-proving program might be able to find the least general form that captures the three given instances.

6.2. Least Common Generalization Demodulators

The demodulators for obtaining the least common generalization will be the most complex set of demodulators presented in this paper. Therefore the reader may wish merely to skim these, then study the demodulators for counting and for bookkeeping in detail, and then return to the present complex set.

The demodulators to be given will demodulate any term of the form

FINALFORM(LEASTCOMGEN("s","t","varlist"))

to a term which is the least common generalization of the two arguments "s" and "t". The third argument, "varlist", will be assumed to be a list of distinct variables, for example,

L(v1,L(v2,L(v3,L(v4,end)))).

Note the use of the constant "end" to end the list. This convention will be followed throughout this section. The number of variables on the list, "varlist", must be at least the number required to find the least common generalization. If the list contains too few variables, a term containing the function, "ERROR", will be derived in place of the sought after generalization.

We divide the main problem into the following four tasks:

1. If "s" and "t" are identical terms, then return this term as the least common generalization.

2. If "s" and "t" are not identical but have the same major function symbol, then do the following.
   a. Pair off the corresponding arguments.
   b. Find the least common generalization of each pair.
   c. Recombine the generalizations of the arguments within the common major function symbol.

3. If neither 1 nor 2 applies to the pair s,t, then the appropriate generalization is a variable.

4. When the processing is complete, return the final form of the generalization with all temporary information removed.

Note that the very nature of the tasks may result in recursion. For example, the accomplishment of b under 2 may call for the accomplishment of 1.

The first task, that of a trivial generalization, is accomplished with the demodulator

EQUAL(LEASTCOMGEN(x,x,varlist),GOTLCG(x,varlist)).

The result is returned along with the list of available variables within the function GOTLCG.

The second task, that of processing two terms with the same major function symbol, is accomplished as follows. We first pair off the corresponding arguments. We have

EQUAL(LEASTCOMGEN(P(x1,x2),P(y1,y2),varlist),
     EVAL1(ARGLIST(P(x1,x2),P(y1,y2)),varlist,P(end,end))).

The function EVAL1 receives three arguments: a list of pairs of arguments, the list of variables, and the function name P with dummy arguments. One such demodulator must be supplied for each function that might occur in the expressions being generalized. For each such function a demodulator must also be supplied to create the argument list:

EQUAL(ARGLIST(P(x1,x2),P(y1,y2)),
     L(PAIR(x1,y1),L(PAIR(x2,y2),end))).

Corresponding arguments are paired and the "end" convention is followed.

We next find the least common generalization of each argument pair. We keep a list of pairs "to do" and a list of generalizations "done". The "done" list is initially the empty list, represented by the constant "end". We have

EQUAL(EVAL1(xtodo,varlist,wfunc),
     EVAL2(xtodo,end,varlist,wfunc)).

We next take one pair from the "to do" list and find its least common generalization. We have

    EQUAL(EVAL2(L(PAIR(xi,yi),xtodo),xdone,varlist,wfunc),
          EVAL3(LEASTCOMGEN(xi,yi,varlist),xtodo,xdone,wfunc)).

Once the generalization has been obtained we place it on the "done" list. We have

    EQUAL(EVAL3(GOTLCG(zi,varlist),xtodo,xdone,wfunc),
          EVAL2(xtodo,L(zi,xdone),varlist,wfunc)).

We repeat these operations until we reach the end of the "to do" list. At this point we plug the generalized arguments back into the common major function (P in this case), getting

    EQUAL(EVAL2(end,L(z2,L(z1,end)),varlist,P(end,end)),
          GOTLCG(P(z1,z2),varlist)).

One such demodulator must be supplied for each function that appears in the expressions being generalized. Note that the list of generalized arguments

    L(z2,L(z1,end))

actually appears in reversed order. The order is corrected in the GOTLCG term. This concludes the second task.

(Note that the following demodulator would not accomplish the desired result in all cases.

    EQUAL(LEASTCOMGEN(P(x1,x2),P(y1,y2),varlist),
          P(LEASTCOMGEN(x1,y1,varlist),LEASTCOMGEN(x2,y2,varlist))).

The reason is as follows. The first LEASTCOMGEN operation may reserve one or more variables in the "varlist". The varlist thus altered must then be passed to the second LEASTCOMGEN operation. The demodulator just given uses the original form rather than the altered form in the second operation.)

We proceed now to the third task, that of generalizing two unrelated terms to yield a variable. We would simply choose one variable from the "varlist" if it weren't for the following consideration. In generalizing the pair of expressions $P(a,a)$ and $P(b,b)$, the generalization of the pair a,b is requested twice: once in the first argument position, and once in the second. We wish to return the same variable both times the request is made. In order to do this we keep a list of the requests that have been made already as well as a list of variables. On the first request we simply choose the first variable from the "varlist" and record the request:

    EQUAL(LEASTCOMGEN(xi,yi,L(var,varlist)),
          GOTLCG(var,REQSVARS(L(REQ(xi,yi,var),end),varlist))).

The function REQSVARS contains both the list of requests and the list of variables still available. On a subsequent request we first check whether an identical request was made previously. We have

```
    EQUAL(LEASTCOMGEN(xi,yi,REQSVARS(zreqs,varlist)),
          EVAL4(LOOKUP(xi,yi,zreqs),xi,yi,REQSVARS(zreqs,varlist))).
```

Here is how we look up a request. (Note that again the order of the demodulators is vital to obtaining the correct information.) We check whether the request matches the first entry in the request list:

```
    EQUAL(LOOKUP(xi,yi,L(REQ(xi,yi,var),zreqs)),
          GOTVAR(var)).
```

If the first entry matches, we return the corresponding variable. Otherwise we search further in the list:

```
    EQUAL(LOOKUP(xi,yi,L(REQ(xj,yj,var),zreqs)),
          LOOKUP(xi,yi,zreqs)).
```

If we reach the end of the list without finding a match, we return the constant "nomatch":

```
    EQUAL(LOOKUP(xi,yi,end),nomatch).
```

We now make use of the information (variable or "nomatch") thus obtained. If the present request merely duplicates a previous request, then the corresponding variable is the desired generalization:

```
    EQUAL(EVAL4(GOTVAR(var),xi,yi,REQSVARS(zreqs,varlist)),
          GOTLCG(var,REQSVARS(zreqs,varlist))).
```

If the request is new, on the other hand, we generalize it to yield a new variable and add the new request to the list:

```
    EQUAL(EVAL4(nomatch,xi,yi,REQSVARS(zreqs,L(var,varlist))),
          GOTLCG(var,REQSVARS(L(REQ(xi,yi,var),zreqs),varlist))).
```

If no variable is available we signal the error:

```
    EQUAL(EVAL4(nomatch,xi,yi,REQSVARS(zreqs,end)),
          ERROR(outofvars,nomatch,xi,yi,REQSVARS(zreqs,end))).
```

This completes the third task.

The fourth task, that of returning the final generalization without the temporary lists attached, is easily accomplished. We have

```
    EQUAL(FINALFORM(GOTLCG(z,varlist)),z).
```

By recourse to the demodulators just given, the program will find the least common generalization of two terms if such exists.

7. Case Analysis

7.1. The Basic Problem

In conducting a case analysis with a theorem-proving program [18,20], it is desirable to purge the clause space of those clauses that are only relevant to the subcase under study when that subcase has been completed. Again, with a standard program, the procedure for doing such is obvious. But for theorem proving, perhaps it is not.

Since the clauses relevant to one particular subcase can be harmful if applied to another subcase, the trick is to demodulate the "useless" clauses to some harmless constant and then, by subsumption, discard them. We give below the approach.

However, before presenting the demodulators, we comment that this approach can also be used when the object is that of seeking a number of distinct solutions to a problem. Such is the goal when, for example, one is seeking to design a number of different digital circuits with certain common properties [21]. Thus the "purging" demodulators are more useful than might have been apparent.

7.2. Purging Demodulators

For many problems requiring case analysis, we can make the following assumptions. We assume first that only unit clauses will be derived during the run. This should not be surprising since we emphasize units so heavily in much of our work [22]. Second, we assume that each derived unit is in fact a positive unit. Third, we assume that the problem to be solved does not involve the equality predicate. The equality predicate will be used only for the encoding of the purging demodulators. Means of circumventing certain of these restrictions will be briefly discussed later. At the moment, coping with the relaxation of the last restriction still poses a serious problem and may well be an area for profitable research.

The use of purging demodulators presents three tasks. The first task is that of recording within each derived clause the case to which it applies. The second task is that of deriving, for those clauses that are to be purged at the completion of a case, a demodulator that appropriately transforms all clauses related to that case. The third task is that of providing the mechanism to remove the transformed clauses by subsumption.

We begin with the problem of recording case information within a derived clause. To do this we include a "case code" as one of the arguments of the clause. For example, the clause

   P(a,b,c,CASE(L(3,L(1,2))))

can be used to state that a relation holds among a, b, and c within the case 3.1.2. The argument

CASE(L(3,L(1,2)))

is the case code. There are many alternate ways to encode case codes, such as,

SUBCASE(3,1,2).

But the given encoding has certain advantages. For example, the use of the "list" function avoids the necessity of including a set of "subcase" functions--a set that includes functions with 1, 2, ..., arguments. Two restrictions, discussed later, must be observed in encoding case codes. For reasons of space we omit the discussion of the tricks used to place a case code in a derived clause.

For the second task, that of deriving a purging demodulator, we force the program to derive the demodulator

EQUAL(CASE(L(3,L(1,2))),junk).

Then all clauses relevant to the case designated by the case code,

CASE(L(3,L(1,2))),

will be subject to back demodulation by means of this demodulator. For example, the clause

P(a,b,c,CASE(L(3,L(1,2))))

is back demodulated to

P(a,b,c,junk).

For the third and final task, that of removing the transformed clauses by subsumption, we include in the input the clause,

P(x,y,z,junk).

When a case has been completed, the program will thus delete the then unneeded clauses. For each predicate that is to receive similar treatment, a corresponding subsumer clause is included in the input. We specify in the input that these particular subsumer clauses are not to participate in inferences.

We now give the two restrictions, mentioned earlier, upon the encoding of case codes. The first restriction is that no case code term may appear in any clause other than as a case code. Adherence to this requirement prevents the purging demodulator from transforming clauses not directly germane to the case at hand. The second restriction is that no case code may appear as a subterm of another case code. If violated, the purging that occurs at the completion of a case would incorrectly affect other cases. In the example given, all case codes are enclosed in the function, CASE, in order to satisfy these two restrictions.

We close this section by briefly discussing actions that can be taken that permit relaxation of the given assumptions. For example, if negative units are derived during various subcases, and if one wishes to have access to a parallel treatment to that already given, the clause,

-P(x,y,z,junk),

can be included in the input. Such a clause will then subsume those unneeded clauses after the appropriate back demodulation occurs following the completion of a case. Note that it would then be necessary to instruct the program not to recognize the unit conflict between this clause and

P(x,y,z,junk).

Alternatively, we could use another predicate, NOTP, to encode the negation of P in a positive unit and thus avoid the problem of negative units. We would then give the relationship between the two predicates by inputing appropriate nuclei.

If the assumption that all derived clauses within the case analysis are units does not hold, one need only include a case code in at least one literal of each derived nonunit. The purging demodulator and the subsuming units given above then suffice to purge such nonunit clauses.

Finally, if one does not assume that all derived equalities will be purging demodulators, we simply remark that no excellent solution is known. One could have recourse to an "equality" predicate that allows a third argument for the case code, but this could interfere with certain built-in features. This situation presents an area for potentially valuable research.

8. Set-Theory

8.1. The Objective

Much effort has been devoted to handling set-theory problems in a natural and efficient manner with an automated theorem-proving program. The main focus is that of illustrating the potential value of some new strategy and/or inference rule. Many are acquainted with the disappointing tediousness with which such problems are solved with standard theorem-proving approaches.

Here we present a notation and a set of demodulators that enable many trivial examples to be dispatched trivially by the program.

8.2. Set-Theory Demodulators

We use the predicate EQUALLOG to denote logical equivalence of its two arguments. Since our program treats all predicates that begin with the letters,

EQUAL, as an equality predicate, positive units in EQUALLOG may become demodulators. We have in particular

    -EQUALLOG(true,false).

The function IFF denotes the result--true or false--of testing its two arguments for logical equivalence. The following demodulators encode standard properties of the logical connectives AND, OR, NOT, IM (implication) and IFF.

```
EQUALLOG(AND(x,true),x)
EQUALLOG(AND(x,false),false)
EQUALLOG(AND(true,x),x)
EQUALLOG(AND(false,x),false)
EQUALLOG(OR(x,true),true)
EQUALLOG(OR(x,false),x)
EQUALLOG(OR(true,x),true)
EQUALLOG(OR(false,x),x)
EQUALLOG(NOT(true),false)
EQUALLOG(NOT(false),true)
EQUALLOG(NOT(NOT(x)),x)
EQUALLOG(AND(x,x),x)
EQUALLOG(OR(x,x),x)
EQUALLOG(IM(true,x),x)
EQUALLOG(IM(false,x),true)
EQUALLOG(IM(x,true),true)
EQUALLOG(IM(x,false),NOT(x))
EQUALLOG(IM(x,x),x)
EQUALLOG(IM(x,y),OR(NOT(x),y))
EQUALLOG(IFF(x,true),x)
EQUALLOG(IFF(true,x),x)
EQUALLOG(IFF(x,false),NOT(x))
EQUALLOG(IFF(false,x),NOT(x))
EQUALLOG(IFF(x,x),true)
EQUALLOG(IFF(x,NOT(x)),false)
EQUALLOG(IFF(NOT(x),x),false)
EQUALLOG(IFF(x,y),AND(IM(x,y),IM(y,x)))
EQUALLOG(NOT(OR(x,y)),AND(NOT(x),NOT(y)))
EQUALLOG(NOT(AND(x,y)),OR(NOT(x),NOT(y)))
EQUALLOG(AND(x,y),AND(y,x))
EQUALLOG(AND(x,AND(y,z)),AND(y,AND(x,z)))
EQUALLOG(OR(x,y),OR(y,x))
EQUALLOG(OR(x,OR(y,z)),OR(y,OR(x,z)))
EQUALLOG(AND(AND(x,y),z),AND(x,AND(y,z)))
EQUALLOG(OR(OR(x,y),z),OR(x,OR(y,z)))
EQUALLOG(OR(AND(x,y),z),AND(OR(x,z),OR(y,z)))
EQUALLOG(OR(x,AND(y,z)),AND(OR(x,y),OR(x,z))).
```

For the concepts of union and intersection and the like, the function symbols U, I, CO, IFEQSET, IFELOF, and D are introduced to encode the standard axioms of set theory. IFELOF is that function that returns true if the first argument is an element of the second and false otherwise. The following demodulators define U, I, and CO (union, intersection, and complement within an assumed universe) in terms of IFELOF.

```
EQUALLOG(IFELOF(x,U(y,z)),OR(IFELOF(x,y),IFELOF(x,z)))
EQUALLOG(IFELOF(x,I(y,z)),AND(IFELOF(x,y),IFELOF(x,z)))
EQUALLOG(IFELOF(x,CO(y)),NOT(IFELOF(x,y))).
```

IFEQSET is that function that returns true if its two arguments are equal sets and false otherwise. It is defined in terms of the function IFELOF through the use of the Skolem function D:

```
-EQUALLOG(IFEQSET(x,y),false)
    EQUALLOG(IFF(IFELOF(D(x,y),x),IFELOF(D(x,y),y)),false),
```

which says that, given two sets which are not identical, there exists an element in one but absent from the other.

The following clause denies the theorem that union distributes over intersection:

```
EQUALLOG(IFEQSET(U(a,I(b,c)),I(U(a,b),U(a,c))),false).
```

We place just this clause in the set of support [23]. A single resolution with the preceding nucleus yields a clause that demodulates to

```
EQUALLOG(true,false).
```

Proof by unit conflict is thus obtained in a single resolution step, through the extensive use of demodulation. Similar one-step proofs can be obtained for other theorems of this type.

9. Conclusions

In this paper we have given a number of uses of demodulation outside the familiar ones of simplification and canonicalization. These uses range from simple bookkeeping and straightforward counting to the finding of the least general subsumer of two given terms. By relegating such subsidiary tasks to demodulation, the program is free to concentrate on its main assigned task. Thus, the problem of the scheduling of the various subsidiary tasks and subtasks that occur during the attempt to solve some complex problem is at least in part solved.

Although each of the tasks discussed here could have been handled by means of additional programming, such a move defeats one of the main goals of the field. The goal in question is that of providing a general-purpose program that can be used to solve many and diverse problems and without burdening the user with additional programming. The given examples demonstrate the power of both the general input language of automated theorem proving and also that of the procedure of demodulation. In particular, the set of demodulators given to cope with problems of set theory shows how naturally such problems can be solved. This approach presents a sharp contrast to that so often quoted in earlier attempts with theorem-proving programs.

The notion is that the material contained herein is but a fragment of that which can be done with demodulation. By demonstrating the ability to replace the standard subroutine with a corresponding set of demodulators, we hopefully have provided a needed aid to the use of various existing theorem-proving programs.

References

[1] Allen, J. and Luckham, D., "An interactive theorem-proving program," Machine Intelligence, Vol. 5(1970), Meltzer and Michie (eds), American Elsevier, New York, pp. 321-336.

[2] Kalman, J., "A shortest single axiom for the classical equivalential calculus," Notre Dame Journal of Formal Logic, Vol. 19, No. 1, January 1978, pp. 141-144.

[3] Lukasiewicz, J., "Der Aquivalenzenkalkul," Collectanea Logica, Vol. 1 (1939), pp. 145-169. English translation in [McCall], pp. 88-115 and in [Lukasiewicz/ Borkowski], pp. 250-277.

[4] Lukasiewicz, J., Jan Lukasiewicz: Selected Works, ed. by L. Borkowski, North-Holland Publishing Co., Amsterdam (1970).

[5] Lusk, E., and Overbeek, R., "Data structures and control architecture for implementation of theorem-proving programs," 5th Conference on Automated Deduction, Vol. 87, Lecture Notes in Computer Science, ed. W. Bibel and R. Kowalski, Springer-Verlag, Berlin, 1980, pp. 232-249.

[6] Lusk, E., "Input translator for the environmental theorem prover - user's guide," to be published as an Argonne National Laboratory technical report.

[7] McCall, S., Polish Logic, 1920-1939, Clarendon Press, Oxford (1967).

[8] McCharen, J., Overbeek, R. and Wos, L., "Problems and experiments for and with automated theorem proving programs," IEEE Transactions on Computers, Vol. C-25(1976), pp. 773-782.

[9] McCharen, J., Overbeek, R. and Wos, L., "Complexity and related enhancements for automated theorem-proving programs," Computers and Mathematics with Applications, Vol. 2(1976), pp. 1-16.

[10] Overbeek, R., "An implementation of hyper-resolution," Computers and Mathematics with Applications, Vol. 1(1975), pp. 201-214.

[11] Peterson, J., "Shortest single axioms for the equivalential calculus," Notre Dame Journal of Formal Logic, Vol. 17(1976), pp. 267-271.

[12] Peterson, J., "An automatic theorem prover for substitution and detachment systems," Notre Dame Journal of Formal Logic, Vol. XIX, Jan. 1978, pp. 119-122.

[13] Robinson, J., "A machine-oriented logic based on the resolution principle," J. ACM, Vol. 12(1965), pp. 23-41.

[14] Robinson, J., "Automatic deduction with hyper-resolution," International Journal of Computer Mathematics, Vol. 1(1965), pp. 227-234.

[15] Smith, B., "Reference manual for the environmental theorem prover," to be published as an Argonne National Laboratory technical report.

[16] Veroff, R., "Canonicalization and demodulation," Argonne National Laboratory, Technical Report ANL-81-6, Argonne, Illinois, February 1981.

[17] Winker, S. and Wos, L., "Automated generation of models and counterexamples and its application to open questions in ternary Boolean algebra," Proc. of the Eighth International Symposium on Multiple-valued Logic, Rosemont, Illinois, 1978, IEEE and ACM Publ., pp. 251-256.

[18] Winker, S., Wos, L. and Lusk, E., "Semigroups, antiautomorphisms, and involutions: a computer solution to an open problem, I," Mathematics of Computation, Vol. 37 (1981), pp. 533-545.

[19] Winker, S., "Generation and verification of finite models and counterexamples using an automated theorem prover answering two open questions," to appear in J. ACM.

[20] Winker, S., Wos, L. and Lusk, E., "Semigroups, antiautomorphisms, and involutions: a computer solution to an open problem, II," in preparation.

[21] Wojciechowski, W. and Wojcik, A., "Multiple-valued logic design by theorem proving," Proc. of the Ninth International Symposium on Multiple-valued Logic, Bath, England, 1979.

[22] Wos, L., Carson, D. and Robinson, G., "The unit preference strategy in theorem proving," Proc. of the Fall Joint Computer Conference, 1964, Thompson Book Company, New York, pp. 615-621.

[23] Wos, L., Carson, D and Robinson, G., "Efficiency and completeness of the set-of-support strategy in theorem proving," J. ACM, Vol. 12(1965), pp. 536-541.

[24] Wos, L., Robinson, G., Carson, D. and Shalla, L., "The concept of demodulation in theorem proving," J. ACM, Vol. 14(1967), pp. 698-709.

[25] Wos, L., Winker, S., and Lusk, E., "An automated reasoning system," AFIPS Conference Proceedings, Vol. 50 (1981), National Computer Conference (Chicago, Ill., 1981), AFIPS Press, pp. 697-702.

[26] Wos, L., Winker, S., Veroff, R., Smith, B. and Henschen, L., "Questions concerning possible shortest single axioms in equivalential calculus: an application of automated theorem proving to infinite domains," submitted to the Notre Dame Journal of Formal Logic for consideration for publication, May 1981.

THE APPLICATION OF HOMOGENIZATION TO SIMULTANEOUS EQUATIONS

Bernard Silver

Department of Artificial Intelligence
University of Edinburgh

Abstract

We have been studying the problem of solving small systems of symbolic simultaneous equations, of the type found on A level Mathematics exam papers. We have found that Homogenization, described in [Bundy and Silver 81], can be extended to provide a fairly powerful method for solving these problems. The work described here has been implemented as an extension to PRESS, a computer program, written in PROLOG, [Clocksin and Mellish 81], for solving symbolic, transcendental, non-differential equations, described in [Bundy and Welham 81], and [Sterling et al 82].

We also discuss the technique of Elimination, and suggest how this might be implemented.

Keywords

Homogenization, simultaneous equations, meta level inference, algebraic manipulation, mathematical reasoning, equation solving, rewrite rules.

1. Introduction

We have been examining various techniques used in the solution of sets of simultaneous equations. The example equations we have used are from A level[1] mathematics papers. A typical question is:

$\cosh(x) - 3.\sinh(y) = 0$

$2.\sinh(x) + 6.\cosh(y) = 5$ (A.E.B. 1973)

It was discovered that some of the equations encountered required methods more sophisticated than those used by PRESS. We discovered that an extension of **Homogenization** was sufficient to allow many of these problems, including the above example, to be solved. We also examined the process of Elimination, a technique commonly used in solving sets of simultaneous equations.

2. The Basic Method

In this section we describe the method originally implemented in PRESS, the **basic method**. Given a set of equations, $\{e_i\}$, and a set of unknowns $\{x_i\}$, we are required to find those values of the x_i which satisfy the e_i.

---

[1] The A level exam is taken at the age of 18 in England and Wales, and is used as a criterion in selection for university entrance.

At any point we have three lists, X, E, and S. Initially, X is the complete set of x_i, in the order they appeared in the problem, and E is the e_i, similarly ordered. S is initially empty. The method falls into two parts, reduction and back-substitution. Reduction consists of the following steps.

1. If X contains only one element exit and proceed to back-substitution. Otherwise, consider the first element of X, x_j say. Delete this from X.

2. Find the first member of E which contains x_j. Call this equation the x_j-equation.

3. Try to solve the x_j-equation for x_j. If we fail to solve this equation, backtrack to the previous step and try to find another one. If there are no more possible x_j-equations exit with failure. Otherwise we obtain an expression for x_j in terms of some of the other unknowns, the x_j-expression. Put this expression on S. Delete the x_j-equation from E.

4. Using the x_j-expression, substitute for x_j in the equations remaining in E.

5. Now recursively apply this procedure.

Each application of the above procedure reduces the number of unknowns in X by one, so it is guaranteed to terminate. Every time an unknown is deleted from X, it is substituted for in E, so the number of unknowns in the equations in E also reduces. We will reach a stage where there is just one unknown, x_k say, in X, and there will be an equation in E which contains x_k as the only unknown. (There may be more than one such equation.) We solve this equation for x_k.

We now apply back-substitution. This consists of substituting the value obtained for x_k in all the equations in S. After simplification, this will produce a value for another unknown, which is then substituted into the rest of the equations. We repeat this process of back-substitution until values have been found for all the unknowns.

Let us consider an example:

$3.x + y = 5$ (i)

$x^2 + 2.y^2 - 3.x + 2.y + 2 = 0.$ (Oxford 1976) (ii)

From (i) the x-expression

$x = (5 - y)/3$ (iii)

is obtained by reduction. Using (iii) to substitute for x into (ii) we get:

$(19.y^2 + 17.y - 2)/9 = 0$.

Solving this for y gives $y = -1$ or $y = (2/19)$. Back-substituting these values for y in (iii) gives $x = 2$ or $x = (31/19)$. Therefore the solution of the set of equations is:

    x = 2 and y = -1,   or

    x = (31/19) and y = (2/19).

    The performance of this method can be improved by some heuristics. The method implemented in PRESS is very simple. The variables are solved for in the order they are given, the only choice being which equation should be used to solve for a particular variable. The only heuristic used is to ensure that the equation chosen contains the unknown we are trying to solve for. If there are alternatives the program simply chooses one arbitrarily, and backtracks to make a different choice later if necessary.

    There are other heuristics which would improve performance. For example, it may be the case that we already have an equation which contains only one unknown. If so we should solve this equation first of all, and substitute for this unknown throughout the set of equations. This may then give another equation with only one unknown so we may repeat the process. When we can no longer continue, the current basic method can be applied.

    If we have a choice of equations to solve for a particular variable, we want to select the 'easiest'. A possible heuristic for this would be to choose the equation with the fewest occurrences of that variable. However, there are other considerations as well. Note that in the example above ((i) and (ii)) it would have been better to solve for y first rather than x. We would like to be able to choose the variables in this order because (i) is (slightly) easier to solve for y rather than x. This may be possible if we have some concept of complexity, for example, (i) is easier to solve for y rather than x because the tree-size of the term containing x is greater than that of the term containing y.

<u>Scope and limitations</u>

    The solution of simultaneous equations is a major concern of numerical analysis. Many methods have been devised for solving systems of linear equations, with numerical coefficients, see [Conte and de Boor 72] for example. By comparison, on such systems our method is extremely inefficient. However, our method is applicable to a much wider class of problems, we are not restricted either to linear equations, or to numerical coefficients.

    In theory, the basic method is as good as any other, in that we will not miss any solutions using it that would be found by other methods. In practice, the method often becomes overburdened with complex terms, and fails to find a solution in a reasonable time. For example consider the following question:

    $\cos(x) + \cos(y) = 1$ (iv)

    $\sec(x) + \sec(y) = 4$ (A.E.B. 1976)   (v)

Solving (iv) for x we obtain as one set of solutions[1]

$$x = n.360 + \cos^{-1}(1 - \cos(y)).$$

Substituting this value in (v) produces

$$\sec(n.360 + \cos^{-1}(1 - \cos(y))) + \sec(y) = 4.$$

This equation cannot be solved by PRESS, even if we use the principal value of x (i.e. set n to 0), as we have not provided simplification rules for functions of the type[2]

$$\sec(\cos^{-1}(y)) \rightarrow 1/y.$$

However, even if PRESS did have access to the simplification rules required, it would still require a lot of work to solve the resulting equation. It is clear that far too much effort is used. In particular we do not need to solve (iv) for x, only for cos(x). Then using the fact that sec(x) is 1/cos(x) the equations can be quite easily solved. Presumably, this method is the one that the examiner expected to be used.

How can we tell when the basic method is not the best approach? If it is not we would also like to know a better method! In the following section we will discuss a method we have implemented to help solve these problems.

3. Homogenization

Homogenization is a technique originally developed for the solution of single equations in one variable. Here we describe its application to simultaneous equations. Homogenization is described in detail in [Bundy and Silver 81]. There follows a brief outline of Homogenization.

3.1. Brief description

We first describe the Homogenization method as it applies to single equations. We will refer to this process as **Standard Homogenization**. Homogenization prepares the equation for the Change of Unknown method, which is described in [Bundy and Welham 81]. Briefly, if given an equation of the form $f(g(x)) = 0$, where $g(x)$ occurs more than once, the Change of Unknown method applies the substitution $y = g(x)$ and then solves $f(y) = 0$ to obtain a value for y, $y = k$ say. Finally it solves $g(x) = k$ for

---

[1] n is an arbitrary integer. The other set of solutions is $x = n.360 - \cos^{-1}(1 - \cos(y))$.

[2] There are a large number of rules of this type even if we neglect the hyperbolic cases. We would prefer PRESS to manage without them, human students certainly do.

x.[1]

Usually the equation is not of the required form $f(g(x)) = 0$. Homogenization attempts to transform the equation into this form. For example, consider the equation

$$4.\log_x 2 + \log_2 x = 5 \qquad \text{(London 1978)} \qquad (i)$$

in which the occurrences of the unknown, x, appear within dissimilar subterms, namely $\log_x 2$ and $\log_2 x$. Some preparation of the equation is required before the unknown can be changed. In the case of our example, the subterm $\log_x 2$ must first be converted to $1/\log_2 x$, with the aid of the rewrite rule

$$\log_u v \rightarrow 1/\log_v u.$$

We will refer to the original equation, prior to Homogenization, as the **input equation**, and the resulting equation will be called the **output equation**. The output equation belongs to the class of algebraic equations, i.e. those involving only the functions +, -, *, / and exponentiation to a rational number power. The input equation can always be regarded as an algebraic equation <u>in some set of non-algebraic subterms in x</u>, e.g. equation (i) above can be regarded as an algebraic equation in the set $\{\log_x 2, \log_2 x\}$. These subterms are called the **offending terms** and the set of them is called the **offenders set**. The idea is that this is a set of subterms preventing the equation being algebraic: a type of equation which PRESS knows a lot about.

The essence of the Standard Homogenization method is to replace each of the offending terms by some algebraic function of a single term, called the **reduced term**. In the example above the reduced term is $\log_2 x$.

The Standard Homogenization method is as follows:

(a) The offenders set is found by trying to parse the input equation as an algebraic equation. When the parse is blocked, because the current subterm is x or a non-algebraic subterm containing x, then this is added to the offenders set and the parse forced to continue.

(b) A reduced term is selected. The method of selection depends on a classification of the type of equation, e.g. trigonometric, exponential etc.

(c) Now an attempt is made to rewrite each term in the offenders set as an algebraic function of the reduced term. If the method succeeds a rewrite

---

[1] In general we may obtain a disjunctive solution for y, i.e. $y = k_i$, $i = 1,2...n$. In this case we solve all the equations $g(x) = k_i$, to obtain a disjunctive solution for x.

is found for every term in the offenders set, so we have

$ot_i \rightarrow af_i(rt)$

where ot_i is an offending term, rt is the reduced term and af_i is an algebraic function. If no rewrite rule is applicable, then backtrack to choose a new reduced term if this is possible, otherwise fail.

(d) Substitute the rewrites for the offending terms in the input equation to give the output equation. This equation is now an algebraic equation of the reduced term, i.e. it is homogenized, so exit with success. Change of Unknown can now be successfully applied, substituting y for the reduced term in the output equation.

3.2. Application to Simultaneous Equations

We will now describe the application of Homogenization to simultaneous equations. We will call this application **Extended Homogenization**, to distinguish it from Standard Homogenization.

The method is best illustrated by an example. Consider the equations we discussed above:

$\cos(x) + \cos(y) = 1$

$\sec(x) + \sec(y) = 4$.

We use the Standard Homogenization parser. Parsing the equations with x as the unknown gives the **x-offenders set** $\{\cos(x), \sec(x)\}$, and parsing with y as the unknown gives the y-offenders set, $\{\cos(y), \sec(y)\}$. We choose a reduced term from the x-offenders set as in Standard Homogenization. In this case the **x-reduced term** is $\cos(x)$, the y-reduced-term is $\cos(y)$. (Note that in general the x_i-offenders sets need not contain the same functions.) The terms in the offenders sets are rewritten as algebraic functions of the reduced terms, using the the rewrite rules

$\sec(x) \rightarrow 1/\cos(x)$

$\sec(y) \rightarrow 1/\cos(y)$.

Substituting these rewrites into the equations, and replacing $\cos(x)$ by x1, and $\cos(y)$ by x2 we obtain:

x1 + x2 = 1

1/x1 + 1/x2 = 4.

Solving these equations by the basic method yields x1 = 1/2 and x2 = 1/2. Finally, solving $\cos(x) = 1/2$ and $\cos(y) = 1/2$ gives the solution to the problem.

Let us generalize this. Firstly, we require the concept of Homogenization with **respect to a variable**. In Standard Homogenization this variable is the unknown. In Extended Homogenization, to homogenize an equation Y with respect to x_i means to apply the process of Standard Homogenization to Y treating x_i as the sole unknown.

If we are homogenizing with respect to x_i, a term which does not contain x_i is effectively a constant, and will not appear in the offenders set created by this operation.

We will now describe the process of Extended Homogenization.

We have the list X which consists of the unknowns, the x_i, and the list E of the equations, the e_i. The basic process is to homogenize the equations with respect to each variable. This is done as follows:

- Consider the first element of X, x_j say. We now homogenize the equations in E with respect to x_j.

- To do this we treat the set of equations in E as one entity, a conjunction of equations. We parse this to obtain the x_j-offenders set. This set is the union of the offenders sets produced by parsing each of the e_i separately with respect to x_j. (As noted above the terms which do not contain x_j are treated as constants during this parse, and thus are not put into the offenders set.)

- Using the Standard Homogenization process we rewrite each member of the x_j-offenders set as an algebraic function of the x_j-reduced term, which is chosen in the standard way. These rewrites are then substituted into the conjunction of the equations of E.

- The equations are now homogeneous with respect to x_j.

- We now repeat the process with the next member of X, and continue until every member of X has been used. At this point the equations are homogeneous in every unknown.

- Now Change of Unknown can be performed. We substitute y_i for the x_i-reduced term, for every x_i in X. We record these substitutions in R, the **reduced term list**. R therefore consists of equations of the form:

    $y_j = f_j(x_j)$

    where f_j is some function.

- The equations in E are now algebraic in the y_i, and we can solve them using the basic method. This gives us the values of the y_i, and substituting these values into R gives us a set of independent equations for the x_i.

In some cases the equations may already be homogenized (or algebraic) with respect to some of the variables. This is detected by the x_i-offenders set being a singleton. If this singleton is x_i this variable need no longer be considered for Homogenization, and we proceed with the next variable. If the singleton is some other term, $f(x_i)$ say, the substitution $y_i = f(x_i)$ is made, and we proceed with the next variable.

If Homogenization succeeds, then the solution to the equations is obtained more neatly than by using the basic method on its own. If all the x_i-offenders sets produced contain only the x_i the equations are algebraic, and we can apply the basic method without attempting Homogenization. In other cases however, there seems to be no easy way of telling whether Homogenization should be attempted. Consider the following question:

$$\log_y(x) = 2 \qquad \text{(ii)}$$

$$\log_2(x) + \log_2(y) = 3. \qquad \text{(A.E.B. 1976)} \qquad \text{(iii)}$$

In this case we can solve (ii) for x to obtain $x = y^2$. Substituting this value in (iii) gives an equation for y which PRESS solves easily. Thus the basic method is quite appropriate for this question. Homogenization fails if it is attempted, but this cannot be predicted.

To overcome this problem, the present implementation adopts the following strategy: If the equations are all algebraic the basic method is used. Otherwise Homogenization is attempted. If Homogenization fails the basic method is tried. These tactics allow PRESS to solve examples such as the one above. However, we have wasted time trying to homogenize.

Extended Homogenization increases the range of problems which PRESS is able to solve. However, Homogenization is not taught to A level students, although they may 'rediscover' the method while working on a particular example. A method they <u>are</u> taught is **Elimination**. We will now discuss this method, and compare it with Extended Homogenization.

4. Elimination

The term **Elimination** is often applied to solving systems of simple linear equations. The basic method is fairly satisfactory for equations of this type, but Elimination is in fact applicable to a much wider range of problems.

The process of Elimination consists of transforming some of the equations in some way, and combining these transformed equations so that a variable is eliminated from the set of equations.

We will begin the explanation of Elimination with an example of linear equations.

4.1. The method of Elimination

Consider the set of equations

$$3.x + 2.y = 9 \qquad \text{(i)}$$

$$2.x - 5.y = -13. \qquad \text{(ii)}$$

We decide to eliminate y, say. To do this we multiply (i) by 2, and (ii) by 3, and subtract the resulting equations. We are left with a single equation in y, $19.y = 57$.

Hence y equals 3 and thus we can find the value of x from (i) or (ii).

Obviously, on such a simple example we have gained nothing. Let us consider a more interesting case.

$\cos(x) - 3.\sin(y) = 0$ (iii)

$6.\cos(y) = 5 - \sin(x)$ (iv)

Here, from (iii) we obtain

$\cos(x) = 3.\sin(y)$ (v)

and from (iv)

$\sin(x) = 5 - 6.\cos(y)$.   (vi)

Now square (v) and (vi) and add the two resulting equations to obtain:

$1 = 9.\sin^2(y) + 25 - 60.\cos(y) + 36.\cos^2(y)$.   (vii)

We have eliminated x from the equations. (vii) can be solved to give the value of y, and substituting this value in (iii) yields an equation for the value of x.

4.2. Comparison with Extended Homogenization

How does Elimination compare with Extended Homogenization? Using the same example as above we obtain the offenders sets $\{\cos(x), \sin(x)\}$ and $\{\cos(y), \sin(y)\}$. Letting $x1 = \cos(x)$ and $y1 = \cos(y)$ we obtain the equations

$x1 - 3.(1 - y1^2)^{1/2} = 0$

$(1 - x1^2)^{1/2} + 6.y1 = 5$.

Solving the first equation for x1, and substituting the value obtained into the second gives:

$(9.y1^2 - 8)^{(1/2)} = 5 - 6.y1$.   (viii)

Squaring (viii) and simplifying gives

$27.y^2 - 60.y + 33 = 0$,

a quadratic in y1. This equation is also generated when PRESS solves (vii), obtained by applying Standard Homogenization to (vii).

Comparing the two methods of solution seems to indicate that Elimination offers no particular advantage over Extended Homogenization on this kind of problem. Both involve a squaring operation and the application of Homogenization, the relative order varying. Both arrive at the same quadratic. Thus given an implementation of Extended Homogenization it seems that Elimination is superfluous.

However, as Elimination is such a well known method we will discuss how it could be implemented.

4.3. Planning the elimination

When is Elimination possible, and how do we proceed if it is? We will not discuss the case of linear equations. In this case Elimination is always possible[1] but usually it is unnecessary.

At the time of writing Elimination has not been implemented. Therefore the following should be viewed as a possible method of implementation, rather than a working program.

For Elimination to succeed we require that the equations are of the form:
$$a_i f_i(x) + g_i(y,z,u,v,\ldots) = 0$$
where we distinguish x. Note that the a_i are constants, any of the other variables have been merged into the g_i.

To eliminate x we need a rule which relates the f_i, or a set of such rules. The type of rules required are modified Collection rules. Collection rules are of the form LHS $\rightarrow$ RHS where RHS contains fewer occurrences of the unknown than LHS. (See [Bundy and Welham 81] for more details.)

For Elimination we rearrange the Collection rules to the form of LHS1 $\rightarrow$ RHS1 where RHS1 is free of the unknown. This is always possible, because in the simplest case we can rewrite the Collection rule as LHS - RHS $\rightarrow$ 0, which is of the required form.

In the above example the rule used was
$$\cos^2(x) + \sin^2(x) = 1.$$

We need a method for determining if a suitable collection rule exists. It seems that Homogenization should be applicable. A possible approach could be the following: We have a set of equations $\{e_i\}$, and a set of unknowns, the $\{x_i\}$.

- Using the Homogenization parser we parse all the equations with respect to all the unknowns, as in Extended Homogenization. However, there is one important difference. Each equation produces an offenders set for each unknown. In Extended Homogenization we take the union of all the x_i offenders sets, to produce <u>the</u> x_i-offenders set. In Elimination this union is not performed, the offenders sets are kept separately. Parsing e_j with respect to x_i produces an offenders set we will call $O_{i,j}$.

- We now try to find an i such that for all j, $O_{i,j}$ is a singleton set. If

---

[1] unless all the equations are nearly solved, that is of the form $x_i + a_i = b_i$, where a_i and b_i are constants.

there is such an i, this means that every occurrence of x_i is isolatable within the equation in which it occurs. We will therefore try to eliminate x_i.

- We take the union over k of the $O_{i,k}$ to produce the x_i-offenders set. From this set we choose a reduced term, $f(x_i)$, using the Standard Homogenization method.

- We now try to rewrite each member of the x_i-offenders set in terms of $f(x_i)$. This is of course a step that occurs in homogenizing the equations with $f(x_i)$ as a reduced term.

- If we succeed in the above step each occurrence of x_i can be eliminated using equation e_j, which contains $f(x_i)$. We simply transform the rewrite rule into a modified collection rule, as above.

Let us consider this process on the example above.

$$\cos(x) - 3.\sin(y) = 0$$

$$\sin(x) + 6.\cos(y) = 5$$

The offenders sets produced are $\{\cos(x)\}$, $\{\sin(x)\}$, $\{\sin(y)\}$, $\{\cos(y)\}$. All four sets are singleton so we choose any one, say $\{\cos(x)\}$. The x-offenders set is $\{\cos(x),\sin(x)\}$. We now attempt to rewrite each member of this set in terms of $\cos(x)$. The rewrite of $\cos(x)$ is trivial, and we find that

$$\sin(x) \rightarrow (1 - \cos^2(x))^{1/2}.$$

Transforming this to a modified collection rule produces:

$$\sin(x) - (1 - \cos^2(x))^{1/2} = 0 \qquad \text{(ix)}$$

We transform the equations so that the term containing x is isolated. This gives us:

$$\cos(x) = 3.\sin(y) \qquad \text{(x)}$$

$$\sin(x) = 5 - 6.\cos(y). \qquad \text{(xi)}$$

We now use (ix) to eliminate x. The left hand side of (ix) contains two subterms. The first, $\sin(x)$, occurs in (xi). We use (x) to produce the second subterm, applying transforms to both sides of the equation. Hence

$$(1 - \cos^2(x))^{1/2} = (1 - (3.\sin(y))^2)^{1/2} \qquad \text{(xii)}$$

Now we subtract (xii) from (xi). The LHS of the equation produced matches the LHS of (ix). Hence we can equate the RH sides. This gives

$$0 = 5 - 6.\cos(y) - (1 - (3.\sin(y))^2)^{1/2},$$

we have eliminated x.

The above method is not as neat as we would like. For example, instead of (ix) we would like:

$$\cos^2(x) + \sin^2(x) = 1$$

However, if this rule was used we would have to transform equation (xi), which is not necessary in the above method.

5. Conclusion

We have discussed various methods of solving simultaneous equations. Some of these techniques have been implemented in PRESS, and we have given an outline of how Elimination could be implemented.

We have seen that Extended Homogenization seems to offer at least as much as Elimination for non-linear equations, and for linear equations the basic method is adequate.

Homogenization can also be used to implement Elimination. As is often the case in the algebra domain, the principle problem is to constrain the search space. Homogenization requires very little search, and is thus a promising candidate for use in solving simultaneous equations.

Acknowledgements

We would like to thank Alan Bundy, Lawrence Byrd, Richard O'Keefe and Leon Sterling for their useful ideas and help.

This work is supported by SRC grant GR/B 29252 and an SRC studentship to the author.

References

[Bundy and Silver 81]
   Bundy, A. and Silver, B.
   Homogenization: Preparing Equations for Change of Unknown.
   In Schank, R., editor, IJCAI7. International Joint Conference on Artificial Intelligence, 1981.
   Longer version available from Edinburgh as DAI Research Paper No. 159.

[Bundy and Welham 81]
   Bundy, A. and Welham, B.
   Using meta-level inference for selective application of multiple rewrite rules in algebraic manipulation.
   Artificial Intelligence 16(2), 1981.

[Clocksin and Mellish 81]
   Clocksin, W.F. and Mellish, C.S.
   Programming in Prolog.
   Springer Verlag, 1981.

[Conte and de Boor 72]
   Conte, S.D. and de Boor, C.
   Elementary Numerical Analysis.
   McGraw-Hill Kogakusha, 1972.

[Sterling et al 82]
   Sterling, L., Bundy, A., Byrd, L., O'Keefe, R., and Silver, B.
   Solving Symbolic Equations with PRESS.
   Research Paper 171, Dept. of Artificial Intelligence, Edinburgh, 1982.
   To appear in EUROCAM 1982 Proceedings.

META-LEVEL INFERENCE AND PROGRAM VERIFICATION
Leon Sterling and Alan Bundy

Department of Artificial Intelligence
University of Edinburgh

Abstract

In [Bundy and Sterling 81] we described how meta-level inference was useful for controlling search and deriving control information in the domain of algebra. Similar techniques are applicable to the verification of logic programs. A developing meta-language is described, and an explicit proof plan using this language is given. A program, IMPRESS, is outlined which executes this plan.

Acknowledgments

This work was supported by SERC grant GR/B/29252.

Keywords

meta-level inference, logic programming, program verification

1. Introduction

It is well-known that logic programs have a dual interpretation - a procedural one and a semantic one (see for example [Kowalski 79]). Program statements can be interpreted both as commands to be executed under some control regime and as first-order predicate calculus clauses.

Consider the following 'code' for appending two lists.*

append([],Y,Y) ←
append([H|X],Y,[H|Z]) ← append(X,Y,Z)

where [] denotes the nil list, and [H|T] the constructor function cons(H,T).

The naive semantic interpretation of this 'code' is that two theorems about 'append' are true, namely append([],Y,Y) is true for all Y, and for all X,Y,Z if append(X,Y,Z) is true then append([H|X],Y,[H|Z]) is true. More powerful things can be said moreover. Clark [Clark 79] shows how applying a fixpoint interpretation to the above logic program for append leads to the theorem in first-order predicate calculus

append(X,Y,Z) ⟷ (X=[] & Y=Z) v
 ∃ H,X1,Z1 (X=[H|X1] & Z=[H|Z1] & append(X1,Y,Z1)).

Using the fact that the two cases above are essentially disjoint, he further breaks this down into two theorems. The 'nil' case is

---

*Throughout the paper we will use the notation conventions of DEC-10 Prolog [Pereira et al 79], one implementation of some of the ideas of logic programming. In particular, variables begin with upper-case letters and constants begin with lower-case letters.

$$\text{append}([\,],Y,Z) \longleftrightarrow Y=Z \ .$$

We shall be making implicit use of this sort of inference throughout the paper.

To give a procedural interpretation one needs to distinguish between input and output variables, i.e. decide what use will be made of the program. The most common use of the append program is when X and Y are input variables, both lists, and one wants to compute Z, the result of appending X and Y. This is a determinate program. On the other hand, one could use Z as the input variable, a list, and compute nondeterministically ways of partitioning it into two lists, X and Y.

Given a specific use of a program one can analyse its properties. In [Clark 79] three properties of logic programs are given special attention - namely, correctness, termination, and total correctness. We will concentrate mainly on the first property, correctness, though the techniques to be described seem to have applications to the other properties. If P(X,Y) is a program, where X is a vector of input variables and Y is a vector of output variables then a **correctness property** of P is a theorem of the form:

$$I(X) \ \& \ P(X,Y) \longrightarrow O(X,Y)$$

where I(X) is an **input condition** and O(X,Y) is an **output condition**.

Program verification is basically proving program correctness properties. For example, with the append example above and the use for computing the result of appending two lists, one might like to verify that if you start off with two lists, you end up with a list. As a theorem, expressed in Prolog form, this is

$$\text{list}(Z) \leftarrow \text{list}(X) \ \& \ \text{list}(Y) \ \& \ \text{append}(X,Y,Z).$$

We have built a program which can prove the above theorem among others. The program, IMPRESS, was originally designed for proving properties of an equation solving program written in Prolog [Bundy and Sterling 81]. Its scope has since expanded to general theorems expressed in horn-clause form. This has particular applications for verification of logic programs.

An important aspect of building IMPRESS is developing a suitable meta-language of concepts about proofs and proof plans. These concepts will be described throughout the paper.

In the next section we give an example verification. Then the meta-level concepts are discussed in some detail. A brief comparison to other work in this area follows, and the final section gives conclusions and points to future directions of the research.

2. An Example Verification

As an example of a verification which illustrates the language we are evolving, consider the relationship between the length of the lists involved in the append predicate. That is, if you append two lists together, the length of the resultant list should be the sum of the lengths of the two lists. In verification terms this could be expressed as

$$\text{append}(X,Y,Z) \longrightarrow \{\text{length}(X,N) \ \& \ \text{length}(Y,M) \longrightarrow \text{length}(Z,N+M) \ \}.$$

The form of the theorem as proved by IMPRESS, and as will be described in this paper, is

$$\text{length}(Z,N+M) \longleftarrow \text{length}(X,N) \ \& \ \text{length}(Y,M) \ \& \ \text{append}(X,Y,Z).$$

This is proved by induction on the variable list X. Using an induction schema of append or an induction schema of length would give rise to a virtually identical proof.

Before describing the proof, let us write down the program/axioms for length and append.

    length([],0).
    length([H|T],N+1) ← length(T,N).

    append([],X,X).
    append([H|X],Y,[H|Z]) ← append(X,Y,Z).

The structure of the programs for length and append are essentially identical. Both consist of two clauses, the base clause and the step clause. The step clause has a simple structure, just a recursive call to itself. This recursive call we call the recursant. In general this structure will not be so simple. In [Bundy and Sterling 81] we describe a proof of the correctness of isolation, a method for solving equations. There the step clause has the form

    isolate([N|Tail],Y,Z) ← isolax(N,Y,Y1) & isolate(Tail,Y1,Z).

In this case we distinguished between the isolate term, which we called the recursant, and the isolax term which we called the performant. These distinctions were important in guiding the correctness proof. In this paper we will restrict the proofs to programs whose step clauses only have a recursant. (This can be regarded as a clause with a nil performant).

An induction proof has two parts, the base case and the step case. The appropriate instantiation for the base case when proving a theorem about lists is the nil list, []. The instantiation for the step case is cons(Head,Tail), or in our terms [Head|Tail]. Taking the base case first, the theorem to be proved is

    length(Z,N+M) ← length([],N) & length(Y,M) & append([],Y,Z).

Using the implicit information from the fixpoint semantics, N is instantiated to 0 because of the theorem length([],N) ⟷ N=0 and Z is unified with Y because of the theorem append([],Y,Z) ⟷ Y=Z. This leaves the theorem to be proved as

$$\text{length}(Y,M) \leftarrow \text{length}(Y,M),$$

which is trivially established.

The step case is more interesting. Here an induction hypothesis is asserted as a theorem, and critically used in the proof. The form of the induction hypothesis can be written down immediately. In this example the induction hypothesis is

$$\text{length}(Z,N+M) \leftarrow \text{length}(\text{list},N) \ \& \ \text{length}(Y,M) \ \& \ \text{append}(\text{list},Y,Z).$$

The theorem to be proved is then

length(Z1,N1+M)
    $\leftarrow$ length([H|list],N1) & length(Y,M) & append([H|list],Y,Z1).

Prior to skolemizing the theorem to be proved we use the theorems, length([H|X],N+1) $\longleftrightarrow$ length(X,N) and append([H|X],Y,[H|Z]) $\longleftrightarrow$ append(X,Y,Z) to replace N1 by N+1 and Z1 by [H|Z]. After skolemization, the theorem to be proved becomes

length([h|z],n+m+1)
    $\leftarrow$ length([h|list],n+1) & length(y,m) & append([h|list],y,[h|z]).

The proof is as follows:

1. The three propositions in the body of the theorem are asserted into the database, namely

    length([h|list],n+1) $\leftarrow$
    length(y,m) $\leftarrow$
    append([h|list],y,[h|z]) $\leftarrow$

2. Resolve append([h|list],y,[h|z]) $\leftarrow$ against
append(X,Y,Z) $\leftarrow$ append([H|X],Y,[H|Z]) to get

    append(list,y,z) $\leftarrow$

3. The proof now proceeds backwards by linear search with goal $\leftarrow$ length([h|z],n+m+1). Resolve the goal against length([H|X],N+1) $\leftarrow$ length(X,N) to give

    $\leftarrow$ length(z,n+m)

4. Resolve this against the induction hypothesis to give

    $\leftarrow$ length(list,n) & length(y,m) & append(list,y,z)

5. Use the assertion length(y,m) $\leftarrow$ to remove the central proposition, leaving

    $\leftarrow$ length(list,n) & append(list,y,z)

6. Use the theorem length(X,N) $\leftarrow$ length([H|X],N+1) and the assertion length([h|list],n+1) $\leftarrow$ to leave as the goal

    $\leftarrow$ append(list,y,z)

7. Resolving this against append(list,y,z) $\leftarrow$ produces the empty clause and

hence a proof of the theorem.

3. Meta-level Concepts

The inductive proofs of many other correctness theorems appear to follow the same basic plan as the proof above. Let us try to identify the meta-level concepts involved. We restate the theorem for convenience.

$$length(Z,N+M) \leftarrow length(X,N) \;\&\; length(Y,M) \;\&\; append(X,Y,Z). \qquad (i)$$

This fits the schema for a correctness property with program hypothesis, append(X,Y,Z), input condition, length(X,N) & length(Y,M), and output condition, length(Z,N+M).

We choose an induction scheme and induction variable by analogy with the recursion scheme and recursion variable of the program hypothesis. The predicate append is defined by primitive recursion on the structure of its argument, which is a list. Thus to prove the theorem we use the induction scheme

$$\forall X \; Q(X) \leftarrow \{Q([\,]) \;\&\; \forall X \; \forall H \; Q([H|X]) \leftarrow Q(X)\}$$

where Q is the conjecture (i) and X is the first argument of append.

Using this induction scheme will generate two subgoals: $Q([\,])$, which we call the **base case**; and $Q([H|X]) \leftarrow Q(X)$, which we call the **step case**.

Note that, in this example, had length(X,N) been chosen as the program hypothesis we would have ended up with an identical induction scheme and base and step subgoals.

A specific proof plan can thus be spelt out.

- Locate the program hypothesis of the conjecture.

- Choose an appropriate induction scheme and induction variable by analogy with the recursion scheme and recursion variable of the program hypothesis.

- Prove the base case after the appropriate instantiation.

- Prove the step case after the appropriate instantiation.

In [Bundy and Sterling 81] we outlined a proof plan for the step case, which we repeat here. Note that, since the definition of the program hypothesis has an empty performant, the application of this proof plan is necessarily simplified. Bracketed comments refer to the proof of the last section.

(a) Assert the induction hypothesis and the step versions of the input condition and the program hypothesis as temporary axioms. (step 1)

(b) Unfold the step program hypothesis into its performant and recursant. (step 2)

(c) Proceed to prove the step version of the output condition.

(d) Fold the step output condition into its performant and recursant. (step 3)

(e) Establish the step output condition performant from the program hypothesis performant. (not needed in this proof)

(f) Apply the induction hypothesis to the step output condition recursant. (step 4)

(g) Establish the induction hypothesis input conditions from the corresponding step input conditions and the program hypothesis performant. (steps 5 and 6)

(h) Establish the induction hypothesis program hypothesis from the program hypothesis recursant. (step 7)

In our example steps 5 and 7, where assertions were used to establish subgoals, were single resolutions, whereas step 6 required two resolutions. In general, these steps can be arbitrarily complex, but a large measure of search guidance is provided by specifying those axioms which are and those which are not involved in the search. Currently, IMPRESS does not get involved in this search, but uses the proof plan to print out a lemma to be proved.

4. Related work

The Edinburgh LCF project [Gordon et al 79] built a computer system for doing formal proofs interactively. The environment provided various primitive steps which the user combined together to generate a proof. The emphasis was to provide a flexible tool for investigation of proofs.

IMPRESS, on the other hand, has no such interactive facility. Development is concentrated on generating proofs automatically, according to explicit proof plans. The proofs exhibited using LCF are goals for IMPRESS to prove.

Other important proof strategies are contained in work on program transformations. Darlington [Darlington 81] gives a meta-language vocabulary, which we are extending, for discussing such transformations. His basic manoeuvres, for example fold/unfold, have been incorporated into our proof plan. His program has no a priori representation of proof plans to apply to conjectures, at the level of the IMPRESS plan above.

We have also built on the work of Boyer and Moore [Boyer & Moore 79]. The selection of a suitable induction scheme and variable, by analogy with the recursion scheme and variable, is related to their technique of chosing an induction scheme and variable after the breakdown of symbolic evaluation. (In fact we built a toy Boyer/Moore program in Prolog and used the experience gained when building IMPRESS.)

Much of the knowledge of the Boyer/Moore program, however, is embedded implicitly in code. For example, much implicit inference is done when type checking at an early stage of a proof. Our emphasis is more in developing a language to describe proofs. Using this language we are able to express heuristics about how to undertake a proof. These heuristics are then converted into explicit proof plans, such as the one described above.

5. Future Directions and Conclusions

As suggested above, our aim is to be able to prove a wide range of theorems using meta-level inference to guide the search. There are many directions in which to proceed. For example, to extend the logic program proofs to termination and total correctness. Also, to translate the experiences of other program manipulation systems into a form suitable for IMPRESS to use. This has already been started with respect to Darlington and Boyer and Moore's systems.

In this paper, we have outlined the current state of our ideas. An example program verification proof is described that our program, IMPRESS, is capable of. It should be emphasised that this schema seems to cover a wide number of proofs. Logic programming seems an excellent domain in which to continue this research.

REFERENCES

[Boyer & Moore 79]
    Boyer, R.S. and Moore, J.S.
    ACM monograph series. : A Computational Logic.
    Academic Press, 1979.

[Bundy and Sterling 81]
    Bundy, A. and Sterling L.S.
    Meta-level Inference in Algebra.
    Research Paper 164, Dept. of Artificial Intelligence, Edinburgh, September, 1981.
    Presented at the workshop on logic programming for intelligent systems, Los Angeles, 1981.

[Clark 79]
    Clark, K.L.
    Predicate Logic as a Computational Formalism.
    Report 79/59, Department of Computing, Imperial College, London, December, 1979.

[Darlington 81]
    Darlington J.
    An Experimental Program Transformation and Synthesis System.
    Artificial Intelligence 16(3):1-46, August, 1981.

[Gordon et al 79]
    Gordon M.J., Milner A.J., and Wadsworth C.P.
    Lecture Notes in Computer Science. Volume 78: Edinburgh LCF - A mechanised logic of computation.
    Springer Verlag, 1979.

[Kowalski 79]
    Robert Kowalski.
    Logic for Problem Solving.
    North Holland, 1979.

[Pereira et al 79]
    Pereira, L.M., Pereira, F.C.N. and Warren, D.H.D.
    User's guide to DECsystem-10 PROLOG .
    Occasional Paper 15, Dept. of Artificial Intelligence, Edinburgh, 1979.

AN EXAMPLE OF FOL USING METATHEORY

FORMALIZING REASONING SYSTEMS AND INTRODUCING DERIVED INFERENCE RULES

Richard W. Weyhrauch

Stanford University
Stanford, CA 94305

1. Introduction

This paper shows how FOL, Weyhrauch [1977, 1980], can be used to formalize and implement reasoning systems. The reasoning system we have chosen as an example is the implicational part of the propositional logic P_1 described in Church [1956], p. 72. The formulas of P_1 are either *sentential constants*, SENTCONSTs, or are built up from other formulas using the implication symbol, $\supset$. For any formulas A, B, and C the following are axioms:

AXIOM1(A,B) $A \supset (B \supset A)$
AXIOM2(A,B,C) $(A \supset (B \supset C)) \supset ((A \supset B) \supset (A \supset C))$.

We say that each axiom is a theorem and that, if A is a theorem and $A \supset B$ is a theorem, then B is a theorem. The latter inference rule is called *modus ponens*.

Suppose A is some particular sentential constant. The following is a derivation showing that the particular formula $A \supset A$ is a theorem.

1. $A \supset ((A \supset A) \supset A)$ AXIOM1($A, A \supset A$)
2. $(A \supset ((A \supset A) \supset A)) \supset ((A \supset (A \supset A)) \supset (A \supset A))$ AXIOM2($A, A \supset A, A$)
3. $(A \supset (A \supset A)) \supset (A \supset A)$ ModusPonens 1, 2
4. $A \supset (A \supset A)$ AXIOM1(A, A)
5. $A \supset A$ ModusPonens 4, 3

It is clear from the above proof that we could have used any formula in place of A. This suggests the following *derived rule*.

OBVIOUS: **If A is a formula then $A \supset A$ is a theorem**

The meta-theorem OBVIOUS although simple is a typical example of a derived rule that we might want to add to our system.

In section 2 we give the construction of META, an FOL context which formalizes the syntax and inference rules of P_1. (In Prolegomena [1980], FOL contexts were called L/S pairs.) This formalization follows Church quite closely. Unfortunately Church's description of modus ponens is not in a form that we can use. We discuss the reason for this in section 3 where we also use the theorem-generating features of FOL to derive a more useful statement of the rule. In section 4 we define the FOL context, THEORY, which implements the reasoning system entailed by the **declarative** description of P_1 in META. After demonstrating how to use the FOL reflection principles to generate proofs in THEORY, we return our attention to META, and in section 5 we use the axiomatization of P_1 to prove the theorem OBVIOUS so that it can be used as a new derived rule and thus extends the expressive power of the system.

I am presenting this simple example because it is completely self contained and I do not want the details of the example to obscure the details of how the FOL features work. The important point here is not the complexity of the example, but rather the details of how FOL provides a formal framework for constructing reasoning systems in which we can both prove the correctness of the new reasoning principles and add them to the reasoning machinery available to us. The ability to reason about the inference rules essentially means that we can add provably correct new rules to our system.

The techniques described here have been used in the current FOL system to build a context META that contains an almost complete axiomatization of the FOL formalism itself. This formalism will be described in detail in Weyhrauch[1982]. This axiomatization has been used to do extremely complex examples. Aiello and Weyhrauch[1980] have used it for algebraic simplification. Talcott and Weyhrauch[1982] use it for reasoning about actions in a new formulation of the McCarthy situation calculus (McCarthy and Hayes[1969]). The example of this paper was first run on the FOL system in the summer of 1977.

2. Formalizing Implicational Propositional Logic

We start by describing the propositional language of P_1 to FOL. The text below contains the actual FOL commands necessary to carry out this example. The commands to FOL are preceded by 5 stars, *****. The FOL response appears as lines of text beginning with line numbers. Even though this is a little harder to read than an explanation in English, I want to give some flavor of the complexity of interacting with FOL and to present a complete example.

Well-formed formulas, WFFs, are built by starting with *sentential constants*, SENTCONSTs, and using the function *make implication*, mkimp, to build new formulas out of old ones. The first command directs FOL's attention to the context META.

```
***** CHANGE TO META;
***** DECLARE SORT SENTCONST;
***** DECLARE INDVAR sentconst ε SENTCONST;
***** DECLARE SORT WFF;
***** DECLARE INDVAR wff1, wff2, wff3 ε WFF;
***** DECLARE OPCONST mkimp 2;
```

FOL has no defaults so it must be told everything. In particular since each FOL context has its own language we must tell it what identifiers are used in what way. The first and third FOL commands make SENTCONST and WFF into sorts, i.e., predicate constants of one argument. We also declare that SENTCONST is a variable which ranges over sentential constants and wff1, wff2, and wff3 are variables over well-formed formulas. mkimp is declared to be a function symbol (or operation constant) of two arguments. The axioms

```
***** AXIOM WFF1:    ∀sentconst.WFF(sentconst) ;
            WFF2:    ∀wff1 wff2.WFF(mkimp(wff1,wff2)) ;;
```

state that every thing of sort SENTCONST is of sort WFF and that mkimp maps WFFs onto WFFs. In FOL another way of specifying this is to say that the sort WFF is *more general* than the sort SENTCONST; and to use the *function map*, FMAP, command to specify the sort of the value of a function given the sorts of its arguments. The commands are:

```
***** MOREGENERAL WFF ≥ {SENTCONST};
***** FMAP mkimp(WFF,WFF)=WFF;
```

Now that we know what formulas are, we need some facts about *implications*, IMPs.

```
***** DECLARE SORT IMP;
***** DECLARE OPCONST hypof(IMP)=WFF, concl(IMP)=WFF;

***** AXIOM IMPL1:   ∀wff1 wff2.IMP(mkimp(wff1,wff2)) ;
            IMPL2:   ∀wff1 wff2.hypof(mkimp(wff1,wff2))=wff1 ;
            IMPL3:   ∀wff1 wff2.concl(mkimp(wff1,wff2))=wff2 ;
            IMPL4:   ∀wff1.(IMP(wff1)⊃wff1=mkimp(hypof(wff1),concl(wff1))) ;;
```

This axiom states the syntactic properties of implications: 1) if you make an implication (mkimp) of any two formulas, then it is an implication (IMP); 2) the *hypothesis* (hypof) if an implication is its first component; 3) the *conclusion* (concl) of an implication is its second component; 4) If you have an implication then you can reconstruct it by making an implication out of its hypothesis and its conclusion. These are facts about formulas.

We introduce the sort THEOREM.

```
***** DECLARE SORT THEOREM;
```

The idea that the axioms are theorems is formalized by the axioms

```
***** AXIOM
        HILBERT1:   ∀wff1 wff2.THEOREM(
                        mkimp(wff1,(mkimp(wff2,wff1)))) ;

        HILBERT2:   ∀wff1 wff2 wff3.THEOREM(
                        mkimp(
                            mkimp(wff1,mkimp(wff2,wff3)),
                            mkimp(mkimp(wff1,wff2),mkimp(wff1,wff3)) )) ;;
```

Modus ponens is described by the axiom

```
***** AXIOM MODUSP: ∀wff1 wff2.(THEOREM(mkimp(wff1,wff2))⊃(THEOREM(wff1)⊃THEOREM(wff2))) ;;
```

The above axioms are a direct expression of the description of P_1 found in Church.

In order to complete our formalization of P_1 we must make a distinction among the *facts* of an FOL context, the particular *formula* which is asserted by an FOL fact and a theorem of P_1. The theorems of P_1 are a collection of formulas determined by its inference rules. An FOL fact is a complex data structure which appears in an FOL context and that contains information about what formula is asserted, why it is being asserted in this context and other data. Consider the following fact in META:

```
HILBERT1:  ∀wff1 wff2.THEOREM(mkimp(wff1,(mkimp(wff2,wff1))))    AXIOM
```

We write it this way to emphasize that facts *are not formulas*. This fact is *called* HILBERT1. It *asserts* the formula ∀wff1 wff2.THEOREM(mkimp(wff1,(mkimp(wff2,wff1)))) and is *justified* in being called a fact because it is an AXIOM.

This next axiom states a very general principle about how FOL works. We compute the formula that is asserted by some FOL fact using the function *well-formed formula of*, wffof. For historical reasons facts of an FOL context are called VLs.

```
***** DECLARE SORT VL;
***** DECLARE INDVAR fact, v11, v12 ε VL;
***** DECLARE OPCONST wffof(VL)=WFF;

***** AXIOM THM: ∀fact.THEOREM(wffof(fact)) ;;
```

This very general axiom states the simple idea that the formula asserted by a fact of P_1, (wffof(fact)), is a theorem. A new idea has been introduced here. This axiom, which is a fact of META, mentions a fact of P_1. One question to ask is: where do the facts of P_1 come from?

The reflection principles embedded in the FOL system allow us to use the above axiomatization to create a reasoning system embodying P_1. We begin by creating a new context THEORY.

```
***** MAKELS THEORY;
***** CHANGE TO THEORY;
```

The first command creates the new context; the second focuses our attention on it. The command

```
***** DECLARE SENTCONST A,B;
```

declares A and B to be particular sentential constants in THEORY.

This completes our description of P_1. From this point on there are no new axioms and everything is entirely within the FOL representational framework.

3. Reformulating the statement of modus ponens

The statement of MODUSP in section 1 is simple and appropriate for metatheoretic discussions, but we also want to generate some particular facts in THEORY. For this purpose MODUSP is not very useful. One reason is that it only mentions the *theorems* of P_1. We want to make use of the *facts* of THEORY. Another way of saying this is that MODUSP asserts an implication whose hypothesis, THEOREM(mkimp(wff1,wff2)), cannot be computed simply *by looking* in THEORY.

A more useful formulation of this idea would be to say, "Suppose we have two facts of THEORY and the first asserts an implication and the second asserts the formula which is the hypothesis of the first. Then we can state that the conclusion of the first fact is a theorem." We call this principle MP and it is written in META as

```
MP:  ∀v11 v12.(IMP(wffof(v11)) ∧ hypof(wffof(v11))=wffof(v12) ⊃ THEOREM(concl(wffof(v11))) )
```

This is an easy consequence of the axioms above. Suppose we have two FOL facts, fact1 and fact2, and w1 (w2) is the formula asserted to be true by fact1 (fact2). Then if IMP(w1)∧hypof(w1)=w2, i.e., w1 is of the form w2⊃w3, then we can conclude that both THEOREM(mkimp(w2,w3)) and THEOREM(w2). By MODUSP we have THEOREM(w3).

The proof of MP as a theorem of META using FOL follows.

***** CHANGE TO META;

***** ASSUME IMP(wffof(v11))∧hypof(wffof(v11))=wffof(v12);

1 IMP(wffof(v11))∧hypof(wffof(v11))=wffof(v12)   (1)

***** ∀E IMPL4 wffof(v11);

2 IMP(wffof(v11))⊃wffof(v11)=mkimp(hypof(wffof(v11)),concl(wffof(v11)))

***** TAUT wffof(v11)=mkimp(hypof(wffof(v11)),concl(wffof(v11))) 1,2;

3 wffof(v11)=mkimp(hypof(wffof(v11)),concl(wffof(v11)))   (1)

***** ∀E THM v11;

4 THEOREM(wffof(v11))

***** SUBST 3 IN 4;

5 THEOREM(mkimp(hypof(wffof(v11)),concl(wffof(v11))))   (1)

***** ∀E THM v12;

6 THEOREM(wffof(v12))

***** ∧E 1 2;

7 hypof(wffof(v11))=wffof(v12)   (1)

***** SUBSTR 7 IN 6;

8 THEOREM(hypof(wffof(v11)))   (1)

***** ∀E MODUSP hypof(wffof(v11)),concl(wffof(v11));

9 THEOREM(mkimp(hypof(wffof(v11)),concl(wffof(v11))))⊃
      (THEOREM(hypof(wffof(v11)))⊃THEOREM(concl(wffof(v11))))

***** TAUT THEOREM(concl(wffof(v11))) 5,8,9;

10 THEOREM(concl(wffof(v11)))   (1)

***** ⊃I 1⊃10;

11 (IMP(wffof(v11))∧hypof(wffof(v11))=wffof(v12))⊃
      THEOREM(concl(wffof(v11)))

***** ∀I 11 v11 v12;

12 ∀v11 v12.((IMP(wffof(v11))∧hypof(wffof(v11))=wffof(v12))⊃
      THEOREM(concl(wffof(v11))))

***** THEOREM MP 12;

This is the first example we have given in this paper of a derived rule. It is called a derived rule (Kleene[1952] p. 86.) because we have proved it from the rules postulated in META. This proof guarantees that the additional method of inference MP does not increase the class of theorems of P_1. In the next section we will show how the reflection principles of FOL will use MP to generate some of the facts of THEORY.

4. A proof of A⊃A in THEORY

In this section we use the FOL reflection command to produce a proof of A⊃A.

***** CHANGE TO THEORY;

***** REFLECT HILBERT1 A A⊃A;

1 A⊃((A⊃A)⊃A)

This is for the first time we have used the reflection feature of FOL. The idea is that we want to use the meaning of the META axiom HILBERT1, i.e., ∀wff1 wff2.THEOREM(mkimp(wff1,(mkimp(wff2,wff1)))) as the justification for asserting the formula A⊃((A⊃A)⊃A) in THEORY. The reflection command takes as its first "parameter" the name of a fact in META – in this case HILBERT1. It notices if it is a universally quantified formula, which it is, and determines the sorts of the variables. In this case they are two well-formed formulas. The rest of the "parameters" to the reflection command are then expected to be objects in THEORY which are of these sorts, in this case, the formulas A and A⊃A. These formulas are then given names in META and the appropriate semantic attachments are made. This assigning of names in META to the objects of THEORY is a use of the *reflection principles* described in Prolegomena [1980]. We then instantiate the axiom. This produces the formula THEOREM(mkimp(Wff1,mkimp(Wff2,Wff1))) in META, where Wff1 (Wff2) is an individual constant in META whose semantic attachment is the formula A (A⊃A respectively). In less formal language Wff1 is the *name* in META for the formula A in THEORY. Similarly Wff2 is the name for the formula A⊃A. We then evaluate this formula using the FOL evaluater and use the *reflection principle for theorems* to assert the formula A⊃((A⊃A)⊃A) in THEORY.

***** REFLECT HILBERT2 A A⊃A A;

2 (A⊃((A⊃A)⊃A))⊃((A⊃(A⊃A))⊃(A⊃A))

***** REFLECT MP ↑,↑↑;

3 (A⊃(A⊃A))⊃(A⊃A)

Notice how in the last command we referred to MP – the theorem we proved in section 3, and that this theorem takes facts as "arguments" rather than formulas. In FOL ↑ refers to the most recent fact, ↑↑ the penultimate fact, *etc.*

***** REFLECT HILBERT1 A A;

4 A⊃(A⊃A)

***** REFLECT MP ↑↑,↑;

5 A⊃A

This is the formal proof of A⊃A which was described informally in the introduction. This is a proof of the particular formula A⊃A. We would need to repeat this collection of steps in order to prove the theorem B⊃B. This means that every time we want to conclude $A \supset A$ for some formula A, we need to repeat these five steps. In terms of theorem proving this seems excessive, but without some metatheoretic machinery we are stuck with the built in rules of the system. In the next section we fix this problem.

5. The derived rule: $A \supset A$

Immediately below is a proof in META of the general principle

OBVIOUS: If A is a formula then $A \supset A$ is a theorem.

The notation ↑:#1#2 means, take the last line, ↑, of the proof, select its well-formed formula, :, then take its first subpart, #1, and then take the second subpart, #2, of that. This way of pointing at previous formulas makes commands to FOL substantially shorter. In addition multiple commands can be typed on one line.

***** CHANGE TO META;

***** ∀E HILBERT1 wff1 mkimp(wff1,wff1);

1 THEOREM(mkimp(wff1,mkimp(mkimp(wff1,wff1),wff1)))

```
***** ∀E HILBERT2 wff1,mkimp(wff1,wff1),wff1;

2 THEOREM(mkimp(mkimp(wff1,mkimp(mkimp(wff1,wff1),wff1)),
        mkimp(mkimp(wff1,mkimp(wff1,wff1)),mkimp(wff1,wff1))))

***** ∀E MODUSP ↑:#1#1 ↑:#1#2;   TAUT ↑:#2#2 ↑↑↑:↑;

3 THEOREM(mkimp(
         mkimp(wff1,mkimp(mkimp(wff1,wff1),wff1)),
         mkimp(mkimp(wff1,mkimp(wff1,wff1)),mkimp(wff1,wff1))))⊃
      (THEOREM(mkimp(wff1,mkimp(mkimp(wff1,wff1),wff1)))⊃
        THEOREM(mkimp(mkimp(wff1,mkimp(wff1,wff1)),mkimp(wff1,wff1))))

4 THEOREM(mkimp(mkimp(wff1,mkimp(wff1,wff1)),mkimp(wff1,wff1)))

***** ∀E HILBERT1 wff1,wff1;

5 THEOREM(mkimp(wff1,mkimp(wff1,wff1)))

***** ∀E MODUSP ↑↑:#1#1 ↑↑:#1#2;   TAUT ↑:#2#2 ↑↑↑:↑;

6 THEOREM(mkimp(mkimp(wff1,mkimp(wff1,wff1)),mkimp(wff1,wff1)))⊃
    (THEOREM(mkimp(wff1,mkimp(wff1,wff1)))⊃THEOREM(mkimp(wff1,wff1)))

7 THEOREM(mkimp(wff1,wff1))

***** ∀I 7 wff1;

8 ∀wff1.THEOREM(mkimp(wff1,wff1))

***** THEOREM OBVIOUS 8;
```

This theorem of META can now be used by reflection to produce theorems in THEORY.

```
***** CHANGE TO THEORY;
***** REFLECT OBVIOUS A; REFLECT OBVIOUS B; REFLECT OBVIOUS A⊃B;

6   A⊃A

7   B⊃B

8   (A⊃B)⊃(A⊃B)
```

These three facts of P_1 were created using 11 steps (counting the proof of OBVIOUS) as opposed to the 15 steps necessary to repeat the proof of the preceding section three times. This reduction of the amount of work necessary to generate new facts is where the leverage of having metatheory available pays off.

By comparing the commands for the proof of A⊃A in THEORY to those of OBVIOUS in META, we can see that they have the same form.

```
REFLECT HILBERT1 A A⊃A;
REFLECT HILBERT2 A A⊃A A;
REFLECT MP ↑,↑↑;
REFLECT HILBERT1 A A;
REFLECT MP ↑↑,↑;

∀E HILBERT1 wff1 mkimp(wff1,wff1);
∀E HILBERT2 wff1 mkimp(wff1,wff1) wff1;
∀E MODUSP ↑:#1#1 ↑:#1#2;   TAUT ↑:#2#2 ↑↑↑:↑;
∀E HILBERT1 wff1 wff1;
∀E MODUSP ↑↑:#1#1 ↑↑:#1#2;   TAUT ↑:#2#2 ↑↑↑:↑;
```

The meta-meta-theorem suggested by the above remark would allow us to generalize a proof in the theory to a proof of some general facts in the meta-theory, which will not be explicitly stated or proved or used in this paper. But it is important to realize that the features already described in this paper are enough to formalize this kind of theorem.

6. Conclusion

Now that we have the example before us let me summarize the ideas in this example. We started out by formalizing (in FOL) a reasoning system. That is, the implicational part of the propositional logic, P_1. This should be looked upon as a *typical* reasoning system. The same techniques that we used here to formalize P_1 can and have been used to formalize most of FOL itself. This is a formulation that includes both the syntactic aspects of what it means to be a formula and the inference rules of the system. It is essential to realize here that, in the majority of currently active theorem proving systems, these aspects of the system are *fixed*, embedded in the system and inaccessible to a user. He is caught with them.

The situation here is very different. First, we formalized the reasoning system P_1 in the way that it was presented to us. By this I mean that we formalized it using the description we found in Church. This included the straightforward description of modus ponens, MODUSP. We then used the theorem proving (checking) facilities of FOL to reformulate this rule into the more useful rule MP. This ability to reason about the rules of our system is here used in an essential way. This new theorem, MP, was then added to our store of inference rules by "implementing" it using the FOL reflection principles. This ability to state metatheorems and have them available for "use" by the theory was the motivating force behind the introduction of reflection principles into FOL. In terms of old questions it represents a kind of formal solution to what people in AI used to call the "declarative/procedural controversy". By using reflection principles as we do, a sentence is used "declaratively" when thought of as a metatheorem, and used "procedurally" when it is the object of a reflection command. As pointed out in the main part of this paper, the resulting proof of A⊃A using reflection is very close to the informal one.

The next thing we did was show that

OBVIOUS: If A is a formula then $A \supset A$ is a theorem

was a theorem of META. This simple principle is to represent the typical **theorem proving** strategy we might want to add to our system. Let me explain this more carefully. A theorem prover can be thought of as a function that takes certain facts, $\mathcal{F}$ and tries to determine if some particular fact, F follows from $\mathcal{F}$, if it does then we are justified calling F a theorem. We can imagine that theorem provers when expressed in an appropriate metatheory have the form

$$\forall \mathcal{F}\ F.(\text{theoremprover}(\mathcal{F},F)=\text{OK} \supset \text{THEOREM}(F))$$

Thus theorem provers and the example given above are special cases of a more general schema for subsidiary deduction rules

$$\forall \overline{x}.(P(\overline{x}) \supset \text{THEOREM}(f(\overline{x})))$$

where $\overline{x}$ is any reasonable collection of parameters. This becomes a subsidiary deduction rule when we specify P, $\overline{x}$ and f. In the example of this paper $\overline{x}$ is simply wff1, $P(\overline{x})$ is TRUE, and $f(\overline{x})$ is mkimp(wff1,wff1). In Prolegomena[1980], p. 151, the example of doing an algebraic simplification in arithmetic is another example.

The importance of building a system in this way is that we do not have to decide beforehand what tools we want to allow ourselves to use for constructing proofs in the theory. FOL reflection allows you to add verifiably correct new rules to your system.

Of course the use of metatheoretic ideas is now accepted as a potential tool for extending the power of a representation system. In an interactive system it increases the quality of the conversation that you can have with the system. Recent other work exploring metatheory is found in Boyer and Moore[1979], LCF[1978], Bowen and Kowalski[1981], and Konolige[1981].

Another aspect of this example is that we have not strayed out of first order logic. Instead, we have used two different first order theories (one a metatheory of the other) connected by semantic attachments and reflection principles to form a much more flexible environment for expressing proofs than the simpler "one level" system. Equally important is the fact that from the perspective of the FOL system both theories are constructed using the same principles.

The most important results of the FOL effort is to provide a collection of epistemological ideas which show that certain complicated idea such as the workings of a reasoning system can be formalized in a clear way.

7. References

Aiello and Weyhrauch[1980]
> Aiello, L, and Weyhrauch, R., *Using meta- theoretic reasoning to do algebra*, Proceedings of the 5-th Workshop on Automated Deduction, Springer-Verlag, Lecture Notes in Computer Science, 1980.

Bowen and Kowalski[1981]
> Bowen, K.A., and Kowalski, R., *Amalgamating Language and Metalanguage in Logic Programming*, School of Computer and Information Science, Syracuse University, June 1981.

Boyer and Moore[1979]
> Boyer, R.S. and Moore, J.S., *Metafunctions: Proving Them Correct and Using Them Efficiently as New Proof Procedures*, SRI International Technical Report CSL-108, December 1979.

Church[1956]
> Church, A., *Introduction to mathematical logic, vol 1*, Princeton University Press, Princeton, 1956.

LCF[1978]
> Gordon, M.J., Milner, A.J., and Wadsworth, C.P., *Edinburgh LCF: A Mechanized Logic of Computation*, Springer-Verlag, Lecture Notes in Computer Science 78.

Kleene[1952]
> Kleene, S. C., *Introduction to metamathematics*, Van Nostrand, New York and Toronto, 1952.

Konolige[1981]
> Konolige, K., *A metalanguage representation of relational databases for deductive question-answering systems*, in Proceedings of the 7-th IJCAI, 1981.

McCarthy and Hayes[1969]
> McCarthy, J. and Hayes, P.J., *Some Philosophical Problems from the Viewpoint of Artificial Intelligence*, in (D.Michie,ed.) Machine Intelligence, 7, Edinburgh U.P., Edinburgh, 1969.

Prolegomena[1980]
> see Weyhrauch [1980].

Talcott and Weyhrauch[1982]
> Talcott, C, and Weyhrauch, R., *Reasoning about actions*, in preparation.

Weyhrauch[1977]
> Weyhrauch, R., *FOL: A Proof Checker for First-order Logic*, Stanford Artificial Intelligence Laboratory Memo AIM-235.1, Stanford University, Stanford, 1977.

Weyhrauch[1980]
> Weyhrauch, R., *Prolegomena to a theory of mechanized formal reasoning*, Artificial Intelligence, 13, 1980, pp.133-170.

Weyhrauch[1982]
> Weyhrauch, R., *A Mechanizable Formulation of Logic*, in preparation.

COMPARISON OF NATURAL DEDUCTION AND LOCKING RESOLUTION IMPLEMENTATIONS[1]

S. Greenbaum, A. Nagasaka, P. O'Rorke and D. Plaisted
Department of Computer Science
University of Illinois
Urbana, Illinois 61801

ABSTRACT

Two versions of the simplified problem reduction format are compared with locking resolution on some examples. The simplified problem reduction format is a complete strategy which has a goal-subgoal structure and combines features of resolution and natural deduction. Checking subgoals against previously generated subgoals to avoid solving the same subproblems repeatedly was found to be a significant benefit in the simplified problem reduction format. In general, the simplified problem reduction format with checking for repeated subgoals was found to be a better strategy than locking resolution in terms of total computing time used, although the search spaces for locking resolution were often smaller. The smaller search spaces and longer times for locking resolution persisted when the deletion method was replaced by a faster, less powerful deletion method, indicating that the difference with the problem reduction format is not due to the choice of a deletion method. The use of a faster, less powerful deletion method did lead to better performance in the locking resolution prove, however.

1. INTRODUCTION

The simplified problem reduction format is a natural deduction strategy that was introduced in [13]. This complete strategy accepts input in clause form but looks for solutions in the style of a natural deduction prover, using a goal-subgoal search structure. Also, this strategy permits a natural use of nontrivial semantic information to delete unachievable subgoals, although that feature was not used much in these experiments. One feature of this strategy is that it checks new subgoals against existing subgoals and avoids solving the same subproblem more than once. The rule that permits this is called the "domination rule". We have run some experiments to determine whether the domination rule is useful. Our conclusions are that the domination rule occasionally slows down the search because of the necessity to check new subgoals against existing subgoals, but almost always reduces search space size and search time, sometimes dramatically. Therefore the domination rule appears to be an essential feature of this strategy, and is probably an essential feature for any natural deduction theorem prover. The simplified problem reduction

---

[1]This work was partially supported by the National Science Foundation under Grant-MCS 79-04897 and MCS-81-05896.

format was also compared to two versions of a locking resolution prover. For a
discussion of locking resolution see [3]. One version of the locking prover used a
slow, more exhaustive instance test to delete superfluous clauses; the other version
used a faster, less powerful deletion method. The latter version is more comparable
to the problem reduction format implementation, which also used a simple,
non-exhaustive deletion method. Although the locking prover with fast deletion was
almost always faster in finding proofs than the locking prover with slow deletion,
it could not find any more proofs within the time limit, and neither version of the
locking resolution prover could find as many proofs as the simplified problem
reduction format program. In several cases, the problem reduction format was able
to obtain proofs that caused the locking prover to exceed the time limit. Neither
locking prover was able to obtain any proofs that the simplified problem reduction
format was not able to obtain. One advantage of the simplified problem reduction
format is that it accepts input in clause form, which makes comparisons possible
with conventional resolution theorem proving programs. Clause form sometimes has
disadvantages [1], but we feel this can usually be overcome by nonstandard
translations into clause form.

2. NON-STANDARD CLAUSE FORM TRANSLATION

In the context of a comparison of natural deduction and resolution theorem
provers, it may be appropriate to mention a nonstandard clause form translation
which gives resolution theorem provers some of the advantages of natural deduction
provers. Instead of giving a complete description, we illustrate with an example.
To show a formula W satisfiable, we introduce new predicate symbols P_1, P_2, ..., P_n
and new formulae W_1, W_2, ..., W_k which use these predicate symbols, such that W is
satisfiable iff $W_1 \wedge W_2 \wedge ... \wedge W_k$ is. Then each W_i is translated into clause form in
the usual way. The P_i represent sub-formulae of W. This translation avoids the
exponential increase in size that can result from the standard translation, although
on small examples the standard translation is usually better.

As an example, suppose W is

$$\forall x [[\forall y A(x, y) \vee B(x, y)] \supset [\exists z C(x, z) \wedge \neg B(x, z)]]$$

where A, B, and C are arbitrary formulae. We introduce predicates P_1, P_2, and P_3
and the following formulae W_1, W_2, W_3, and W_4:

W1: $\forall x\ P_1(x) \supset P_2(x)$
W2: $\forall x\ [\forall y\ A(x, y) \vee P_3(x, y)] \equiv P_1(x)$
W3: $\forall x\ [\exists z\ C(x, z) \wedge \neg P_3(x, z)] \equiv P_2(x)$
W4: $\forall x \forall y\ B(x, y) \equiv P_3(x, y)$

The top-level equivalences in the formulae W_i can often be replaced by implications,
depending on which logical connectives surround the appropriate sub-formulae of
W. A similar translation can be applied to A, B, and C if desired. This
translation should probably not be applied to small sub-formulae of W. It appears

that some of the advantages of non-clausal theorem proving [11] can be obtained for resolution theorem provers using this approach.

3. THE STRATEGIES USED

The natural deduction strategy used is presented in detail in [13]. Here we give a forward chaining version of the method for ground clauses, for the sake of clarity. The method of lifting to non-ground clauses and converting to backward chaining is fairly straightforward. In the following description, L and M, possibly with subscripts, refer to positive literals. Also, a clause is represented as an implication $L_1 \wedge \ldots \wedge L_k \supset M_1 \vee \ldots \vee M_n$ in which the order of the literals is significant. The symbol H represents an unordered set of positive literals, regarded as a conjunction. GOAL is a predicate symbol not occurring in the input clauses. In this system we show that a set S of ground clauses is inconsistent by deriving an assertion of the form S |- GOAL. There are four rules, G1, G2, S, and A.

Rule G1 (Subgoal)

$$\frac{\overline{L_1} \vee \ldots \vee \overline{L_k} \in S, \; S \vdash H \supset L_1, \ldots, S \vdash H \supset L_k}{S \vdash H \supset \text{GOAL}}$$

Rule G2 (Subgoal)

$$\frac{L_1 \wedge \ldots \wedge L_k \supset L \in S, \; S \vdash H \supset L_1, \ldots, S \vdash H \supset L_k}{S \vdash H \supset L}$$

Rule S (Split)

$$\frac{L_1 \wedge \ldots \wedge L_k \supset M_1 \vee \ldots \vee M_n \in S, n > 1, S \vdash H \supset L_1, \ldots, S \vdash H \supset L_k,\; S \vdash H \wedge M_1 \supset L, \ldots, S \vdash H \wedge M_n \supset L}{S \vdash H \supset L}$$

Rule A (Assumption)

$$\frac{L \in H}{S \vdash H \supset L}$$

One can show that S |- GOAL is derivable iff S is inconsistent, that is, S |- H ⊃ GOAL is derivable for empty H. The split control parameter limits the size of H. For Horn sets S, H will always be empty. For non-Horn sets, each application of the split rule may increase the number of literals in H by one (when using backward chaining). The above rules are given in a forward chaining format.

The locking resolution prover uses locking resolution with indices chosen so that negative literals resolve away first. The prover uses straightforward breadth-first search and has no parameters for deleting clauses with too many literals or too deep a nesting of function symbols. Two deletion methods were tried. In the first method, clauses are deleted which are instances of other clauses, ignoring the order of the literals. This is an expensive test because many permutations of the literals may have to be examined. In the second method, clauses are deleted that are equal to existing clauses, taking into account the order of the literals. This second deletion method can be performed using a simple equality test. The locking prover computes lock factors of a clause each time the clause is resolved. Therefore the locking prover is slowed down by unnecessary repetitions of work, although this may partially be offset by the fact that the factors are not added to the search space, making the deletion test faster.

Since the problem reduction format uses backward chaining and the locking prover as tested simulates forward chaining, this comparison may not seem valid. However, the simplified problem reduction format performs backward chaining in such a way that the size of the search space is proportional to the number of distinct subgoals generated. This is not true of all backward chaining systems, because the same subgoal can occur repeatedly and thereby contribute more than one element to the search space. For example, Prolog [12] solves a subgoal again each time it is generated. In forward chaining for Horn sets, the number of positive unit clauses generated is the number of solutions to subgoals found, in a sense. Also, due to theoretical considerations mentioned in [13], it appeared that forward chaining would be a better choice for locking resolution and would therefore provide a better comparison with the natural deduction prover. Most of the sets of input clauses were Horn sets or nearly Horn sets. Using backward chaining, the sizes of the clauses produced grows quickly, but not when forward chaining is used. We hope to experiment soon with a backward chaining approach to locking resolution on these examples.

4. COMPARISON RUNS

All strategies were run compiled in LISP on the CYBER 175. The simplified problem reduction format was run with straightforward breadth-first search and no heuristics or semantics; adding these might significantly increase its usefulness. The problem reduction format was run both with and without the domination rule. This rule requires subproblems to be checked to see if they are instances of any other subproblems. A subproblem has the structure of a Horn clause. Instead of implementing a complete instance test for Horn clauses, we simply check if one clause is an instance of another if the order of the literals is fixed; that is, no permutations of the literals are examined. However, in the first version of the locking prover, an exhaustive instance checker is used to see if a clause can be deleted; this essentially requires all permutations of the clause to be examined. The second version of the locking prover simply tests if two clauses are identical. For details of the simplified problem reduction format, see [13]; for a similar strategy see [9]. The problems run were the nine examples from Appendix A.4 of

Chang and Lee [5], plus three examples from [13], and two additional examples. The last five examples are given in the Appendix. Although Chang and Lee's prover was able to find proofs of all nine of their examples, their prover was provided with guidance from the user concerning sets of support for the non-unit clauses, how many times a non-unit clause may be selected before a function-depth test is used, and a maximum function depth that a clause may have. In addition, the strategy they used is unit resolution with a specialized control structure, and unit resolution by itself is not complete. Our provers received no user guidance except a "split control" parameter which is a non-negative integer telling how many negative literals a subgoal may contain; this parameter had a value of 0,1, or 2 in all cases but one; in that case (PSET), the value was 3, and no proof was found.

The nine examples from Chang and Lee are all group theory or number theory problems in which equality is treated relationally: that is, $x*y = z$ is written $P(x, y, z)$ and so on. We do not feel that this is the best way to treat equality; strategies based on term-rewriting systems [7, 8] or paramodulation [14] seem preferable. The modification method [4] is another possibility. However, these problems do provide a readily accessible basis for comparison. These nine examples are referred to by CLEX1, CLEX2, ..., CLEX9 in the following diagrams. Also, EX5 is a verification condition from Hoare's FIND program, taken from [2], EX6 is a database query, adapted from [16], and EX7 is a planning problem from [13] using the situation calculus formalism from artificial intelligence research. The PSET example is from [2]; the theorem is that the powerset of $A \cap B$ is equal to the intersection of the powersets of A and B. Finally, EX0 is from Manna [10]. The theorem is to show that a symmetric transitive relation R is reflexive if $\forall x \exists y R(x,y)$. The following table gives running times in seconds for SPRF (simplified problem reduction format) with DTEST = T (domination test used) and DTEST = NIL (no domination test performed), and for locking resolution with slow and fast instance tests. For all provers, signs of literals were chosen to make the clauses as nearly as possible Horn sets [6]. In CLEX8 and CLEX9, the sign of the P predicate was changed for this reason. For the locking resolution provers, indices were chosen so that negative literals resolve away first, thereby simulating P_1 deduction[15] to some extent. "No proof" means that the theorem was not proved in 50 seconds of CPU time, or that the prover ran out of space.

Note that SPRF obtains proofs for at least 5 of the examples that locking resolution could not prove in the time limit. Furthermore, the domination test led to a significant improvement in CLEX3 and EX7, in one case making the difference between finding a proof and not finding a proof. On two examples, CLEX9 and EX6, the proof was found faster without the domination test. On nine examples, the proof was found faster with the domination test. The locking provers were much slower on examples CLEX5 and EX6; they were much faster on examples CLEX7 and CLEX9. Also,

Figure 4-1: CPU TIME USED

Example	SPRF, DTEST = T	SPRF, DTEST = NIL	Locking slow deletion	Locking, fast deletion
CLEX1	2.75 sec.	3.60 sec.	3.52 sec.	1.16 sec.
CLEX2	no proof	no proof	no proof	no proof
CLEX3	5.25	25.11	no proof	no proof
CLEX4	2.76	4.11	no proof	no proof
CLEX5	0.21	0.26	17.84	6.34
CLEX6	1.72	3.22	no proof	no proof
CLEX7	2.60	3.89	1.15	1.19
CLEX8	1.55	3.33	3.69	3.15
CLEX9	5.48	2.75	1.18	1.06
EX5	18.80	28.93	no proof	no proof
EX6	1.53	1.39	6.47	5.2
EX7	5.12	no proof	no proof	no proof
PSET	no proof	no proof	no proof	no proof
EX0	0.49	0.48	0.57	0.31

locking with fast deletion was almost always faster in finding proofs than locking with slow deletion. Comparing search space sizes between locking resolution and the simplified problem reduction format, the search space sizes are actually much closer than these figures indicate. The following table shows the search size for these strategies. For SPRF, the number of chains generated is given; for locking resolution, the number of clauses generated is given. If no proof was found, the number of clauses generated before running out of time or space is given, followed by an asterisk.

In cases in which both SPRF and the locking provers found proofs, the search space was much smaller for locking in CLEX1, CLEX7 and CLEX9. The search space was much smaller for SPRF in CLEX5. Comparing SPRF with and without the domination test, the domination test almost always reduced the search space, and in CLEX3 and EX7 it did so dramatically. The domination test sometimes increases the search space size; this seems to be because the order of the breadth-first search is slightly changed. Comparing the two locking provers, the fast deletion method led to slightly larger search spaces in almost all cases in which proofs were found, but this was more than offset by the decrease in computing time. Possibly a fast, simple deletion test is an advantage in general for resolution theorem provers. It is interesting that when locking resolution finds a proof, the search spaces seem to

Figure 4-2: SEARCH SPACE SIZES

Example	SPRF, DTEST = T	SPRF, DTEST = NIL	Locking, slow detection	Locking, fast detection
CLEX1	67	97	19	19
CLEX2	426*	499*	66*	183*
CLEX3	100	306	78*	222*
CLEX4	61	98	78*	222*
CLEX5	15	19	60	76
CLEX6	52	84	76*	340*
CLEX7	79	94	15	18
CLEX8	60	114	37	39
CLEX9	97	95	19	19
EX5	223	416	137*	249*
EX6	50	50	30	30
EX7	84	373*	93*	377*
PSET	314*	367*	72*	346*
EX0	24	28	11	12

be smaller for locking resolution than for the simplified problem reduction format. Search space size may not be a proper method for comparison, since a different quantity is being computed (clauses versus chains). Also, it could be that locking resolution is a much more complicated operation than the operations of the simplified problem reduction format. Perhaps the problem reduction format was programmed better.

5. CONCLUSIONS

These experiments tend to indicate that the simplified problem reduction format has some advantages over locking resolution. Also significant is the fact that the domination rule led to a dramatic reduction in search space sizes and search times on two example, and almost always reduced the search time and search space size. This has implications for other natural deduction provers as well. In addition, a fast, simple deletion method led to shorter search times for the locking resolution prover than a slower, more exhaustive deletion method. Finally, the ability of the simplified problem reduction format to use heuristics and semantic information needs to be explored. Actually, we did test EX5 with some simple semantics on an interpretive version of the problem reduction format prover, and the semantics made the difference between finding a proof and not finding a proof. However, much more needs to be done.

6. EXAMPLES

EX5

C1 $x<y \lor y<x$

C2 $j<i$

C3 $m<p$

C4 $p<q$

C5 $q<n$

C6 $m<x \land x<i \land j<y \land y<n \supset A[x]<A[y]$

C7 $m<x \land x<y \land y<j \supset A[x]<A[y]$

C8 $i<x \land x<y \land y<n \supset A[x]<A[y]$

C9 $\neg(A[p] < A[q])$

Here x and y are variables and i, j, m, n, p, and q are constants.

EX6

C1 $\neg$answer(x)

C2   ocean(x) $\land$ borders(x, y) $\land$ african(y) $\land$ borders(x, z) $\land$ asian(z) $\supset$ answer(x)

(This clause formalizes the query "Which oceans border African and Asian countries?")

C3	ocean(Atlantic)	C22	african(Angola)
C4	ocean(Indian)	C23	african(Somalia)
C5	borders(Atlantic, Brazil)	C24	african(Kenya)
C6	borders(Atlantic, Uruguay)	C25	african(Tanzania)
C7	borders(Atlantic, Venezuela)	C26	asian(India)
C8	borders(Atlantic, Zaire)	C27	asian(Pakistan)
C9	borders(Atlantic, Nigeria)	C28	asian(Iran)
C10	borders(Atlantic, Angola)		
C11	borders(Indian, India)		
C12	borders(Indian, Pakistan)		
C13	borders(Indian, Iran)		
C14	borders(Indian, Somalia)		
C15	borders(Indian, Kenya)		
C16	borders(Indian, Tanzania)		
C17	south-american(Brazil)		
C18	south-american(Uruguay)		
C19	south-american(Venezuela)		
C20	african(Zaire)		
C21	african(Nigeria)		

EX7

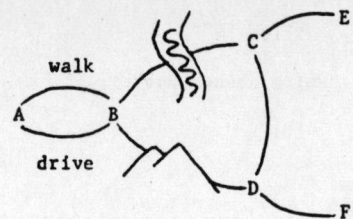

C1 ¬AT(A, s)

C2 AT(F, s0)

C3 COLD(s) ∨ WARM(t)

C4 AT(A, s) ⊃ AT(B, WALK(B, s))

C5 AT(A, s) ⊃ AT(B, DRIVE(B, s))

C6 AT(B, s) ⊃ AT(A, WALK(A, s))

C7 AT(B, s) ⊃ AT(A, DRIVE(A, s))

C8 COLD(s) ∧ AT(B, s) ⊃ AT(C, SKATE(C, s))

C9 COLD(s) ∧ AT(C, s) ⊃ AT(B, SKATE(B, s))

C10 WARM(s) ∧ AT(B, s) ⊃ AT(D, CLIMB(D, s))

C11 WARM(s) ∧ AT(D, s) ⊃ AT(B, CLIMB(B, s))

C12 AT(C, s) ⊃ AT(D, GO(D, s))

C13 AT(D, s) ⊃ AT(C, GO(C, s))

C14 AT(C, s) ⊃ AT(E, GO(E, s))

C15 AT(E, s) ⊃ AT(C, GO(C, s))

C16 AT(D, s) ⊃ AT(F, GO(F, s))

C17 AT(F, s) ⊃ AT(D, GO(D, s))

PSET

A1 $x = y \supset (x \subset y)$
A2 $x = y \supset (y \subset x)$
A3 $(x \subset y) \wedge (y \subset x) \supset x = y$
A4 $(x \subset y) \wedge (z \in x) \supset z \in y$
A5 $g(x,y) \in x \vee (x \subset y)$
A6 $g(x,y) \notin y \vee (x \subset y)$
A7 $x \in P(y) \supset (x \subset y)$
A8 $(x \subset y) \supset x \in P(y)$
A9 $x \in y \cap z \supset x \in y$
A10 $x \in y \cap z \supset x \in z$
A11 $x \in y \wedge x \in z \supset x \in y \cap z$
A12 $P(A \cap B) \neq P(A) \cap P(B)$

EXO

C1 $R(x,y) \supset R(y,x)$
C2 $R(x,y) \wedge R(y,z) \supset R(x,z)$
C3 $R(x, f(x))$
C4 $\neg R(c, c)$

REFERENCES

[1] Andrews, P.B.
Theorem proving via general matings.
J. ACM 28:193-214, 1981.

[2] Bledsoe, W.W.
Non-resolution theorem proving.
Artificial Intelligence 9:1-35, 1977.

[3] Boyer, R.S.
Locking, a restriction of resolution.
PhD thesis, University of Texas at Austin, TX, 1971.

[4] Brand, D.
Proving theorems with the modification method.
SIAM Journal of Computing 4:412-430, 1975.

[5] Chang, C.L. and Lee, R.C.
Symbolic Logic and Mechanical Theorem Proving.
Academic Press, 1973.

[6] Henschen, L. and Wos, L.
Unit refutations and Horn sets.
J. ACM 21:590-605, 1974.

[7] Knuth, D.E. and Bendix, P.B.
Simple word problems in universal algebras.
Pergamon Press, Oxford, 1970, .

[8] Lankford, D.S.
Canonical algebraic simplification in computational logic.
Technical Report ATP-25, University of Texas at Austin, Texas, 1975.
Automatic Theorem Proving Project.

[9] Loveland, D.W. and Reddy, C.R.
Deleting repeated goals in the problem reduction format.
J. ACM 28:646-661, 1981.

[10] Manna, Z.
Mathematical Theory of Computation.
McGraw-Hill, New York, 1974.

[11] Murray, N.V.
Completely non-clausal theorem proving.
Artificial Intelligence 18:67-86, 1982.

[12] Pereira, L. Pereira, F. and Warren, D.
User's guide to DEC System-10 PROLOG.
Technical Report 03/13/5570, Labortorio Nacional De Engenharia Civil, Lisbon, 1978.

[13] Plaisted, D.
A simplified problem reduction format.
Artificial Intelligence, to appear.

[14]
  Robinson, G.A. and Wos, L.
  Paramodulation and theorem proving in first-order theories with equality.
  American Elsevier, New York, 1969, .

[15]
  Robinson, J.A.
  Automatic deduction with hyper-resolution.
  Int. J. Comput. Math. 1:227-234, 1965.

[16]
  Warren, D.H.D.
  Efficient processing of interactive relational database queries expressed in logic.
  Technical Report, Dept. of Artificial Intelligence, University of Edinburgh, 1981.

Derived Preconditions
and Their Use in Program Synthesis[*]

Douglas R. Smith
Department of Computer Science
Naval Postgraduate School
Monterey, California 93940, USA

Abstract

In this paper we pose and begin to explore a deductive problem more general than that of finding a proof that a given goal formula logically follows from a given set of hypotheses. The problem is most simply stated in the propositional calculus: given a goal A and hypothesis H we wish to find a formula P, called a precondition, such that A logically follows from $P \wedge H$. A precondition provides any additional conditions under which A can be shown to follow from H. A slightly more complex definition of preconditions in a first-order theory is given and used throughout the paper. A formal system based on natural deduction is presented in which preconditions can be derived. A number of examples are then given which show how derived preconditions are used in a program synthesis method we are developing. These uses include theorem proving, formula simplification, simple code generation, the completion of partial specifications for a subalgorithm, and other tasks of a deductive nature.

0. Introduction

Traditionally, the subject of automatic theorem proving has dealt with the problem of finding a proof that a given goal formula A logically follows from a given hypothesis H. In this paper we pose a more general deductive problem and suggest that systems for solving this more general problem can extend the utility of deductive mechanisms, and provide a framework for overcoming some problematic features of current theorem provers. The problem is most simply stated in the propositional calculus: given a goal A and hypothesis H we wish to find a formula P, called a precondition, such that A logically follows from $P \wedge H$. In other words a precondition provides any additional conditions under which A can be shown to follow from H.

---
[*] The work reported herein was supported by the Foundation Research Program of the Naval Postgraduate School with funds provided by the Chief of Naval Research.

A formal system in which preconditions can be derived is described in section 2. Each rule in this natural deduction-like system has a reduction component which reduces a goal A_0 to subgoals $A_1, A_2, \ldots, A_k$ and a composition component which composes preconditions of subgoals $A_1, A_2, \ldots, A_k$ to form a precondition of A_0.

After presenting basic terminology in section 1 a formal system for deriving preconditions is given in section 2. A number of examples are presented in section 3 which show how derived preconditions are used in a program synthesis method we are developing [9,10]. These uses include theorem proving, formula simplification, simple code generation, the completion of partial specifications for a subalgorithm, and other tasks of a deductive nature.

1. Terminology

The examples given below are drawn from a program synthesis system which works within a many-sorted first-order theory TT. The theory includes data types such as $\mathbb{N}$ (natural numbers), LIST($\mathbb{N}$) (linear lists of natural numbers), and BAGS($\mathbb{N}$) (multisets of natural numbers). We will use the (possibly subscripted) symbols i,j,k for variables ranging over $\mathbb{N}$, x,y,z for variables over LIST($\mathbb{N}$), and B as a variable over BAGS($\mathbb{N}$). The theory also includes a number of functions and predicates defined on these types and axiomatic specifications of their interactions. The notions of term, atomic formula, literal, and (well-formed) formula have their usual definitions [5]. Let T and F be propositional constants which have the values true and false respectively in all models of TT. We make use of a distinguished subset of the theorems of TT called <u>known theorems</u> which are assumed to be immediately available to the deductive system. The set of known theorems may change over time but initially includes all axioms of TT. All of the known theorems required by the examples are listed in the appendix.

Let $Q_1 x_1 Q_2 x_2 \ldots Q_n x_n$ G be a closed formula not necessarily in prenex form where Q_i is either $\exists$ or $\forall$ for $i=1,2,\ldots,n$. A $x_1 x_2 \ldots x_n$-<u>precondition</u> of $Q_1 x_1 Q_2 x_2 \ldots Q_n x_n$ G is a quantifier-free formula P dependent only on variables $x_1, x_2, \ldots, x_n$ such that

$$Q_1 x_1 Q_2 x_2 \ldots Q_n x_n [\ P \Rightarrow G\ ]$$

is valid in TT. P is also a <u>weakest</u> $x_1 x_2 \ldots x_n$-<u>precondition</u> if

$$Q_1 x_1 Q_2 x_2 \ldots Q_n x_n [\ P \equiv G\ ]$$

is valid in TT.

Two well-known special cases of these concepts can be given. First, if T can be derived as a $x_1 x_2 \ldots x_n$-precondition of a goal $Q_1 x_1 Q_2 x_2 \ldots Q_n x_n$ G then the derivation is in fact a proof of the validity of $Q_1 x_1 Q_2 x_2 \ldots Q_n x_n$ G since

$$Q_1x_1Q_2x_2\ldots Q_nx_n\ [\ T \Rightarrow G\ ] \equiv Q_1x_1Q_2x_2\ldots Q_nx_n\ G$$

Therefore any system for deriving preconditions can also be used for theorem proving. Second, Dijkstra's concept [3] of a "weakest pre-condition" WP(S,R) of a program S with respect to post-condition R may be defined as a weakest q-precondition of

$$\forall q \exists k \exists p[\ \text{TERMINATE}(S,q,k,p)\ \wedge\ R(p))\ ]$$

where TERMINATE(S,q,k,p) holds iff program S activated in initial state q terminates within k steps (assuming a suitable definition of a program step) in a final state p. I.e.,

$$\forall q[\ \text{WP}(S,R)[q]\ \equiv\ \exists k \exists p\ \text{TERMINATE}(S,q,k,p)\ \wedge\ R(p)\ ]$$

Our program synthesis method is not directly related to Dijkstra's approach to algorithm design [3].

In general a given goal may have many preconditions. Characteristics of a useful precondition seem to depend on the application domain. In program synthesis we generally want preconditions which are a) easily computable, b) in as simple a form as possible, and c) as weak as possible. (Criterion (c) prevents the boolean constant F from being an acceptable precondition for all goals.) Clearly there is a tradeoff between these criteria. We are currently investigating the possibility of measuring each criterion by a separate heuristic function, then combining the results to form a net complexity measure on preconditions. For reasons to be discussed later we assume that such a complexity measure ranges over a well-founded set (such as IN under the usual < relation) and that we seek to minimize complexity over all preconditions. In this paper however we are mostly concerned with setting up a formal system within which preconditions can be derived, and showing how to solve some program synthesis problems using it.

2. A Formal System for Deriving Preconditions

2.1 Goal Preparation

In presenting a set of rules which allow us to derive preconditions we use the notation $\frac{A}{H}$ to denote the statement that well-formed formula A logically follows from the set of hypotheses H in TT, i.e., $h_1 \wedge h_2 \wedge \ldots \wedge h_k \Rightarrow A$ is valid in TT where $H = \{h_1, h_2, \ldots h_k\}$.

A goal statement $\frac{A}{H}$ and the known theorems of TT are prepared as follows. First, all occurences of equivalence ($\equiv$) and implication ($\Rightarrow$) signs are eliminated and negation signs are moved in as far as possible. H and the known theorems of TT are then skolemized in the usual way [5], i.e., existentially quantified variables are replaced by skolem functions of the universally quantified variables on which they depend. Quantifiers are then dropped with the understanding that all remaining variables are universally quantified. The goal A is skolemized in a dual manner with universally quantified variables replaced by skolem functions of the existential variables on which they depend. All quantifiers are then dropped with the understanding that all variables in A which remain are existentially quantified. The preparation of A is equivalent (via duality of goals and assertions) to preparing ~A as an hypothesis then taking the negation of the result as our prepared goal.

2.2 Reduction/Composition Rules

Rules which reduce a goal statement to two subgoal statements are expressed in the following form:

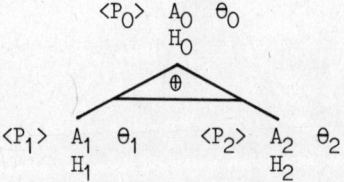

where A_0, A_1, and A_2 are goal formulas, H_0, H_1, and H_2 are sets of hypotheses, θ_0, θ_1, and θ_2 are substitutions, P_0, P_1, and P_2 are formulas (the derived preconditions), and $\oplus$ is either $\vee$ or $\wedge$. A rule of this form asserts that if P_i is a (weakest) precondition of $H_i\theta_i \Rightarrow A_i\theta_i$ where i=1,2 then P_0 is a (weakest) precondition of $H_0\theta_0 \Rightarrow A_0\theta_0$. P_0 generally is $P_1 \oplus P_2$. Substitution θ_0 is formed from substitutions θ_1 and θ_2 in ways that depend on $\oplus$.

If $\oplus$ is $\wedge$ then θ_0 is the unifying composition of θ_1 and θ_2, denoted $uc(\theta_1, \theta_2)$ [7]. If $\theta_0 = uc(\theta_1, \theta_2)$ then θ_0 is a most general substitution such that for any literal L

$$(L\theta_1)\theta_0 = (L\theta_0)\theta_1 = L\theta_0 = (L\theta_2)\theta_0 = (L\theta_0)\theta_2.$$

$uc(\theta_1, \theta_2)$ may be computed by finding the most general unifier of

$$(t_1, \ldots, t_n, t_{n+1}, \ldots, t_{n+m})$$
$$(v_1, \ldots, v_n, v_{n+1}, \ldots, v_{n+m})$$

where

$$\theta_1 = \{t_1/v_1, \ldots, t_n/v_n\}$$
$$\theta_2 = \{t_{n+1}/v_{n+1}, \ldots, t_{n+m}/v_{n+m}\}.$$

If these expressions cannot be unified then the result is a special atom NIL. For example,

$$uc(\{a/z\},\{b/z\}) = \text{NIL}$$

$$uc(\{\},\{a/z\}) = \{a/z\}$$

$$uc(\{f(x)/z\},\{f(a)/z\}) = \{f(a)/z, a/x\}$$

If θ is $\vee$ then θ_0 is formed by the <u>disjunctive composition</u> of P_1, θ_1, P_2 and θ_2, which is denoted $dc(P_1,\theta_1,P_2,\theta_2)$. The disjunctive composition may be computed as follows assuming that the derived preconditions P_1 and P_2 contain no variables. Let $\{S_1,S_2,\ldots,S_m\}$ be the set of skolem function names in P_1 which come from the top level goal in the current deduction. For example if the top level goal is $Q(u,f_1(u)) \Rightarrow R(x,f_2(x),f_3)$ and P_1 is $W(f_1(f_3),g_2(f_3))$ then $\{f_1,f_3\}$ is the set of skolem function names in P_1 which comes from the top level goal. Let $P_1(y_1,\ldots,y_k)$ be the formula resulting from the replacement of each occurence of skolem function S_j by variable y_j in P_1. In the above example $P_1(y_1,y_2)$ denotes $W(y_1,g_2(y_2))$. Function dc is defined as follows.

$dc(P_1,\theta_1,P_2,\theta_2) =$ if θ_1=NIL and θ_2=NIL then NIL
                    else if P_1=T or θ_2=NIL then θ_1
                    else if P_2=T or θ_1=NIL then θ_2
                    else if $\theta_1=\{\}$ then θ_2
                    else $\{h_x(S_1,S_2,\ldots,S_m)/x \mid t/x \in \theta_1$ or $t/x \in \theta_2\}$

where

$$h_x(y_1,\ldots,y_m) = \text{if } P_1(y_1,\ldots,y_m) \text{ then } x\theta_1 \text{ else } x\theta_2.$$

Loosely speaking, the disjunctive composition of P_1,θ_1,P_2, and θ_2 behaves like θ_1 when P_1 holds and behaves like θ_2 otherwise. Some examples:

$$dc(a_0>3, \{f_1(a_0)/x\}, T, \{a_0/x\}) = \{a_0/x\}$$

$$dc(f_1>f_2(f_1), \{f_1/z, f_2(f_3)/x\}, f_1<f_2(f_3), \{f_2(f_1)/z, f_3/x\})$$
$$= \{h_z(f_1,f_2,f_3)/z, h_x(f_1,f_2,f_3)/x\}$$

where

$$h_z(y_1,y_2,y_3) = \text{if } y_1>y_2 \text{ then } y_1 \text{ else } y_2$$

$$h_x(y_1,y_2,y_3) = \text{if } y_1>y_2 \text{ then } y_2 \text{ else } y_3$$

A complete deduction involving a disjunctive composition is given in section 2.5.

Rules which reduce a goal statement to one subgoal are notated

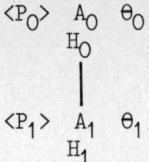

Occasionally, as in the application of known theorems which are implications, the relation between goal and subgoals is not one of equivalence but implication. Rules of this kind are notated

$$
\begin{array}{l}
\langle P_0 \rangle \; A_0 \; \theta_0 \\
\quad\;\; H_0 \\
\quad\;\; \uparrow \\
\langle P_1 \rangle \; A_1 \; \theta_1 \\
\quad\;\; H_1
\end{array}
$$

which asserts that if P_1 is a precondition of $H_1\theta_1 \Rightarrow A_1\theta_1$ then P_0 is a precondition of $H_0\theta_0 \Rightarrow A_0\theta$. For rules of this kind we cannot assert that P_0 is a weakest precondition of $H_0\theta_0 \Rightarrow A_0\theta_0$ even if P_1 is known to be a weakest precondition of $H_1\theta_1 \Rightarrow A_1\theta_1$.

The following rules are for the most part extensions of typical goal reduction rules [2,5,8].

R1. Reduction of Conjunctive Goals

R2. Reduction of Disjunctive Goals

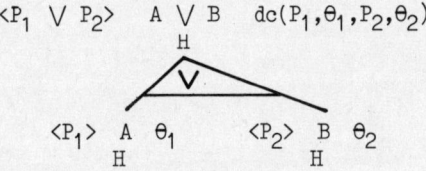

R3. Reduction of Conjunctive Hypotheses

$$\frac{\langle P \rangle \quad A \quad \theta}{\{B \wedge C\} \cup H}$$
$$\Big|$$
$$\frac{\langle P \rangle \quad A \quad \theta}{\{B,C\} \cup H}$$

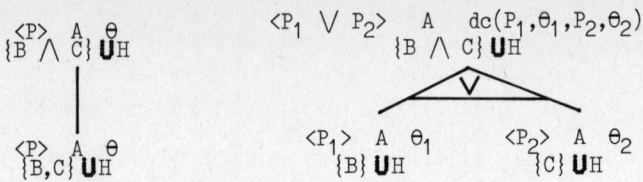

R4. Reduction of Disjunctive Hypotheses

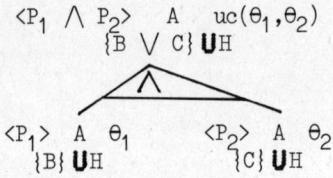

R5. Application of an Equivalence Formula

$$\begin{array}{c} \langle P \rangle \quad A \quad \theta\theta_1 \\ H \\ \Big| \\ \langle P \rangle \quad B\theta \quad \theta_1 \\ H \end{array}$$

if $C \equiv B$ is a known theorem of TT
or an hypothesis in H and θ unifies $\{A,B\}$

R6. Application of an Implicational Formula

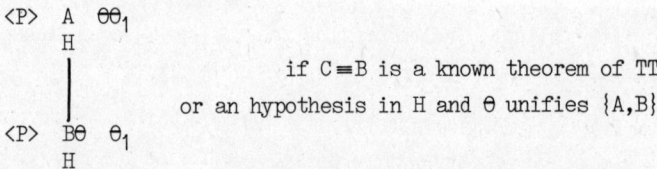

if $C \Rightarrow B$ is a known theorem of TT or hypothesis in H,
and D is $C\theta$ where θ unifies $\{A,B\}$
or D is $\sim B\theta$ where θ unifies $\{A,\sim C\}$ or $\{\sim A,C\}$

R7. Forward Inference from an Hypothesis

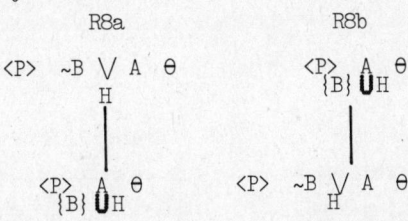

$$\frac{\langle P \rangle \quad A \quad \theta}{\{B\} \cup H}$$

if $D \Rightarrow E$ or $D \equiv E$ is a known theorem of TT
or hypothesis in H and θ_1 unifies $\{B,D\}$

$$\frac{\langle P \rangle \quad A \quad \theta}{\{B, E\theta_1\} \cup H}$$

R8. Goal/Hypothesis Duality rules

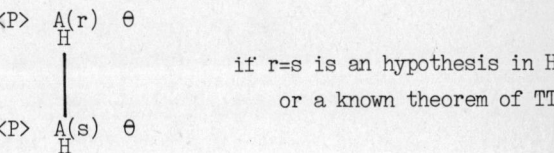

R8a

$$\langle P \rangle \quad \underset{H}{\sim B \vee A} \quad \theta$$

$$\langle P \rangle \quad \underset{\{B\} \cup H}{A} \quad \theta$$

R8b

$$\langle P \rangle \quad \underset{\{B\} \cup H}{A} \quad \theta$$

$$\langle P \rangle \quad \underset{H}{\sim B \vee A} \quad \theta$$

R9. Substitution of Equal Terms

$$\langle P \rangle \quad \underset{H}{A(r)} \quad \theta$$

$$\langle P \rangle \quad \underset{H}{A(s)} \quad \theta$$

if $r=s$ is an hypothesis in H
or a known theorem of TT

R10. Conditional Equality Substitution

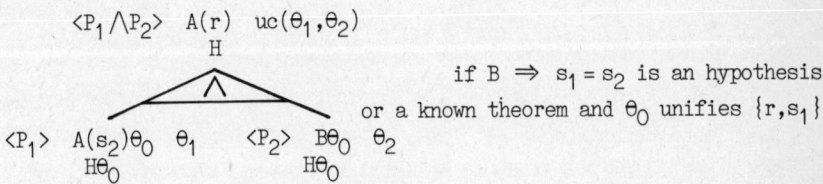

$$\langle P_1 \wedge P_2 \rangle \quad \underset{H}{A(r)} \quad uc(\theta_1, \theta_2)$$

$$\langle P_1 \rangle \quad \underset{H\theta_0}{A(s_2)\theta_0} \quad \theta_1 \qquad \langle P_2 \rangle \quad \underset{H\theta_0}{B\theta_0} \quad \theta_2$$

if $B \Rightarrow s_1 = s_2$ is an hypothesis
or a known theorem and θ_0 unifies $\{r, s_1\}$

2.3 Primitive Goals

There are several types of primitive goal statements in our system. Each are described by notations of the form $\langle P \rangle \underset{H}{A} \theta$ which assert that P is a precondition of $H\theta \Rightarrow A\theta$ if the associated condition holds.

P1. $\langle T \rangle \underset{H}{A} \theta$ if θ unifies $\{A,B\}$ where B is a known theorem of TT or $B \in H$

P2. $\langle F \rangle \underset{H}{A}$ NIL if θ unifies $\{A, \sim B\}$ or $\{\sim A, B\}$, where B is a known theorem of TT

In addition to P1 and P2 any goal with a null hypothesis may be taken as primitive:

P3. $\langle A' \rangle \, {}^{A}_{\{\}} \, \{\}$ if A has the form $\bigvee_{i=1}^{k} A_i$ and A' has the form $\bigvee_{j=1}^{m} A_{i_j}$ where $\{A_{i_j}\}_{j=1,m} \subseteq \{A_i\}_{i=1,k}$ and for each j, $1 \leq j \leq m$, A_{i_j} depends on the variables $x_1, x_2, \ldots, x_n$ only when we seek a $x_1, x_2, \ldots, x_n$-precondition.

Primitive goals of type P1 and P2 yield weakest preconditions but in general primitive goals of type P3 do not. Note that any goal statement can be converted to an equivalent goal with a null hypothesis by repeated applications of rule R8b.

2.4 The Deduction Process

The derivation of a precondition of goal statement $\frac{A}{H}$ can be described by a two stage process. In the first phase rules are repeatedly applied to goals reducing them to subgoals and generating a goal tree. Rules are not applied to a goal satisfying the primitive goal tests P1 and P2 or if the goal has been specially converted to satisfy P3. If for some reason, such as limits on computational resource, it is desired to terminate the reduction process before all subgoals have been reduced to primitive goals of type P1 or P2, then any subgoals waiting for rule application can be converted to a primitive goal of type P3. The result of this reduction process is a goal tree with primitive goals as leaf nodes.

The second phase involves the bottom-up composition of preconditions and substitutions. Initially each primitive goal yields a precondition and a substitution. Subsequently whenever a precondition or substitution has been found for each subgoal of a goal $\frac{A}{H}$ then a precondition and substitution is composed for $\frac{A}{H}$ according to the reduction/composition rule employed. Each newly composed precondition is then run through a simplification process to be described later.

Usually several rules can be applied to a given goal and each rule will generate a precondition. In an computer implementation of this system we would make use of a complexity measuring function and select that precondition of least complexity among the alternatives.

2.5 An Example

As an example of the use of this system suppose that we wish to show that

$$\forall i_0 \forall i_1 \exists i_2 [(i_0 < i_1 \land i_2 = 0) \lor (i_0 \geq i_1 \land i_2 = 1)] \tag{1}$$

is valid in TT where i_0, i_1, i_2 are variables over $\mathbb{N}$ (natural numbers). We do so by trying to derive T as a $i_0 i_1 i_2$-precondition of (1). The goal after preparation is:

$$(r_0 < r_1 \land i_2 = 0) \lor (r_0 \geq r_1 \land i_2 = 1)$$

where r_0 and r_1 are skolem constants of type $\mathbb{N}$. The derivation is depicted below in figure 1. Initially (1) is reduced via rule R2 to two subgoals then each of these subgoals are reduced via rule R1 to two other subgoals. Subgoals $i_2 = 0$ and $i_2 = 1$ match axiom $i = i$ (theorem n0 in the Appendix) with substitution $\{0/i_2\}$ and $\{1/i_2\}$ respectively and thus are primitive goals of type P1. Suppose that goals $r_0 < r_1$ and $r_0 \geq r_1$ are taken as primitive goals of type P3. The composition phase now begins. Subgoals $r_0 < r_1 \wedge i_2=0$ and $r_0 \geq r_1 \wedge i_2=1$ yield preconditions ($T \wedge r_0 < r_1$) and ($T \wedge r_0 \geq r_1$) respectively. A simplification process reduces these preconditions to $r_0 < r_1$ and $r_0 \geq r_1$ respectively. The composed substitutions for the immediate subgoals of (1) are just the unifying compositions $uc(\{0/i_2\},\{\}) = \{0/i_2\}$, and $uc(\{1/i_2\},\{\}) = \{1/i_2\}$ respectively. The derived precondition of goal (1) is ($r_0 < r_1 \vee r_0 \geq r_1$) which simplifies (via theorem n4) to T. The composed substitution is the disjunctive composition $\{f_{i_2}(r_0,r_1)/i_2\}$ where

$$f_{i_2}(j_1,j_2) = \text{if } j_1 < j_2 \text{ then } 0 \text{ else } 1.$$

The derivation shows that T is a precondition of

$$(r_0 < r_1 \wedge f_{i_2}(r_0,r_1)=0) \vee (r_0 \geq r_1 \wedge f_{i_2}(r_0,r_1)=1)$$

i.e., that our original goal is valid. Furthermore we have obtained a substitution term for the one existentially quantified variable in (1). After requantifying we obtain the valid formula:

$$\forall i_0 \forall i_1 [(i_0 < i_1 \wedge f_{i_2}(i_0,i_1)=0) \vee (i_0 \geq i_1 \wedge f_{i_2}(i_0,i_1)=1)].$$

In this example and all that follow we annotate the arcs with the name of the rule and theorem used and note the primitive goal type of each leaf node. Also in

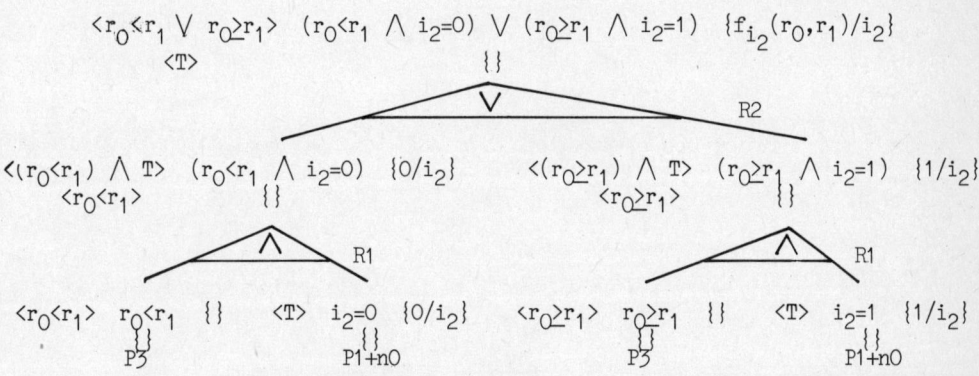

Figure 1.

this example we write the simplified form of the composed precondition P immediately under P. Hereafter in examples we will simply omit the composed precondition in favor of its simplified form. Also we omit substitutions when they are inessential to an understanding of a derivation.

2.6 Formula Simplification

Any deductive mechanism needs a means to simplify formulas which are generated during the deductive process. Simplification can be usefully viewed as the task of finding a weakest precondition (in all variables) of formula A. The search for a simple weakest precondition is kept short by using only a few of the known theorems of TT. The strategy followed in the examples is to repeat the following sequence of rule applications until the goal has been reduced to literals:
a) simplify the goal as much as possible using known equivalence theorems of TT,
b) multiply subexpressions out using p9 and p10 (DeMorgan's Laws),
c) break the result of (b) down to subexpressions using R1 or R2.
The multiplication step allows us to mix preconditions which were returned from different branches of the goal tree.

A precondition generating mechanism used for simplification purposes must be carefully controlled in order to avoid infinite regress. One way around this problem is to prohibit simplification of preconditions generated during the simplification process. Instead we check whether the final derived precondition P is simpler than the initial goal formula A. If not then A is returned otherwise we attempt to simplify P. If our complexity measuring function ranges over a well-founded set then this simplification process will terminate.

Suppose that we need to simplify the expression

$$(i>j \lor i=0) \land (i<j \lor j=0) \qquad (2)$$

where i and j vary over $\mathbb{N}$. The derivation in figure 2a yields

$$(i>0 \land j=0) \lor i=0$$

as a weakest precondition (i.e. equivalent form) of (2). The derivation in figure 2b yields

$$(i=0 \lor j=0) \qquad (3)$$

as a weakest precondition.   The result is that (2) has been simplified to (3).

3. The Use of Derived Preconditions in Program Synthesis

In this section we show how derived preconditions can play a central role in the design of algorithms [9,10]. Many of the key steps in the design process involve

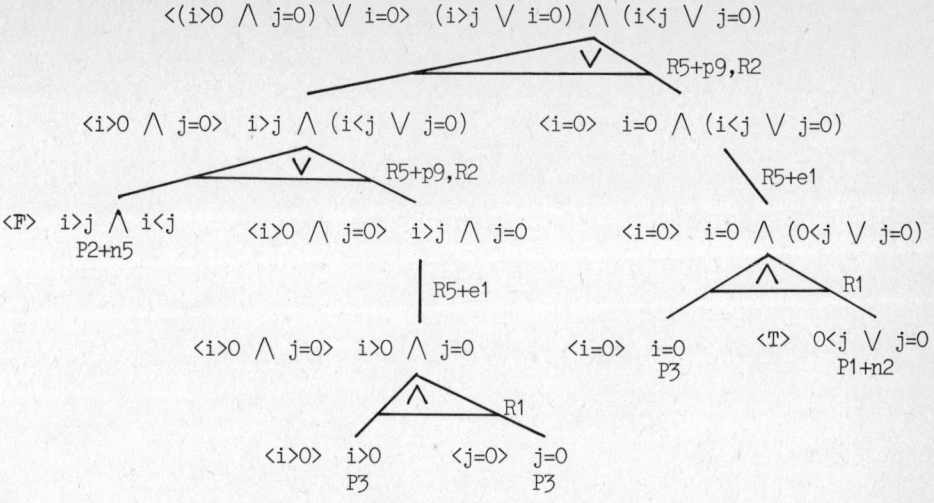

Figure 2a. First pass at simplifying goal formula (2).

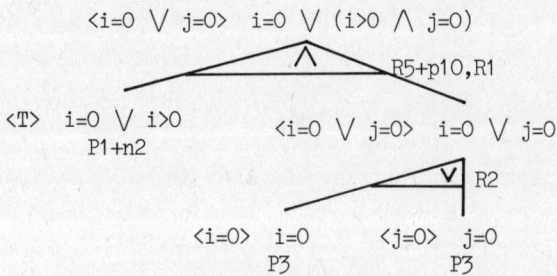

Figure 2b. Second pass: simplifying the result of figure 2a.

finding a precondition of a formula constructed by instantiation of a formula schema with functions, predicates and types from the specification and the partially designed algorithm. The resulting derived precondition is used to either strengthen or complete some aspect of the target algorithm.

Initially a user supplies a complete formal specification of a problem which he desires to solve. The specification consists of a naming of the input and output data types, and two formulas called the input and output conditions. The types, functions and predicates involved in the specification must be part of the language of TT. For example, the problem of sorting a list of natural numbers may be specified as follows:

$$QSORT(x) = z \text{ such that } ORD(z) \wedge BAG(x) = BAG(z)$$
$$\text{where } QSORT: LIST(\mathbb{N}) \rightarrow LIST(\mathbb{N}).$$

Here the input and output types are LIST($\mathbb{N}$) (lists of natural numbers). There is no input condition (except the implicit condition of the input type) and the output condition is ORD(z) $\wedge$ BAG(x)=BAG(z) where ORD(z) holds iff the list z is in nondecreasing order, and BAG(x)=BAG(z) holds iff the multiset (bag) of elements in x and z is the same.

We will construct a divide and conquer algorithm (quicksort) of the form:

QSORT(x) = if
$\quad\quad\quad$ PRIM(x) $\rightarrow$ QSORT := $f(x)$ ▯
$\quad\quad\quad$ ~PRIM(x) $\rightarrow$ (x_1,x_2) := DECOMPOSE(x);
$\quad\quad\quad\quad\quad\quad\quad\quad\quad$ (z_1,z_2) := (QSORT(x_1),QSORT(x_2));
$\quad\quad\quad\quad\quad\quad\quad\quad\quad$ QSORT := COMPOSE(z_1,z_2)
$\quad$ fi

where PRIM is a predicate which determines when to terminate recursion, f is a function which provides a solution for primitive inputs, DECOMPOSE and COMPOSE are decomposition and composition functions respectively. In this program schema PRIM, f, DECOMPOSE, and COMPOSE are uninterpreted functions whose value we have to determine. The if-fi construct is Dijkstra's nondeterministic conditional statement [3]. Associated with the algorithm schema is a correctness schema which will be introduced later.

The first step in the synthesis process involves the representation of the users problem by a problem reduction model [10]. This format extends the specification of a problem and restricts the type of algorithms which can be used to solve the problem to one of a small number of algorithms which work by problem reduction. For present purposes the relevant parts of the representation for QSORT are:

a) a relation IDR, called the input decomposition relation, which constrains the way in which input x_0 can be decomposed into objects x_1 and x_2 and serves as a partial output condition on subalgorithm DECOMPOSE in the divide and conquer schema:

$$\text{IDR}(x_0,x_1,x_2) \equiv \text{BAG}(x_0)=\text{BAG}(x_1) \cup \text{BAG}(x_2)$$

where $B_1 \cup B_2$ denotes the bag-union of bags B_1 and B_2.

b) a relation OCR, called the output composition relation, which constrains the way in which output object z_0 can be formed from objects z_1 and z_2 and serves as a partial output condition on the subalgorithm COMPOSE:

$$\text{OCR}(z_0,z_1,z_2) \equiv \text{BAG}(z_0)=\text{BAG}(z_1) \cup \text{BAG}(z_2)$$

c) a well-founded ordering relation $\succ$ on LIST($\mathbb{N}$) is used to ensure that the target program terminates on all inputs:

$$x_0 \succ x_1 \equiv LG(x_0) > LG(x_1)$$

where the function $LG(x)$ returns the length of the list x.

3.1 Checking and Enforcing Compatibility in the Representation

The representation of the user's problem by a problem reduction model is constructed by heuristic means. A formula expressing the mutual compatibility of various parts of the model is constructed and an attempt is made to verify it. If the derived precondition P is T then the parts are compatible otherwise we use P to modify the model to ensure compatibility. For example we want the input decomposition relation IDR to be compatible with the well-founded ordering $\succ$, in the sense that

$$\forall x_0 \forall x_1 \forall x_2 \, [IDR(x_0,x_1,x_2) \Rightarrow x_0 \succ x_1 \wedge x_0 \succ x_2]$$

i.e., if x_0 can decompose into lists x_1 and x_2 then x_1 and x_2 must both be smaller than x_0 under the $\succ$ relation. After substituting in the form of IDR and the well-founded ordering for the QSORT example, and preparing the formula we obtain the goal:

$$BAG(a_0)=BAG(a_1) \cup BAG(a_2) \Rightarrow LG(a_0)>LG(a_1) \wedge LG(a_0)>LG(a_2) \qquad (4)$$

where a_0, a_1, and a_2 are skolem constants for the (universally quantified) variables x_0, x_1, x_2. The derivation of a $x_0 x_1 x_2$-precondition of (4) is given in figure 3. The resulting precondition is

$$BAG(x_0)=BAG(x_1) \cup BAG(x_2) \Rightarrow LG(x_1)>0 \wedge LG(x_2)>0$$

which means that IDR is not strong enough to imply the consequent of the original goal. From the definition of preconditions it follows that the conjunction of IDR and the derived precondition will in fact imply the consequent of (4). Thus we can form a new strengthened input decomposition relation IDR' where

$$IDR'(x_0,x_1,x_2) \equiv IDR(x_0,x_1,x_2) \wedge [BAG(x_0)=BAG(x_1) \cup BAG(x_2) \Rightarrow LG(x_1)>0 \wedge LG(x_2)>0]$$

The derivation in figure 3 guarantees that IDR' is compatible with the well-founded ordering. After simplifying IDR' we have

$$IDR'(x_0,x_1,x_2) \equiv BAG(x_0)=BAG(x_1) \cup BAG(x_2) \wedge LG(x_1)>0 \wedge LG(x_2)>0.$$

3.2 Reducing a Quantified Predicate to a Target Language Expression

The predicate $PRIM(x)$ in the divide and conquer schema is intended to distinguish nondecomposable from decomposable inputs. In the QSORT example it is sufficient for $\sim PRIM(x_0)$ to be a x_0-precondition of

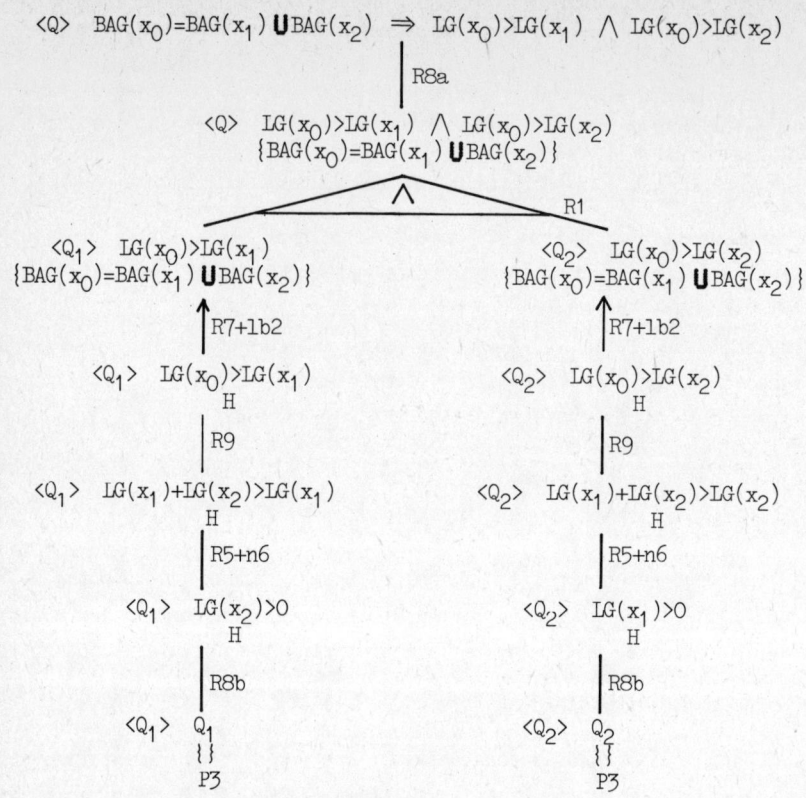

where
Q_1 is $BAG(x_0)=BAG(x_1) \cup BAG(x_2) \Rightarrow LG(x_2)>0$
Q_2 is $BAG(x_0)=BAG(x_1) \cup BAG(x_1) \Rightarrow LG(x_2)>0$
Q is $BAG(x_0)=BAG(x_1) \cup BAG(x_2) \Rightarrow (LG(x_2)>0 \wedge LG(x_1)>0)$
$H = \{BAG(x_0)=BAG(x_1) \cup BAG(x_2), LG(x_0)=LG(x_1)+LG(x_2)\}$

Figure 3. Checking Compatibility of IDR and ⊢

$$\forall x_0 \exists x_1 \exists x_2 \ IDR'(x_0,x_1,x_2)$$

i.e. a list is decomposable only if there are lists into which it can decompose. The deduction in figure 4 yields the precondition $LG(a_0)>1$ and after some simple manipulations $LG(x) \leq 1$ and $LG(x)>1$ can be substituted for $PRIM(x)$ and $\sim PRIM(x)$ respectively in QSORT. One additional mechanism is needed to correctly handle this example. The reduction/composition rule R1 treats each subgoal independently and combines the returned substitutions into their unifying composition. This treatment does not work well when the subgoals have common variables. Most theorem proving systems allow substitutions in one subgoal to be applied to the other (since different substitutions may be found independently for the same variable) and we follow this practice here.

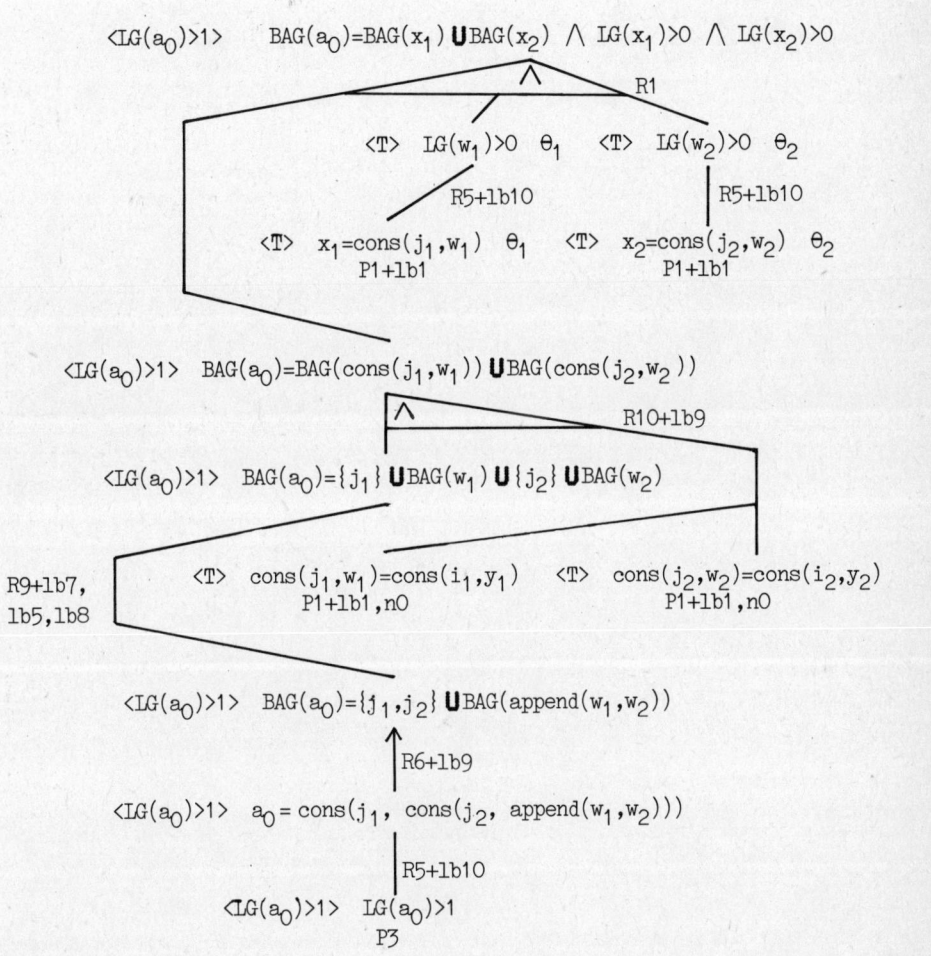

where $\theta_1 = \{cons(j_1,w_1)/x_1\}$
and
$\theta_2 = \{cons(j_2,w_2)/x_2\}$

Figure 4. Generating a target language expression for ~PRIM

3.3 Simple Code Generation through Substitution of a Term for an Output Variable.

With the PRIM predicate in hand the synthesis process can proceed to the task of finding a target language expression to handle primitive inputs in the quicksort algorithm. A correctness formula for the primitive branch of the quicksort algorithm is:

$$\forall x \, \exists z [LG(x) \leq 1 \Rightarrow ORD(z) \land PERM(x,z)].$$

The deduction in figure 5 shows that T is a xz-precondition of this formula thus proving its validity in TT. The substitution gives us a value for z for any x, namely x itself. Thus the primitive branch of our quicksort is completed since x is the desired output value. The target algorithm now has the form

$$QSORT(x) = \text{if}$$
$$\quad LG(x) \leq 1 \rightarrow QSORT := x \; []$$
$$\quad LG(x) > 1 \rightarrow \ldots$$
$$\text{fi}$$

3.4 Completion of the Partial Specification of a Subalgorithm

The next step in the synthesis provides our final example and completes the construction of the top level algorithm for QSORT. The nonprimitive branch of QSORT has two uninterpreted functions COMPOSE and DECOMPOSE which have partial specifications based on OCR and IDR respectively. We look for a known target language function

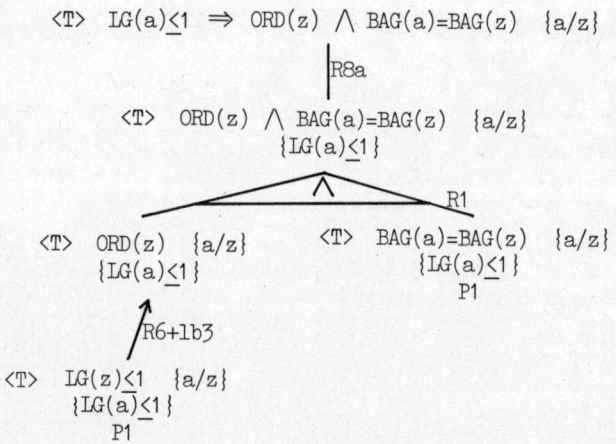

Figure 5. Finding a target language term

satisfying either partial specification and find that the function APPEND, which appends one list onto the end of another, satisfies the (partial) specification for COMPOSE. The algorithm schema then becomes:

$$\text{QSORT}(x) = \text{if}$$
$$\quad LG(x) \leq 1 \rightarrow \text{QSORT} := x \;[\,]$$
$$\quad LG(x) > 1 \rightarrow (x_1, x_2) := \text{DECOMPOSE}(x);$$
$$\quad\quad\quad\quad\quad\quad (z_1, z_2) := (\text{QSORT}(x_1), \text{QSORT}(x_2));$$
$$\quad\quad\quad\quad\quad\quad \text{QSORT} := \text{APPEND}(z_1, z_2)$$
$$\text{fi}$$

where subalgorithm DECOMPOSE remains to be synthesized and has partial specification

$$\text{DECOMPOSE}(x) = (x_1, x_2) \text{ such that } [LG(x) > 1 \Rightarrow (BAG(x) = BAG(x_1) \cup BAG(x_2) \land$$
$$LG(x_1) > 0 \land LG(x_2) > 0)]$$
where DECOMPOSE: $LIST(\mathbb{N}) \rightarrow LIST(\mathbb{N})^2$.

The concern now is to find any additional output conditions needed by DECOMPOSE in order to make QSORT satisfy its formal specifications. A sufficient condition for the total correctness of QSORT [10] is:

$$\forall x_0 \forall x_1 \forall x_2 \forall z_0 \forall z_1 \forall z_2 \; [[\; BAG(x_0) = BAG(x_1) \cup BAG(x_2) \land$$
$$LG(x_1) > 0 \land LG(x_2) > 0 \land$$
$$BAG(x_1) = BAG(z_1) \land ORD(z_1) \land$$
$$BAG(x_2) = BAG(z_2) \land ORD(z_2) \land$$
$$z_0 = APPEND(z_1, z_2)] \Rightarrow (BAG(x_0) = BAG(z_0) \land ORD(z_0))] \quad (6)$$

If (6) is not valid it is because the specification of DECOMPOSE is too weak. We seek therefore a $x_0 x_1 x_2$-precondition of (6) and add it to the output specification of DECOMPOSE. Preparing (6) results in the substitution of skolem constants $a_0, b_1, b_2, c_0, c_1, c_2$ for $x_0, x_1, x_2, z_0, z_1, z_2$ respectively. Let H denote the set of conjuncts in the antecedent of the prepared correctness formula and A the consequent. An expression of the form P(ALL(B)) will be used to abbreviate $\forall x \in B \; P(x)$ where B is a bag variable. The derivations given in figures 6a and 6b yield

$$ALL(BAG(x_1)) \leq ALL(BAG(x_2)).$$

Strengthening DECOMPOSE with this precondition we obtain the complete specification

$$\text{DECOMPOSE}(x) = (x_1, x_2) \text{ such that } [LG(x) > 1 \Rightarrow (BAG(x) = BAG(x_1) \cup BAG(x_2) \land$$
$$LG(x_1) > 0 \land LG(x_2) > 0 \land ALL(BAG(x_1)) \leq ALL(BAG(x_2))]$$
where DECOMPOSE: $LIST(\mathbb{N}) \rightarrow LIST(N)^2$.

The synthesis process is then recursively invoked to design an algorithm meeting these specifications.

<T> $BAG(a_0)=BAG(c_0)$
$\{c_0=APPEND(c_1,c_2)\}\,\cup\,H$

| R9

<T> $BAG(a_0)=BAG(APPEND(c_1,c_2))$
H

| R9+1b5

<T> $BAG(a_0)=BAG(c_1)\cup BAG(c_2)$
$\{BAG(b_1)=BAG(c_1),BAG(b_2)=BAG(c_2)\}\,\cup\,H$

| R9

<T> $BAG(a_0)=BAG(b_1)\cup BAG(b_2)$
$\{BAG(a_0)=BAG(b_1)\cup BAG(b_2)\}$
P1

Figure 6a. Nonprimitive branch of QSORT

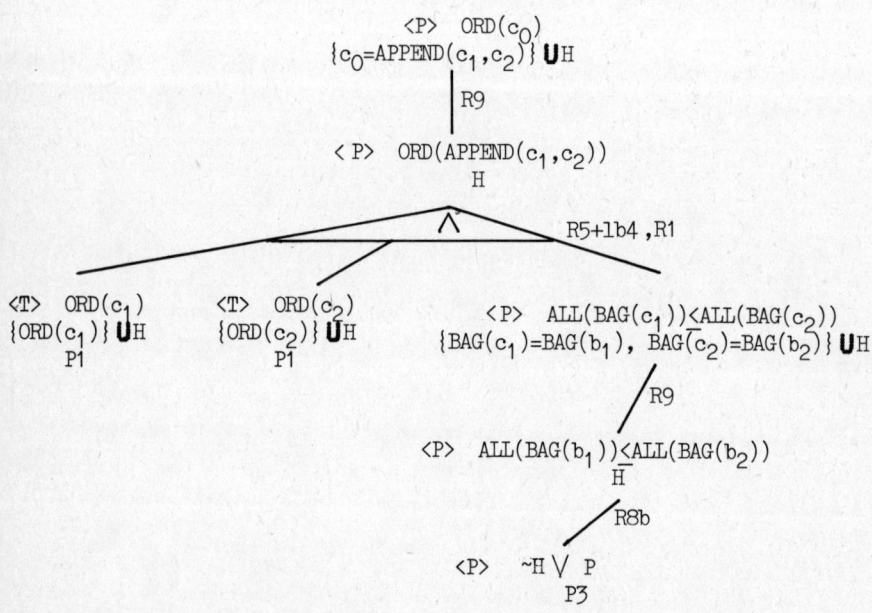

where P is $ALL(BAG(b_1))\leq ALL(BAG(b_2))$

Figure 6b. Completing the Specification of DECOMPOSE

The synthesis system from which we've drawn the examples is an attempt to obtain increased synthesis performance by 1) dividing the synthesis task into a number of relatively small deductive tasks, and 2) using large amounts of knowledge about programming. The system makes use of two types of programming knowledge: 1) control strategy knowledge encoded by program schemas (such as the schema for divide and conquer used above) and their associated correctness schemas, and 2) data structure knowledge represented in part by the known theorems of TT. Other recent deductive approaches to program synthesis [1,4,6] also make use of data structure knowledge, but have different approaches to representing control knowledge and tend to construct programs on the basis of a single large deductive task.

4. Conclusion

In this paper we have defined a new deductive problem, that of finding a precondition of a given formula, and presented a formal system within which preconditions can be derived. We have tried to convey a sense of the flexibility and usefulness of such a system through a number of examples drawn from the domain of program synthesis. We are currently implementing a system based on the one described here and hope to report on such issues as formula complexity measures and control, which we have largely ignored here, in a future paper.

APPENDIX

Listed below are the known theorems used in the examples of this paper. It is important that these assertions are expressed in their strongest form (i.e., as equivalences rather than implications) whenever possible, so that it can be determined whether a weakest precondition has been derived or not. Often a theorem is used in one direction only although it may be stated as an equivalence.

Propositional theorems
p1. $A \vee \sim A$
p2. $\sim(A \wedge \sim A)$
p3. $T \wedge A \equiv A$
p4. $T \vee A \equiv T$
p5. $F \wedge A \equiv F$
p6. $F \vee A \equiv A$
p7. $\sim(A \wedge B) \equiv \sim A \vee \sim B$
p8. $\sim(A \vee B) \equiv \sim A \wedge \sim B$
p9. $A \wedge (B \vee C) \equiv (A \wedge B) \vee (A \wedge C)$
p10. $A \vee (B \wedge C) \equiv (A \vee B) \wedge (A \vee C)$

p11. $(A \Rightarrow B) \equiv (\sim A \lor B)$
p12. $A \lor (A \land B) \equiv A$
p12. $A \land (A \lor B) \equiv A$

Equality theorems

e1. $P(x) \land x=y \equiv P(y) \land x=y$ where $P(x)$ is a formula depending on term x.

Natural number theorems
Let i,j,k denote variables of type $\mathbb{N}$.

n0. $i=i$
n1. $i \geq 0$
n2. $i=0 \lor i>0$
n3. $i \leq j \lor i \geq j$
n4. $i<j \lor i \geq j$
n5. $\sim(i<j \land i>j)$
n6. $i+j>i \equiv j>0$
n7. $\sim(i>k) \equiv i \leq k$
n8. $\sim(i<k) \equiv i \geq k$
n9. $i>k_1 \land j>k_2 \Rightarrow i+j>k_1+k_2+1$

List and Bag theorems

Let w_0, w_1, w_2 vary over LIST($\mathbb{N}$), and let B_1, B_2 vary over BAGS($\mathbb{N}$).

lb1. $w_0 = w_0$
lb2. $BAG(w_0) = BAG(w_1) \cup BAG(w_2) \Rightarrow LG(w_0) = LG(w_1) + LG(w_2)$
lb3. $LG(w_0) \leq 1 \Rightarrow ORD(w_0)$
lb4. $[ORD(w_1) \land ORD(w_2) \land ALL(BAG(w_1)) \leq ALL(BAG(w_2))] \equiv ORD(APPEND(w_1, w_2))$
lb5. $BAG(APPEND(w_0, w_1)) = BAG(w_0) \cup BAG(w_1)$
lb6. $B_1 = B_1$
lb7. $\{i_1\} \cup \{i_2\} = \{i_1, i_2\}$
lb8. $B_1 \cup B_2 = B_2 \cup B_1$
lb9. $w_1 = cons(i_0, cons(i_1, \ldots cons(i_n, w_2) \ldots)) \Rightarrow BAG(w_1) = \{i_0, i_1, \ldots, i_n\} \cup BAG(w_2)$
lb10. $w_0 = cons(i_0, cons(i_1, \ldots cons(i_n, w_1) \ldots)) \equiv LG(w_0) > n$

REFERENCES

1. Bibel, W. Syntax-directed, Semantics-Supported Program Synthesis, Art. Intell. 14(3), 1980, pp 243-262.

2. Bledsoe, W. Nonresolution theorem proving, Art. Intell. 9(1), 1977, pp 1-35.

3. Dijkstra, E.W. A Discipline of Programming, Prentice-Hall Inc., Englewood Cliffs, NJ, 1976.

4. Guiho, G., and Gresse, C., Program Synthesis from Incomplete Specifications, Proc. Fifth Conf. on Automated Deduction, Lecture Notes in Computer Science, Eds. W. Bibel and R. Kowalski, Springer-Verlag, Berlin, 1980.

5. Loveland, D.W. (1978), Automated Theorem Proving: A Logical Basis. North Holland Pub. Co., New York, 1978.

6. Manna, Z., and Waldinger, R. (1980), A Deductive Approach to Program Synthesis, ACM Trans. on Prog. Lang. 2(1), Jan. 1980, pp 90-121.

7. Nilsson, N. Principles of Artificial Intelligence, Tioga Pub. Co., Palo Alto, CA, 1980.

8. Reiter, R. (1976), A Semantically Guided Deductive System for Automatic Theorem Proving, IEEE Trans. on Computers C-25(4), 1976, pp 328-334.

9. Smith, D.R. A Design for an Automatic Programming System, Proc. of IJCAI-7, Vancouver, B.C, Canada, 1981, pp 1024-1027.

10. Smith, D.R. The Synthesis of Divide and Conquer Algorithms, Forthcoming technical report.

Automatic Construction of Special Purpose Programs

Chris Goad
Stanford University

1. Introduction

The automatic programming problem as traditionally formulated is that of passing automatically from specifications of programs (expressed in some particular formal language) to programs which satisfy those specifications. We will be concerned here with a different style of automatic programming. Rather than defining the class of programming problems to be dealt with by the language in which those problems are formulated, we will instead consider smaller and more manageable classes of programming problems; classes of problems defined in ordinary mathematical terms. Also, our aims are different from the traditional aims of automatic programming in that we are interested primarily in increasing the efficiency of computations, rather than in transfering the burden of programming from human to computer. (There is of course some overlap between these aims.)

We will introduce the kind of automatic programming which we have in mind by means of an example - an example which will be treated at length later in the paper. The example concerns an important computational problem from three dimensional graphics, namely, the hidden surface elimination problem. The hidden surface elimination problem is that of determining, given a formal description of one or more opaque objects in three dimensional space, which surfaces of those objects are visible from a given point of view, and which are hidden behind other objects in the scene. In many applications of three dimensional computer graphics, such as computer animation and flight simulation, the appearance of the same scene must be computed repeatedly for many different positions of the viewer. Several algorithms for hidden surface elimination have been devised which exploit this property of an application (which is conventionally referred to as "object coherence"). In each case, the algorithms proceed by first constructing a suitable data structure describing the scene; this is done "off-line". Then the data structure is used at runtime for performing the hidden surface computation for each position of the viewer.

Another way of exploiting this kind of coherence is to automatically construct a different special purpose program for each scene treated. This scheme involves devising an automatic *synthesis method* M : M is a program which takes a scene s as input, and yields a special purpose program Q_s as output. Q_s in turn takes the position of the viewer as input and produces a suitable representation of the scene with hidden surfaces removed as output. Since Q_s has a very limited

This research was supported in part by the National Science Foundation under Grant MCS81-04873 and in part by the Advanced Research Projects Agency of the Department of Defense under Contract MDA903-80-C-0102 task to perform, it is potentially much faster than any general purpose algorithm for hidden surface elimination would be for the same scene. We have applied the automatic programming scheme to hidden surface elimination, with good results. The special purpose programs which we have produced are indeed much faster than any general purpose algorithm to be found in the computer graphics literature.

The hidden surface example illustrates a very general method for improving the efficiency of computation by use of automatic programming. The essential property of the example from the standpoint of automatic programmming is this: a computation taking two inputs (here, the scene s and the viewer position p) is performed repeatedly in a context where the first input changes only occasionally whereas the second input changes rapidly. Any situation which exhibits this property is a candidate for special purpose automatic programming; we may attempt to develop a synthesis method which takes the slowly changing input and generates a fast special purpose program; the special purpose program in turn takes the second input and completes the computation. An overall gain in efficiency is realized if the time saved by the use of the special purpose programs on the rapidly changing input exceeds the time lost in the occasional generation of a new special purpose program upon presentation of a new value of the slowly changing input.

Let us restate the points of the last paragraph in more formal terms. Let $\alpha(p, x, y)$ be a ternary predicate. Suppose that in the course of some large computation we are obliged to repeatedly compute values of y with $\alpha(p, x, y)$ from given values of p and x, and that in the sequence of p's and x's to be treated, p changes slowly, and x rapidly. The automatic programming problem for α is just that of automatically passing from any value for p to a program for $\lambda x\, y.\alpha(p, x, y)$. That is, we want a program S_α such that for each p, and each x, $\alpha(p, x, S_\alpha(p)(x))$ holds. The utility of a solution to the automatic programming problem defined by α depends on the factors mentioned in the last paragraph.

In the situation described above, where y with $\alpha(p, x, y)$ is to be computed from slowly changing p and rapidly changing x, we will say the sequence of inputs exhibits *sweep coherence*. (The picture behind this term is that of a ray performing a circular sweep about a point, with x "attached" farther out than p, so that the position of x changes more rapidly than that of p. Evidently, object coherence in three dimensional graphics is an instance of sweep coherence.) There is a conventional scheme for exploiting sweep coherence which was alluded to earlier in connection with hidden surface elimination. Namely, one devises a data structure which encodes the relevant information about p in such a way that the computation of y can be completed rapidly once a value of x is given. To take the archetypical computer science example, if one needs to repeatedly determine whether a number x belongs to a list p of numbers, for many values of x but only a few of p, one proceeds by sorting each new p, and then using binary search to determine whether each successive x belongs to the current p. Note that there is no convincing way to *formally* distinguish between the "data structure scheme" and the "automatic programming scheme". The reason for this is that automatically constructed special purpose programs may be regarded as data structures which, together with their arguments, are passed to the interpreter for the language in which the programs are expressed. However, the fact that the distinction between a program and a data structure which is not a program cannot be convincely formalized does not make it a useless distinction. For the purposes of this paper, the word "program" is to be read in its ordinary informal sense: a program is a description of a method of computation which is expressible in a programming language of the usual kind.

The principal purpose of this paper is to present evidence that special purpose automatic programming is a useful enterprise - or, more precisely, to present new evidence, since some such evidence exists already in the computer science literature (see section 6). Of necessity, individual concrete examples are the only kind of evidence which can be expected, since at the outset we are supposing that for each programming task (given by a ternary predicate α) a different synthesis program S_α will be required. As already indicated, we will consider one example in detail, namely, the hidden surface elimination problem.

Another purpose of the paper is to indicate that although different synthesis programs are needed for different automatic programming problems, there are nonetheless general - and, in fact, quite primitive - techniques which can be used for a variety of problems. These generally useful techniques are for the most part already in use in automatic programming and automatic program transformation of the traditional kind. Our example shows that these general techniques, although not always very effective alone, can produce good results when combined with considerations which are particular to the (narrow) class of programming problems treated in an instance of special purpose automatic programming - that is, particular to the mathematical specification which we have designated α.

One generally useful technique for special purpose automatic programming is the application of automatic deduction to the simplification of decision trees. Thus, among other things, this paper describes a promising class of practical applications for automatic deduction. It so happens that the requirements for automatic deduction which arise from the particular application which we discuss in detail are satisfied by an old and well known decision procedure, namely, the simplex algorithm. However, there are closely related applications which present more interesting decison problems.

Before closing the introduction, we wish to emphasize the following point. The character of the problems which are treated in special purpose automatic programming is very different from the character of those dealt with in automatic programming of the traditional kind. The traditional style of automatic programming work deals with universal or metamathematically defined classes of problems in that the collection of problems to be solved by a real or ideal automatic programming system is defined by giving the language in which the problems are to be formulated, rather than by reference to particular objects and operations to be carried out on those objects. In computing practice as a whole, this metamathematical or universal style of defining the class of problems to be solved by a particular program is very unusual. The reason for this is simply that such methods as have been found for attacking universal classes of problems are rarely efficient enough to compete with special purpose methods designed for the particular situations which come up in practice. One prefers to solve broad classes of problems at one blow if possible, but such possibilities do not often arise.

In special purpose automatic programming, on the other hand, the class of problems to be solved by any particular synthesis program is defined by an ordinary mathematical predicate α which *will* in general make reference to specific objects and operations. In this sense, the style of automatic programming with which we are concerned is much closer to ordinary computing practice - and much farther away from the universal aims popular in artifical intelligence - than the traditional style. All of this shows, in any case, that automatic programming - in the sense of writing programs which write programs - is not intrinsically bound to universal aims.

For a general discussion concerning choices of problem classes in computing and mathematics, see Kreisel[1981].

2. A simple example of special purpose automatic programming: list membership

The method which we use for synthesis of special purpose hidden surface elimination programs was developed and will be presented in several stages. We will start by describing a simple scheme for automatically generating special purpose programs, namely the scheme of specializing general purpose programs. If this scheme is applied in a direct way to the hidden surface problem, the results are unsatisfactory in that the special purpose programs P_s which are generated for any but very small scenes s are intractably large. However, having formed a picture of what the programs P_s are like, we are able analyse their defects. The result of this analysis is a series of modifications to the P_s which finally yield usable programs Q_s. More precisely, the modifications apply not to the synthesized programs P_s, but to the method by which they are synthesized, so that we end up with a direct method for synthesis of usable special purpose programs. We have chosen to present the synthesis method in a step-by-step manner in order to increase clarity, but also for a more important reason. Namely, the method by which the P_s are produced is directly applicable to a wide variety of problems, whereas the subsequent modifications are particular to the hidden surface problem. Thus the multi-stage presentation has the virtue of exhibiting the method which we use for the hidden surface problem as an instance of a general scheme for exploiting sweep coherence; the scheme is: "take a general purpose algorithm and specialize it, analyse the results, and introduce appropriate modifications."

The general side of our presentation, that is, the side concerned with the specialization of programs, will begin with an example which lacks the technical complexities of the hidden surface problem, but which illustrates several techniques which are used later. The example is essentially a reformulation of the construction of a binary search tree by random insertion [Robson, 1979]. Let u be a list of elements from a linearly ordered set, and let n be an individual member of that set. The computational problem which we wish to solve is that of determining whether n is a member of u. We may set this up as a problem of the kind described in the introduction by defining $M(u, n, r)$ as "$r =$ TRUE if $n \in u$; $r =$ FALSE otherwise". We assume that the sweep coherence criterion for the utility of performing a computation in two stages is met; that is, we assume that membership of n in u must be determined for many different values of n, but few values of u. What is wanted is a synthesis program for M - a program S_M which will take a list u as input and produce as output a fast program $S_M(u)$ for determining membership in that list.

The appropriate "data stucture scheme" for exploiting sweep coherence in the current example involves sorting and binary search. Since this method is provably optimal, no improvements by the use of automatic programming are possible in this case.

With this warning in mind, let us proceed. The synthesis method for M consists of taking a slightly peculiar algorithm for computing list membership and specializing it to the list at hand. The algorithm to be specialized is given by the recursive definition:

$f(u, n) =$ if $empty(u)$ then FALSE else
    if $n < head(u)$ then $f(tail(u), n)$ else
    if $n > head(u)$ then $f(tail(u), n)$ else
    TRUE

The specialization to a concrete list L is carried out by a conventional scheme: first symbolically execute f as applied to L, then simplify the resulting decision tree by removing redundant decision nodes. (Other terms for symbolic execution include partial evaluation [Beckeman et al, 1976], and repeated unfolding [Burstall and Darlington, 1977]). A redundant decision node is one whose outcome is predetermined by the outcomes at nodes along the path leading from the root of the decision tree to the node in question. More precisely, if $P_1 \cdots P_k$ are the decision predicates appearing on the path leading to node N, and Q is the predicate at node N, then N is redundant if $A_1 \wedge A_2 \cdots A_k \supset Q$ is valid, or if $A_1 \wedge A_2 \cdots A_k \supset \neg Q$ is valid, where $A_i = P_i$ if the true branch is taken out of P_i's node in the path leading to N, and $A_i = \neg P_i$ if the false branch is taken. The removal of a redundant decision node is carried out by replacing the subtree of which it is a root by the subtree rooted at its left or its right son, depending on whether the predicate at the node is predetermined to be true or false. Note that determining whether a node in a decision tree is redundant involves deciding whether certain formulas are valid. In the current instance, all such formulas are implications between conjunctions of inequalities containing n as the only variable, so that the decision problem has an easy automatic solution.

Here is an example of the behavior of S_M, the synthesis method for M just described, in a concrete case. Suppose that the underlying ordered set for the membership computation is the integers, and that we wish to compute $S_M(\langle 2, 5\rangle)$. We proceed by symbolically executing $f(\langle 2, 5\rangle, n)$, arriving at the decision tree:

if $n < 2$ then
  (if $n < 5$ then FALSE else
  if $n > 5$ then FALSE else
  TRUE) else
if $n > 2$ then
  (if $n < 5$ then FALSE else
  if $n > 5$ then FALSE else
  TRUE) else
TRUE

The decision node represented by the second line of the above program is redundant, since $n < 2 \supset n < 5$. Its removal yields:

if $n < 2$ then FALSE else
if $n > 2$ then
  (if $n < 5$ then FALSE else
  if $n > 5$ then FALSE else
  TRUE) else
TRUE

No redundant nodes remain. Hence the above program represents the final result of computing $S_M(\langle 2, 5\rangle)$. By the analysis given in [Robson 1979], the behavior of S_M exhibits the following general properties. First, the expected maximum depth of the decision tree generated by S_M when applied to a list L of length n (assuming that L is given in random order with equal probability assigned to each ordering) is $k \, log_2(n)$ for a constant k between 7.26 and 8.62 (the exact value of k is not known). Second, the number of decision nodes in $S_M(L)$ is exactly $2n$, where again n is the length of L. (If we adopt a formulation in which the decision nodes have three branches [for $<$, $=$, and $>$], then the number of nodes and depth are halved.) Thus the worst case running time of the programs generated by S_M, measured by the maximum number of decision nodes to be traversed on the path to the result, may be expected to exceed the optimal worst case running time obtained by sorting and binary search by only a small constant factor. Also the size of the programs generated is linear in the length of L, so that in this respect as well S_M is only a small constant factor worse than optimal. Finally, note that the expected worst case running time of $S_M(L)$ is immensely better than that of the one stage algorithm given by f; the former is logarithmic in the length of L, while the latter is linear.

This example illustrates the following points.

(1) Exploiting sweep coherence is by no means a marginal matter in computer science. The sorted list is just one of many data structures devised for this purpose, but even taken alone sorting has been the target of a large amount of effort.

(2) In the case of list membership, the automatic programming approach failed to produce results which were as good as those produced by the right data structure. But on the other hand, our approach did almost as well, and in a sense required less in the way of intellectual resources than the conventional scheme. For, in constructing special purpose programs for individual lists L, all that was required was the unwinding and optimization of a slight variant of the ordinary one stage program for computing membership. On the other hand, the construction of the sort and binary search method requires additional ideas - ideas which are not already implicit in the one stage program. There are many computational problems exhibiting sweep coherence where a one stage program exists, but where no fully satisfactory data structure for exploiting the coherence is available - the required additional ideas have not been found. Such problems are promising candidates for the automatic programming approach. One such problem, which as we will see has fulfilled its promise, is that of hidden surface elimination.

3. The synthesis method for hidden surface elimination

Now we turn to the our principal subject, the synthesis of special purpose programs for hidden surface elimination.

Recall that the hidden surface elimination problem is that of taking a scene s and a viewer position p, and computing the appearance of s from the viewpoint p with hidden surfaces removed. Many approaches to the hidden surface elimination problem have been developed [Sutherland et al,1974]. The general approach which will concern us here has the following attributes:

The scene to be displayed is represented by a set F of faces, where a face is a convex polygon with some particular position and orientation in three dimensional space. A face is oriented also in the sense that it has a front and a back side. (A face in turn may be given by the ordered list of its vertices, and the vertices by their coordinates in three space.) The faces in F fit together to form (approximations to) the surfaces of the objects in the scene. This representation scheme is currently the one most commonly used in three dimensional computer graphics.

In order to display the scene from a particular point of view, the faces are first sorted into what is called "priority order". A list L of faces is in priority order with respect to a given viewpoint p if whenever face i occludes face j from p, i appears after j in L. (A face i is said to occlude a face j from p if *some* part of j is hidden behind i as viewed from p.) Once a priority ordered list has been computed, the generation of a picture of the scene with hidden surfaces removed can be carried out by a "painting" process, in which the faces are written onto the picture (eg by writing onto the memory of a bit-mapped CRT) in the order in which they are given in L. Then, if one face partially hides another, the hiding face will be written (or painted) after the hidden face; thus elimination of hidden surfaces occurs by overwriting. The priority list approach appears to be the best available for real time applications such as flight simulation; in such applications the painting process is carried out largely by special purpose hardware. (Notes: (1) Priority sorting is used in a number of schemes for hidden surface elimination other than the simple painting method described above. (2) the faces which are oriented away from the viewer - that is, the "back faces" must be eliminated before the painting is done. (3) For some scenes and viewpoints, no priority order exists; see [Sutherland et al, 1974] for an example. However, this happens rarely for naturally occuring scenes. The priority sorting methods which we will consider will perform the priority sort if possible, and will indicate its impossibility otherwise.)

Let $P(F, p, L)$ denote the predicate, "L is a priority sorted list of the faces F from viewpoint p". Our aim is to devise a synthesis program for P, that is, a program S_P which will take F and generate a special purpose priority sorting program $S_P(F)$; $S_P(F)$ then takes the position p as input, and generates a priority sorted list L for p.

We will proceed by first considering the result of applying the direct approach described in section 2 - that of specializing a simple one stage algorithm and then removing redundancies from the resulting decision tree. Then modifications of the direct method will be introduced one by one until a usable final result is obtained.

Here is the brute force algorithm for priority sorting:

(1) Compute the entire occlusion relation; that is, determine for each pair of faces i, j with $i \neq j$ whether or not i occludes j from p.

(2) Topologically sort the faces according to the occlusion relation computed in step (1). If there is a cycle in the occlusion relation, then no topological sort is possible, and in this case the outcome of the computation is an indication of failure. (Priority sorting consists exactly of finding a linear order which is consistent with the occlusion relation; the task of extending an acyclic binary relation to a linear order is the topological sorting problem. Algorithms for topological sorting may be found in standard references such as Knuth[1968]).

Now, consider the decision tree which results from unwinding this algorithm for a particular set F of faces. The decision tree T_0 which we have in mind may be described in precise terms as follows. Let $n = |F|$, and let $\langle i_1, j_1 \rangle, \langle i_2, j_2 \rangle \cdots \langle i_{n(n-1)}, j_{n(n-1)} \rangle$ be an enumeration of the set of pairs of faces from F given in the order in which they are considered in step (1) of the brute force algorithm. T_0, then, is a full binary tree of depth $n(n-1)$ (with $2^{n(n-1)}$ nodes!). Let $occ(p, i, j)$ denote the occlusion predicate: $occ(p, i, j)$ holds iff i occludes j from p. Then the predicate appearing at each of the 2^{k-1} decision nodes at the kth level of the tree is $occ(p, i_k, j_k)$. Evidently, each leaf of the tree corresponds to a particular truth assignment to the occlusion predicates, that is, to a particular occlusion relation on the faces. The computational result appearing at each leaf is the result of topologically sorting the faces according to the occlusion relation associated with that leaf, or an indication of failure if that relation contains a cycle.

The next stage in the process consists of the removal of redundant nodes from the decision tree. However, this requires that we automatically decide whether assertions of the form $occ(p, i, j)$ or $\neg occ(p, i, j)$ follow from sets of other assertions of the same form. As they stand these decision problems do not have any efficient solution. This difficulty can be overcome by simplifying the predicates which appear in the decision tree in the following way. For particular faces i and j, the predicate $occ(p,i,j)$ can be expressed as a conjunction of simpler predicates $occ_1(p,i,j) \cdots occ_r(p,i,j)$. The number r of such predicates is just the sum of the number of vertices in i and the number of vertices in j. Each predicate $occ_m(p,i,j)$ has the form, " the kth vertex of j (or i) lies below (or above) the 'horizon' defined by the lth edge of i (or j), as seen from p", or, equivalently, "p lies above (or below) the plane defined by the endpoints of the lth edge of i (or j) and the kth vertex of j (or i)". Thus, each predicate $occ_m(p,i,j)$ can be written as a linear inequality in the coordinates p_x, p_y, p_z of the viewer position p, with coefficients depending on i,j. Now, let us rewrite the decision tree T_0 to get a decision tree T_1 in which the predicate at each decision node is a linear inequality. This is done by expanding out the occlusion decision nodes according to the scheme:

if $occ(p,i,j)$ then t_1 else $t_2 \rightarrow$

if $occ_1(p,i,j)$ then
  if $occ_2(p,i,j)$ then
.
.
.
      if $occ_r(p,i,j)$ then t_1 else
      t_2 else
.
.
.
  t_2 else
t_2

We are in a good position to automatically remove redundant nodes from T_1, since (1) the predicate at each of its nodes is a linear inequality, (2) the negation of a linear inequality is a linear inequality, and (3) the question of whether a given inequality follows from a set of inequalities can be efficiently decided by use of the simplex algorithm. Further, on intuitive grounds, it is to be expected that the removal of redundant nodes will reduce the size of T_1 by a very large factor, since, once the, say, fifth level of the decision tree has been reached along a given path, the tests met so far along the path will have severely constrained the position of the viewer, and accordingly it is likely that the outcomes at the great majority of nodes below that level will in fact be predecided.

Of course, the problem with T_1 from the practical standpoint is that, even for very small numbers of faces, its size is intractably large. In practice, however, one will not proceed by first generating the decision tree T_1, and then optimizing it, but instead will optimize the tree while it is being generated. This can be done by constructing the tree from top down. When a new decision node is to be generated, one asks first whether its outcome is predetermined. If so, then one need not add the node to the tree; instead, one proceeds directly with the generation of its left or right subtree, depending on whether the predicate at the node is predetermined to be true or false. If this scheme is used, then no redundant nodes are ever generated; instead, the optimized tree is produced in one pass.

So, we have shown how to automatically construct an optimized decision tree T_2 for doing the priority sort for a fixed set of faces, but variable viewpoint. Although we expect that T_2 will be much faster than the brute force method from which it sprang, it is still too large to be of practical use.

The following observations will allow us to do much better. (1) The entire occlusion relation need not be determined in order to do the priority sort. (2) A partial determination of the occlusion relation which is insufficient to do the entire sort may still allow a part of the sort to be carried out. The former observation will allow us to shorten the tree, while the latter will make it possible to diminish the size of the results appearing at the leaves, by moving as much information as possible about the output of the computation to nodes closer to the root of the tree.

Let I be a set $\{I_1 \cdots I_k\}$ of linear inequalities in the the coordinates p_x, p_y, p_z of the viewer position. We may think of I as representing the simplex of points in three dimensional space which satisfy all of the inequalities $I_1 \cdots I_k$. Let $canocc(I, i, j)$ denote the predicate, "there is some point satisfying each of the inequalities in I from which the face i occludes the face j". Now, consider an arbitrary node N in T_2. Let I_N be the set of inequalities which are assumed to hold at N - that is to say, the set of inequalities which must hold if N is to be reached in the course of executing the decision tree. Let G be the graph of $canocc(I_N, i, j)$ viewed as a binary relation on the set of faces. If G is acyclic, then a topological sort of G will yield a priority order which is valid for all view points in the simplex I_N. In this case, the subtree rooted at N is not needed at all; a correct priority order can be generated without further case analysis. Suppose, on the other hand, that G contains cycles. Let $S_1 \cdots S_k$ be the strongly connected components of G. (Recall that a strongly connected component of a directed graph is a maximal set S of points from the graph having the property that, for any two points $p, q \in S$, there is a path from p to q.) As long as there is more than one strongly connected component - that is, as long as G is not itself strongly connected - a part of the priority order can be determined at the current stage. Let G' be the (acyclic) graph which results from collapsing the strongly connected components of G into single vertices. (Formally: the vertices of G' are the strongly connected components $S_1 \cdots S_k$ of G; $[\![S_i \to S_j]\!]$ is an edge in G' iff for some $p \in S_i$, $q \in S_j$, $[\![p \to q]\!]$ is an edge in G.) The result of topologically sorting G' gives an ordering to the $S_1 \cdots S_k$ which constitutes a partial priority sort, in the following sense. Let G'' be the graph of any occlusion relation which is consistent with inequalities I_N. If G'' has a priority ordering at all, then it has one of the form $append(p_1(S_{i_1}), p_2(S_{i_2}) \cdots p_k(S_{i_k}))$ where $S_{i_1} \cdots S_{i_k}$ is the topologically sorted ordering of G', and where $p_j(S_{i_j})$ is some permutation of S_{i_j}. Thus, the set of faces has been partitioned into subsets

$S_1 \cdots S_k$ such that the members of each S_i may be listed consecutively in any final ordering, and the order in which the S_i appear has also been decided. It is only the orderings within the S_i that remain to be determined. All of this has two consequences: (1) the occlusion or lack of occlusion between faces in different strongly connected components need not be considered in the decision tree rooted at N. Further, the strongly connected components may be considered separately; if desired, a different decision tree may be generated for each. (2) The decision tree rooted at N needs to specify only the orderings within, and not between the strongly connected components, provided that the ordering $S_{i_1} \cdots S_{i_k}$ is stored in one way or another at the node N.

The observations of the last paragraph indicate in a fairly direct way how T_2 may be improved on. The result T_3 of this improvement has a structure which is a bit more complicated than that of an ordinary decision tree, in that it has internal nodes which are not decision nodes. One way of describing T_3 is as a simple loop-free program which is built up from constants denoting lists of faces by use of (1) the conditional operator: "if P then t_1 else t_2" where P is a linear inequality in p_x, p_y, p_z, and (2) the *append* operator: "$append(t_1, \cdots t_k)$". Thus, T_3 differs from a decision tree only in that T_3 makes use of two operators ("if" and "*append*"), rather than just one ("if") in constructing its result.

The method by which T_3 is built follows closely the method used to build T_2. The difference is that the *canocc* graph is employed to guide the selection of face pairs for case analysis, and also to split the computation of the priority ordering into separate computations for separate strongly connected components. The method for synthesizing T_3 is given below by the recursive program $\mathcal{R}(I, F)$. Here, I is a set of inequalities, and F a set of faces. The result returned by $\mathcal{R}(I, F)$ is a program which computes an ordering for the faces in F; under the assumption that the inequalities in I hold, this ordering will be a correct priority ordering of the faces. $\mathcal{R}$ makes use of a subroutine $\mathcal{R}_1$ which takes care of generating the individual tests $occ_1(i,j) \cdots occ_r(i,j)$ which together determine whether i occludes j. $\mathcal{R}$ and $\mathcal{R}_1$, then, are as follows: (For the sake of clarity, we make the simplifying assumption here that the set of faces F has a priority ordering from all viewpoints; the other case is not difficult to handle.)

$\mathcal{R}(I, F)$:

(1) Compute the graph G of $canocc(I, i, j)$ for $i \neq j \in F$.

(2) If G is acyclic, then topologically sort G, and return the constant representing this sorted list of faces.

If G is strongly connected then:

    (a) choose i, j such that $occ(i, j)$ is not decided by I.

    (b) return $\mathcal{R}_1(I, F, i, j, 1)$ (This does a case analysis according to whether i occludes j)

Otherwise:

    (a) Let $S_1 \cdots S_k$ be the strongly connected components of G. Topologically sort the graph G' gotten by collapsing the S_j, getting an ordering $S_{i_1} \cdots S_{i_k}$ of the S_j. Compute the programs $P_j = \mathcal{R}(I, S_j)$ for priority sorting the S_j. Return the program, "append"$(P_{i_1}, \cdots P_{i_k})$

$\mathcal{R}_1(I, F, i, j, n)$:

(Here $\langle i, j \rangle$ is the pair of faces to be dealt with; n is the index of the occlusion test to be generated)

(1) If n is greater than the number of occlusion tests needed for the faces i, j - that is, if all the tests have already been generated - then return $\mathcal{R}(I, F)$

Otherwise:

    (a) If I implies that $occ_n(i, j)$ holds then return $\mathcal{R}_1(I, F, i, j, n+1)$.

    (b) If I implies that $occ_n(i, j)$ does not hold then return $\mathcal{R}(I, F)$.

    (c) Otherwise, generate the test; return the program:

"if" $occ_n(i, j)$ "then" $\mathcal{R}_1(I \cup \{occ_n(i, j)\}, F, i, j, n+1)$ "else" $\mathcal{R}(I \cup \{\neg occ_n(i, j)\}, F)$

4. Results of experiments

A program for synthesis of special purpose priority sorting programs has been implemented on the Stanford computer science department PDP-10/KL-10 computer in MacLisp. The program has been tested on one large scale example so far, namely, a description of a hilly landscape derived from a data base provided to the author by the Link division of Singer corporation (Link is a manufacturer of flight simulators). The description consisted of a set L of 1135 faces making up, roughly speaking, a triangulation of the landscape.

The implemented program is based directly on the method described in the last section, but includes the following important refinement. (There are also a number of less important refinements and implementation details whose description is beyond the scope of the current general presentation of the method).

In almost all three dimensional computer graphics applications, the field of view covered by the image to be generated is limited. For example, in several of the flight simulators manufactured by Link, the field of view or "window" spans 48 degrees horizontally, and less than 40 degrees vertically. We exploit this fact by performing an initial case analysis according to the direction in which the viewer is looking. Specifically, this is what is done: Consider the projection of the viewing direction v onto the x,y plane. We divide the "pie" consisting of the set of all possible such projections into ten equal "slices", each 36 degrees wide. Now, one face can *visibly occlude* another only if there is a ray from the viewer's eye which passes through face i and face j *and* whose direction lies within a certain angular distance the viewing direction v. So, the assumption that the x,y projection of v lies within a given 36 degree slice reduces the number of possible occlusions between faces, since it places limitations on the relative positions of visibly occluding pairs of faces. In any case, the ten slices of the pie are considered separately, with field of view parameters of 48 degrees horizontal and 40 degrees vertical, and with the additional assumption that the angle of the viewing direction to the x,y plane (that is, the angular deviation from horizontal flight) is less than 30 degrees. The other cases, where the angle of view is steeply up or steeply down, are handled separately. For each slice, a separate initial *canocc* relation is computed; it is this restricted canocc relation which forms the starting point for the method R described in the last section.

In the experiments performed so far, the synthesis method has been applied to the landscape L for only one of the ten pie slices. The features of the landscape (mountains and valleys) are oriented in more or less random directions, so there is reason to believe that similar results would be obtained for each of the pie slices. Also, the steeply up and steeply down orientations should yield results which are better, and not worse, than the orientations which are close to horizontal.

Some specifics on the synthetic program T_3 produced for the one slice are as follows: worst case number of decision nodes encountered during any execution of T_3: 53; expected number of decision nodes encountered during an execution of T_3, assuming that at each decision node the two possible outcomes of the decision are equally likely: 27; total number of decision nodes in T_3: 85.

For the landscape and slice treated, we estimate that T_3 is about ten times as fast as the fastest and most widely used of the previously known algorithms for doing the priority computation, namely, that of Schumacker and his colleagues [Sutherland et al, 1974] (this is the algorithm used by Link, and by several other flight simulation manufacturers, in their current systems). In addition to its speed, our scheme has the more important advantage of flexibility. The Schumacker method imposes the rather severe restriction that it be possible to hierarchically decompose the scene into clusters by means of a collection of separating planes, where by a "cluster" is meant a set of faces for which one fixed priority order is valid. In contrast, our method imposes no restriction at all on the structure of the scenes to which it may be applied.

The reader is referred to an extended version of this report [Goad, 1982] for a more detailed description of the experimental results, and for further information about the relationship between our method and other methods for priority sorting.

5. Other applications to geometrical computation

Evidently, the general scheme which we have used for hidden surface elimination, namely that of symbolically executing a brute force algorithm, and then introducing appropriate modifications, can be tried on any geometrical problem in which the shape of objects is fixed, but their positions are not. Examples include other problems in three dimensional graphics and simulation, such as hidden line (rather than surface) elimination, and collision detection between moving bodies of known shape. In applications where there is more than one moving object (here, the viewer counts as an object as well), the decision problems which arise will involve non-linear as well as linear inequalities. Devising efficient methods for attacking these more complex decision problems constitutes an interesting class of problems in the field of automatic deduction.

6. Related work on program synthesis and manipulation

As indicated earlier, the techniques which we have used for producing and optimizing decision trees are adapted from standard methods which have been employed in many fields, including program transformation [Burstall and Darlington, 1977], planning [Sproul 1977], program synthesis of the traditional kind [Manna and Waldinger, 1980] and program specialization [Beckeman et al 1976; Emmanuelson, 1980]. The work reported here differs from work in all of these fields in more or less the same way, namely, in that we treat classes of problems defined in a mathematical rather than universal style. For example, in program specialization, one looks for methods which can be usefully applied to any program in a given language; thus, the goal of the enterprise is the development of *universal* methods which will apply to all problems formulated in a particular way. Another example is that of "knowledge based" program synthesis of the kind developed by Barstow[1977]. The aims are again universal, both in the sense that the class of problems to be treated, although restricted to the program's "domain of expertise", is still much wider than the class which we treat, and in the sense that the mechanism for synthesis is intended to have universal application; only the "knowledge" encoded in a set of production rules is particular to the domain of expertise. In contrast, the results which we obtained for the hidden surface elimination problem depended on adapting our methods specifically to the problem at hand.

There are many algorithms from the mainstream of computer science which may be seen as synthesizers of special purpose programs. For example, there is a wide and useful class of algorithms for recognition of patterns in strings [Aho, Hopcroft, and Ullman, 1974] which, when given the pattern, proceed by constructing a finite automaton which in turn performs the search. A finite automaton, like a decision tree, is a simple variety of program, and in this sense, these algorithms construct programs for the special purpose of finding an instance of a pattern in a string. The hidden surface elimination algorithm described in this paper should be regarded as a new piece of work in this tradition.

References

Aho, A.V., Hopcroft, J.E., and Ullman, J.D.[1974], *The design and analysis of computer algorithms*,Addison-Wesley, Reading Mass., 1974, see Chapter 9: pp. 317-361

Barstow, D.[1977], *A knowledge based system for automatic program construction*,Fifth International Joint Conference on Artificial Intelligence, Cambridge, August, 1977

Beckeman,L., Haraldsson, A., Oskarsson,O., and Sandewall, E.[1976], *A partial evaluator and its use as a programming tool*,Artificial Intelligence Journal 7,1976, pp. 319-357

Burstall R.M., and Darlington, J.[1977], *A transformation system for developing recursive programs*,JACM, Vol. 24, No. 1, January 1977

Emmanuelson, P.[1980], *Performance enhancement in a well-structured pattern matcher through partial evaluation*,Ph.D. Thesis, Software Systems Research Center, Linköping University, Linköping, Sweden, 1980

Goad, C.[1982], *Automatic construction of special purpose programs*,Stanford University Computer Science Dept. Report no. STAN-CS-82-897

Knuth, D.E.[1968], *The art of computer programming, vol 1: Fundamental algorithms*,Addison-Wesley, Reading Mass., 1968, pp. 258-268

Kreisel, G.[1981], *Neglected possibilities for processing assertions and proofs mechanically: choice of problems and data*,in: P. Suppes [ed.], University-level computer-assisted instruction at Stanford: 1968-1980. Stanford Calif.; Stanford University, Institute for Mathematical Studies in the Social Sciences, 1981

Manna, Z., and Waldinger, R.[1980], *A deductive approach to program synthesis*,ACM Transactions on programming languages and systems, Vol. 2, No 1., January 1980

Robson, J.[1979], *The height of binary search trees*,The Australian Computer Journal, 11(1979), pp 151-153

Sproull, R.F.[1977], *Strategy construction using a synthesis of heuristic and decision-theoretic methods*,Xerox PARC technical report CSL-77-2, July, 1977

Sutherland, I.E.,Sproull, R.F., and Schumacker, R.A.[1974], *A characterization of ten hidden-surface algorithms*,Computing Surveys, Vol. 6, No. 1, March 1974

Deciding Combinations of Theories

Robert E. Shostak

Computer Science Laboratory
SRI International
333 Ravenswood Avenue
Menlo Park, CA 94025 USA

Abstract

A method is given for deciding formulas in combinations of unquantified first-order theories. Rather than coupling separate decision procedures for the contributing theories, it makes use of a single, uniform procedure that minimizes the code needed to accommodate each additional theory. The method is applicable to theories whose semantics can be encoded within a certain class of purely equational canonical form theories that is closed under combination. Examples are given from the equational theories of integer and real arithmetic, a subtheory of monadic set theory, the theory of *cons, car*, and *cdr*, and others. A discussion of the speed performance of the procedure and a proof of the theorem that underlies its completeness are also given. The procedure has been used extensively as the deductive core of a system for program specification and verification.

1 Introduction

Much attention has been given in the last few years to the problem of developing decision procedures for applications in program verification and mechanical proof of theorems in mathematics. It often happens, in these applications, that the formulas to be proved involve diverse semantic constructs. Verification conditions for programs, for example, often mix arithmetic with data structure semantics. Accordingly, a number of procedures have been developed (such as those of Jefferson [Jef 78] and the author [Sho 79]) for particular *mixes* of theories. One of the most general results along these lines is given by Nelson and Oppen [NeO 79] in their work on simplification using cooperating decision procedures. The central idea, here, is that procedures for diverse unquantified theories in a certain class can be hooked together so as collectively to decide a formula involving constructs from each. The coupled procedures communicate by passing back and forth information they individually derive about equality among terms.

The method described in this paper also addresses the problem of deciding combinations of theories. Rather than connecting separate procedures for the constituent theories, however, it is based on a single, uniform procedure that localizes the mechanism particular to each constituent to a relatively small amount of code. As a result, new semantic constructs can be incorporated quite easily. The method also obviates the explicit computation of entailed equality information required by the distributed approach, and the redundant representation of this information in the separate procedures.

The method is applicable to theories whose semantics can be expressed inside a certain class of purely equational canonical form theories. The class has the property that the canonizing functions of arbitrary members can be combined to obtain a canonizing function for their union. The decision procedure extends the congruence closure methods ([DoS 78],[NeO 80],[Sho 78]) that have been used for the pure theory of equality. The extension depends upon a notion of *semantic signature* that generalizes the earlier concept of signature to accommodate the semantics of the canonical form theories.

This research was supported in part by NSF grant MCS-7904081, and by AFOSR contract F49620-79-C-0099.

The method is useful for a surprisingly rich class of constructs. We give examples from integer and real arithmetic, the theory of recursive data structures, a subtheory of monadic set theory, and the first-order theory of equality. The procedure has been used intensively in a major verification effort [ShS 82] over the course of several months, and has been found to give good speed performance.

The next two sections characterize the class of theories to which the method is applicable, and prove that it is closed under combination. Section 4 gives the decision procedure and exemplifies its operation. Section 5 discusses performance issues, and the last section gives a proof of the theorem on which the procedure is based.

2 σ-Theories and Algebraic Solvability

The procedure described in this paper operates over a subclass of certain unquantified first-order equational theories called σ-theories. Such theories have both *interpreted* and *uninterpreted* function symbols. Terms whose outermost function symbol is interpreted are themselves said to be interpreted, and all other terms (including variables) are said to be uninterpreted. The distinguishing feature of σ-theories is that each has an effective canonizer, i.e., a computable function σ from terms to terms such that a purely interpreted (i.e., containing no uninterpreted function symbols) equation $t = u$ is valid in the theory iff $\sigma(t) = \sigma(u)$. The canonizer σ must act as an identity on uninterpreted terms, must be idempotent (i.e., $\sigma(\sigma(t)) = \sigma(t)$), and must have the property that the arguments of each interpreted canonical term must themselves be canonical.

The semantics of σ-theories are defined with respect to σ-*interpretations*, i.e., Herbrand interpretations that respect σ in the sense that for each interpreted term $t = f(t_1, \ldots, t_n)$, $v(t) = \sigma(f(v(t_1), \ldots, v(t_n)))$, where $v(t)$ is the value of t in the interpretation. Note that these semantics say more than that two terms must be equal in all interpretations iff they have identical canonical forms; they also say that certain terms (for example, distinct canonical constants) must be *unequal* in all interpretations.

A simple example of a σ-theory is the equational theory of addition and multiplication by constants over the reals. Each term in this theory can, of course, be reduced to a linear expression of the form $a_1 x_1 + a_2 x_2 + \ldots + a_n x_n + c$ $(n \geq 0)$, where each a_i is a nonzero real constant, each x_i is a variable, and c is a real constant. By specifying some criterion for ordering the x_i's (alphabetic ordering, for example), removing unity coefficients, and discarding c if $c = 0$, a suitable σ can be defined. Note that the theory can trivially be enriched with uninterpreted function symbols.

The decision procedure we will describe operates on σ-theories with a special property we call *algebraic solvability*. Such theories have a *solver*- a computable function that takes a purely interpreted equation (no uninterpreted symbols other than variables) e (say $t = u$) as input, and returns either **true**, **false**, or a conjunction of equations. The returned formula must be equivalent to e and must be either **true** or **false** if e has no variables. Moreover, the equations (if any) in the returned formula must all be of the form $x_i = t_i$, where the x_i's are distinct variables of t (or u if t has none) that do not occur in any of the t_i's.

Returning to the example of real linear arithmetic, a suitable solver is trivially provided by conventional algebraic manipulation; solving $3x + 2y = 2x + 4$, for example, gives $x = -2y + 4$.

At first glance, one might expect that algebraic solvability is such a strong criterion that there are no nontrivial theories other than, say, theories over fields, that satisfy it. Surprisingly, this is not so. As the following examples suggest, algebraic solvability is characteristic of many constructs one encounters in practice.

Example 1. The Theory of *cons*, *car*, and *cdr*.

First consider the theory of *cons*, *car*, *cdr*, and *nil* with the usual initial algebra semantics. Using the rewrites

$$car(cons(x,y)) \mapsto x, \quad cdr(cons(x,y)) \mapsto y,$$

one can always reduce a term in the theory to a canonical form in which *cons* is not an argument of either *car* or *cdr*.

A *solve* function for this theory is given by the algorithm shown in Figure 1. Note that the algorithm introduces new (existential) variables. The chief function of the main routine *solve* is to grind the left side of the given equation (represented as an ordered pair) down to a variable; the auxilliary function *solve*1 (which takes two terms and returns a term) takes over from there, and attempts to eliminate that variable from the right side. The routine *resolve*, called by *solve*, sees to it that no equations in the returned conjunction have the same left side.

Consider, for example, the action of the algorithm on the equation

$$cons(car(x), cdr(car(y))) = cdr(cons(y, x)).$$

The two sides are first canonized to obtain

$$cons(car(x), cdr(car(y))) = x.$$

This equation is then reduced to the two equations

$$car(x) = car(x), \quad cdr(car(y)) = cdr(x).$$

The first of these reduces to **true**, and is, in effect, discarded. The second becomes

$$car(y) = cons(a, cdr(x)),$$

where a is a new variable, and then (recursing again)

$$y = cons(cons(a, cdr(x)), d),$$

which is then returned.

It is not difficult to show that the algorithm satisfies the criteria for a solver. The reader might find it amusing to prove that it always terminates.

The more general theory of recursive data structures can be treated with a straightforward generalization of σ and the *solve* algorithm just given.

Example 2. Integer Linear Arithmetic

The σ described earlier for real linear arithmetic can be used here as well. A *solve* function is provided by a method for reducing linear Diophantine equations that is based on the Euclidean algorithm. For example, the equation

$$17x - 49y = 30$$

is solved by this method to obtain

$$x = 49z - 4, \quad y = 17z - 2,$$

where z is a new variable. The theory of linear congruences can be treated in a similar way.

$solve(<t, u>)$
    $t \leftarrow \sigma(t)$;
    $u \leftarrow \sigma(u)$;
    if $t = u$ return **true**;
    if t contains no variable swap t and u;
    if t still contains no variable return **false**;
    let a and d be fresh variables (new on each call);
    if t is of the form:
        $car(t_1)$ then return $solve(<t_1, cons(u, d)>)$;
        $cdr(t_1)$ then return $solve(<t_1, cons(a, u)>)$;
        $cons(t_1, t_2)$ then
            return $resolve(solve(<t_1, car(u)>) \wedge solve(<t_2, cdr(u)>))$;
    else return $<t, solve1(t, u)>$;

$solve1(x, u)$
    $u \leftarrow \sigma(u)$;
    if $x = u$ or x does not occur in u return u;
    if x occurs in u as an argument of $cons$ or if u is not a $cons$
        return from $solve$ with **false**;
    let a, d be fresh variables (new on each call);
    replace $car(x)$ by a and $cdr(x)$ by d in u;
    return $cons(solve1(a, car(u)), solve1(d, cdr(u)))$;

$resolve(c)$
    until c contains no two equations of the form $v = t_1, v = t_2$ do
        remove $v = t_2$ from c; $c \leftarrow c \wedge solve(<t_1, t_2>)$;
    return c;

Figure 1. Solver for *car*, *cdr*, and *cons*

Example 3. "Venn Diagram" Monadic Set Theory

Formulas in this theory are constructed from set variables and the empty set under set inclusion, union, intersection, and complement. As is well known, such formulas can be encoded in a purely equational theory with only set variables, the empty set, and the union operator +. The encoding represents each set variable in the given formula by the union of its disjoint constituent pieces in the Venn Diagram corresponding to the variables of the formula. Referring to Figure 2, for example,

$$A \cap B = B \cup C$$

becomes

$$s_2 + s_3 = (s_2 + s_3 + s_4 + s_6) + (s_3 + s_4 + s_5 + s_7).$$

Similarly, $A \subseteq B$ becomes $s_1 + s_5 = \Phi$.

It is easy to see that terms in the encoding theory can be canonized simply by ordering the set variables in some way and eliminating redundancies.

A solver for the encoding theory is also easy to construct. One simply replaces each side of an equation with its σ-form and cancels variables occurring on both sides.

If the two sides thereby become identical, **true** is returned; otherwise, the conjunction

$$\bigwedge v_i = \Phi$$

is returned, where the v_i's are the remaining variables.

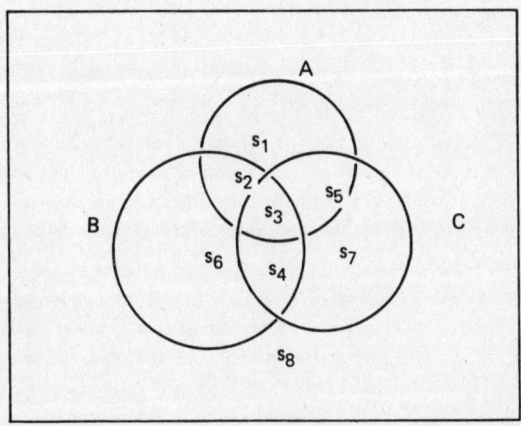

FIGURE 2 EXAMPLE FROM "VENN DIAGRAM" THEORY

Example 4. The Unquantified First-Order Theory of Equality

This theory has no interpreted function symbols at all. σ is therefore just the identity function. The *solve* function merely returns **true** if the given equation is an identity, and acts as an identity otherwise.

Up until now, we have been concerned with solve functions for equations with only purely interpreted terms. In the next section and in the procedure itself, it will be necessary to deal with equations that may have uninterpreted terms other than variables. The notion of solver is extended to deal with such equations in the following way.

We will say that a *solver* for a σ-theory T is a computable function that takes an arbitrary equation e (say $t = u$) over T as input and returns either **true**, **false**, or a conjunction of equations. The returned formula must be equivalent to e and must be one of **true** or **false** if e contains no uninterpreted terms. Moreover, the equations (if any) in the returned formula must all be of the form $u_i = t_i$, where the u_i's are distinct uninterpreted subterms of t (or u if t has none) that do not occur in the t_i's *except possibly as proper subterms of uninterpreted terms*.

Note that all the solvers we have talked about so far can be trivially extended to satisfy the new definition simply by treating the maximally uninterpreted subterms of a submitted equation as if they were distinct variables. For example, the equation

$$f(x) + g(x + 2, f(x)) = 0,$$

when solved in linear arithmetic, produces

$$f(x) = -1 \times g(x + 2, f(x)).$$

Note that although $f(x)$ occurs on the right-hand side of the result, it occurs only as a proper subterm of an uninterpreted term.

3 Combining Theories

Algebraically solvable σ-theories have the elegant and useful property that they can be combined (provided they are compatible in a certain sense) without giving up algebraic solvablility.

One needs to be careful, of course, about what it *means* to combine theories. Simply to say that the interpreted symbols of the combined theory are the union of those of the constituents does not suffice. The semantics of mixed terms (terms with symbols from different constituents) must be explicated, and conflicts that may arise when the constituents share interpreted symbols must be resolved.

The most natural resolution (and the one we espouse) of the issue of mixed terms is simply to assume that interpreted symbols from different theories have no semantic interaction in the combined theory; if a function interpreted in T_1 but not in T_2 is given an argument that is interpreted in T_2 but not in T_1, the effect is to treat the argument *as if it were uninterpreted*. For example, in the combination of real linear arithmetic with the theory of *cons*, *car*, and *cdr*, we will have

$$\sigma(car(cons(1 + 1, nil))) = 1 + 1$$

We will consider that $+$ is *uninterpreted with respect to cons, car, and cdr, (and vice versa)* in the combined theory, and that $1 + 1$ is therefore an uninterpreted subterm of $car(cons(1 + 1, nil))$.

It still remains to resolve the question of what happens if the constituent theories share interpreted symbols. Here we will take the conservative stand of requiring that the shared symbols have exactly the same semantics. More precisely, we will say that T_1 and T_2 are *compatible* if for each term t in their combination whose function symbol is interpreted in both, $\sigma_1(t) = \sigma_2(t)$. We will consider only combinations of compatible theories.

Note that as a consequence of the view that uninterpretedness is relative, the purely interpreted terms of the combined theory are exactly the union of those from the constituent theories; the action of the combined σ on each such term is the action of the σ of the appropriate constituent. It is easy to check that the combined σ thus defined satisfies the requirements for a canonizer.

Given solvers for each of the constituent theories, one can be constructed for the combination in the following way. Let T be the combination of $T_1, \ldots, T_r$, and let $t = u$ be an equation of T to be solved. If neither t nor u contains uninterpreted terms, we return **true** if they have identical canonical forms and **false** if not. If t has no uninterpreted subterms, t and u are exchanged. $solve_i$ is now applied to $t = u$, where T_i is a constituent theory from which the function symbol of t derives. Each equality in the resulting conjunction is then recursively replaced by its solution until no left side of an equality occurs in a right side except as a proper subterm of an uninterpreted term.

The idea is easily understood in the context of an example. Suppose the equation to be solved is

$$5 + car(x+1) = cdr(x+2) + 3.$$

First solving in linear arithmetic, we obtain

$$car(x+1) = cdr(x+2) - 2.$$

Note that whereas $car(x+1)$ is (relatively) uninterpreted in the original equation, it is interpreted here. The solver for car, cdr, and $cons$ is now applied to obtain

$$x + 1 = cons(cdr(x+2) - 2, d),$$

where d is a new variable. Solving once more in linear arithmetic, we have

$$x = cons(cdr(x+2) - 2, d) - 1.$$

As the left side x is an uninterpreted term that does not occur in the right side except as a proper subterm of an uninterpreted term $(x + 2)$, we are done.

4 The Procedure

We now give a refutation procedure for conjunctions of equalities and disequalities in an algebraically solvable σ-theory T. Formulas with arbitrary propositional structure can be decided for validity by placing their negations into disjunctive normal form and applying the refutation procedure to each conjunction; the original formula is valid iff all conjunctions are unsatisfiable.

The procedure extends the congruence closure methods of Nelson and Oppen [NeO 80], Downey, Sethi and Tarjan [DoS], and the author [Sho 78] for deciding purely uninterpreted formulas. These methods all depend on computing a congruence relation that provides a canonical form for each term represented in the relation. The procedure we describe can be viewed as combining this canonical form for purely uninterpreted terms with the σ-form for interpreted terms to obtain a universal canonical form for all terms. Crucial to the extension is a generalization of the notion of *signature* that permits the two kinds of canonical forms to be merged.

The main algorithms of the procedure are *process* and *canon*, shown in Figure 3a. *process* is called on the set of equalities (each represented by an ordered pair) occurring in the conjunction S whose satisfiability is to be determined; if *process* returns **unsatisfiable**, the procedure halts with that conclusion. Otherwise, for each disequality $t \neq u$ of S, $canon(t)$ and $canon(u)$ are computed. If $canon(t) = canon(u)$ for some such disequality, **unsatisfiable** is returned; otherwise, **satisfiable** is returned.

As in the procedure described by Nelson and Oppen, a congruence relation is developed (by *process*) using Tarjan's *union* and *find* primitives ([Tar 75]). The *canon* routine returns the canonical representation, in this relation, of a term to which it is applied. Note from the specification of *canon* that two terms have the same *canon* if they have the same signature. If a term is already represented in the relation (i.e., *canon* has previously been called on a term with the same signature), the existing canonical form is returned; otherwise, the signature of the term is returned as its *canon*.

The *solve* routine is assumed to return a set (representing a conjunction). In accordance with the definition given in the last section, the set must either be a singleton containing one of **true, false**, or a set of equations whose left hand sides are uninterpreted. The right hand sides are assumed to be reduced with respect to σ. Note that *process* uses the convention that e_1 refers to the left side of an equation e, and that e_2 refers to the right side. The order of sides is important.

Note also that the *union* function (called in *merge*) is assumed always to change *find* of its first argument. Initially, $find(t)$ is assumed to be t for each term t.

Aside from the tree structure used to implement *union* and *find*, two global data structures are used: *use* and *sig*. For each canonical term t, $use(t)$ is a list of terms whose signatures have t as an argument. $use(t)$ is maintained so that the signatures of its members are distinct and include the signatures of all represented terms whose signatures have t as an argument. For each u in $use(t)$, $sig(u)$ gives the current signature of u.

Figure 3b. shows auxiliary routines *canon1* (called by *merge*) and *signature1* (called by *canon1*.) *canon1* differs from *canon* only in that it returns the *find* of uninterpreted terms.

Figure 4 shows various stages in the application of the procedure to the conjunction

$$z = f(x-y) \wedge x = z + y \wedge -y \neq -(x - f(f(z)))$$

in the combination of the theory of real linear arithmetic and the pure theory of equality. The solid arrows in each diagram represent the $union - find$ structure (for example, $find(z) = f(x + -1 \times y)$ in Figure 4(a)). The dotted arrows represent the *use* structure; a dotted arrow from t to u indicates that u is a member of $use(t)$.

The procedure first calls *process* on the equalities of the conjunction. The equality $z = f(x-y)$ is selected, canonized (giving $z = f(x + -1 \times y)$), and submitted to *solve*. The canonized equality is its own solution, and so $merge(z, f(x + -1 \times y))$ is invoked. Figure 4(a) shows the state after this call. The dotted structure results from the call of *canon* on $f(x - y)$.

The second equality $x = z + y$ is next selected from *eqset*, canonized to obtain $x = f(x + -1 \times y) + y$, and submitted to *solve*. *solve* again has no effect, and $merge(x, f(x + -1 \times y) + y)$ is called. Note that inside the call, *canon1* is invoked on $x + -1 \times y$, yielding (as the reader should check) $f(x + -1 \times y)$.

$process(eqset)$
 while $eqset \neq \Phi$ do
  select and delete e from $eqset$;
  $s \leftarrow solve(canon(e_1) = canon(e_2));$ /* e_1 refers to the left side of e, e_2 to the right side */
  while $s \neq \Phi$ do
   select and delete e from s;
   if $e = $ **true** continue while;
   if $e = $ **false** return **unsatisfiable**;
   $merge(e_1, e_2);$

$merge(t_1, t_2)$
 $t_1 \leftarrow find(t_1); t_2 \leftarrow find(t_2);$
 if $t_1 = t_2$ return;
 $union(t_1, t_2);$ /* $find(t_1)$ is now t_2 */
 for $u \in use(t_1)$ do
  if u is uninterpreted
   replace t_1 with t_2 in the argument list of $sig(u)$;
   for $v \in use(t_2)$ when $sig(v) = sig(u)$ do
    remove v from $use(t_2)$;
    $s \leftarrow s \cup solve(find(u) = find(v));$
   add u to $use(t_2)$;
  else add $u = canon1(u)$ to s;

$canon(t)$
 $w \leftarrow signature(t);$
 if w is a constant return $find(w)$;
 else
  Say $w = f(w_1, \ldots, w_n);$
  If $w = sig(u)$ for some $u \in use(w_1)$ return $find(u)$;
  else
   for i from 1 to n do add w to $use(w_i)$;
   $sig(w) \leftarrow w;$
   $use(w) \leftarrow \Phi;$
   return(w);

$signature(t)$
 if t is a constant return t;
 else return $\sigma(f(canon(t_1), \ldots, canon(t_n)))$ where $t = f(t_1, \ldots, t_n);$

Figure 3a. Main Routines of Procedure

$canon1(t)$
    if t is uninterpreted return $find(t)$;
    $w \leftarrow signature1(t)$;
    if w is a constant return $find(w)$;
    else
        Say $w = f(w_1, \ldots, w_n)$;
        If $w = sig(u)$ for some $u \in use(w_1)$ return $find(u)$;
        else
            for i from 1 to n do add w to $use(w_i)$;
            $sig(w) \leftarrow w$;
            $use(w) \leftarrow \Phi$;
            return(w);

$signature1(t)$
    if t is a constant return t;
    else return $\sigma(f(canon1(t_1), \ldots, canon1(t_n)))$ where $t = f(t_1, \ldots, t_n)$;

Figure 3b. Routines *canon1* and *signature1*

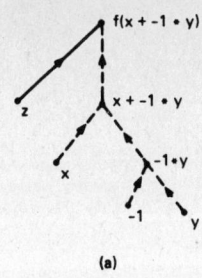

(a)

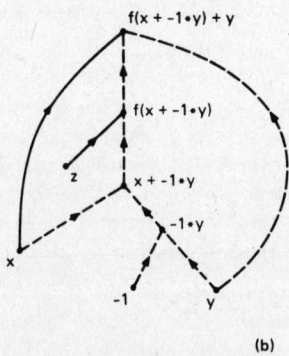

(b)

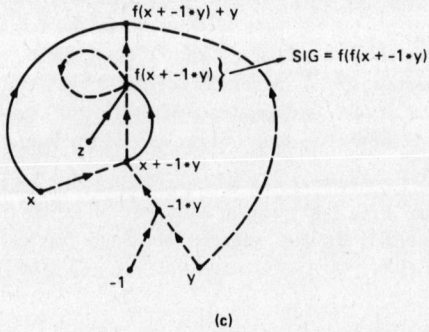

(c)

FIGURE 4 APPLICATION OF THE PROCEDURE TO $z = f(x-y) \wedge x = z + y \wedge -y \neq -(x-f(f(z)))$

The equation

$$x + -1 \times y = f(x + -1 \times y)$$

is then added to s. Figure 4(b) depicts the situation after this call.

$merge(x + -1 \times y, f(x + -1 \times y))$ is now invoked as a result of the addition to s. The effects of this call are shown in Figure 4c. Note that $f(x + -1 \times y)$ is added to its own use list, and that $sig(f(x + -1 \times y))$ is updated to reflect the new signature of $f(x + -1 \times y)$, namely $f(f(x + -1 \times y))$. No new pairs are added to s, so the call to $process$ is complete.

The procedure next computes $canon$ of $-y$ and $-(x - f(f(z)))$. Both calls produce $-1 \times y$, so the conjunction is unsatisfiable.

5 Performance Issues

The procedure has been implemented in INTERLISP-10, and forms the deductive core of a specification and verification system that provides mechanical support for an extension of multisorted first-order logic [ShS 82]. This system has been used intensively over the last year in a successful design proof effort for the SIFT [Wen 78, MeS 82] fault-tolerant operating system. The effort involved the proof of literally hundreds of formulas. We found that formulas occupying several pages of syntax would usually be proved in less than a minute (provided they were true- otherwise this could take substantially longer!)

The time complexity of the procedure naturally depends on the cost of invoking the algorithms for σ and $solve$. Since these costs typically tend to dominate, it is impossible to analyze the performance of the algorithm without making specific assumptions about the particular combination of theories to which it is applied. In the purely uninterpreted case, the procedure becomes very similar to the E-procedure mentioned earlier, whose worst-case time was determined by Nelson and Oppen to be $O(m^2)$, where m is the number of edges in the graph of the developed congruence relation. The differences are that use lists are not kept sorted in our algorithm and that we use a special data structure sig to store the current signatures of terms that represent signature classes. We have found that because use lists are usually quite short, the time required to maintain sorted versions is not worth the theoretical advantage in doing so. We have also found that the use of the sig data structure provides a significant practical improvement, because it eliminates having to compute signatures in the second **for** loop of the $merge$ routine.

We should point out that analytic measures can be quite misleading indicators of performance in real applications. Our experience has been that such measures are often less informative than more precise and objective characterizations of performance, such as "blindingly fast".

6 Correctness of the Procedure

It is easy to check by inspection that the procedure is sound, i.e., that two terms with the same $canon$ must be equal in any σ-interpretation for S. This section gives a proof of the theorem that underlies the procedure's completeness. The theorem extends the result given in [Sho 78] for the pure theory of equality to σ-theories.

Definition. We say that a congruence relation $\simeq$ over a set $\mathcal{R}$ of terms in over a σ-theory T is a σ-congruence relation if each equivalence class $[t]$ in $\simeq$ has a representative $\rho(t)$ such that

(*) For each interpreted term $t = f(t_1, \ldots, t_n)$ in T,

$$\rho(t) = \sigma(f(\rho(t_1), \ldots, \rho(t_n)))$$

Note that, interpreting the *canon* procedure as ρ and the set of represented terms as R, the (*)-property just says that the canon of each interpreted term t must be the semantic signature of t.

Theorem. Let S be a σ-unsatisfiable set of equalities and disequalities over a σ-theory T, and let $\simeq$ be a σ-congruence relation over a set R of terms in T such that $t \simeq u$ for each equation $t = u \in S$. Then there exists a disequality $t \neq u \in S$ such that $t \simeq u$.

Pf. We will suppose that there is no such disequality, and construct a σ-model for S. First, divide the Herbrand Universe of T into layers $H_0, H_1, \ldots$, where $H_0 = R$, and
$$H_{i+1} = H_i \cup \{f(t_1, \ldots, t_n) \mid t_1, \ldots, t_n \in H_i\}$$
for each function symbol f. We say that the R-*height* of a term t is the smallest i for which $t \in H_i$.

The model M is constructed inductively by height. For terms t of height 0, we let $M(t) = \rho(t)$. For terms $t = f(t_1, \ldots, t_n)$, $n \geq 0$, of height $i > 0$, we let

$$M(t) = \begin{cases} M(v) & \text{if there exists a term } v = f(v_1, \ldots, v_n) \\ & \text{of height } < i \text{ such that } M(v_j) = M(t_j), \\ & 1 \leq j \leq n. \\ \sigma(f(M(t_1), \ldots, M(t_n))) & \text{otherwise.} \end{cases}$$

It must be shown that M is functionally reflexive, i.e.,
$$\bigwedge M(t_i) = M(u_i) \supset M(f(t_1, \ldots, t_n)) = M(f(u_1, \ldots, u_n))$$
and that M respects σ. Both these properties are straightforwardly proven by induction on R-height. The first follows directly from the manner in which the model is constructed, the second from the (*)-property. ∎

7 Acknowledgement

The author is grateful to D. Hare, P. M. Melliar-Smith and Richard Schwartz, all of whom helped to correct lacunae in the procedure.

References

[DoS 78]	Downey, P.J., R. Sethi, R. Tarjan. "Variations on the common subexpression problem." To appear in *J.ACM*.
[Jef 78]	Jefferson, D. R., "Type reduction and program verification", Ph.D. Thesis, Carnegie Mellon University, August 1978
[MeS 82]	Melliar-Smith, P. M., R. L. Schwartz, "Formal specification and mechanical verification of SIFT: a fault-tolerant flight control system", *IEEE Trans. on Computers*, July 1982.
[NeO 79]	Nelson, G., D. Oppen, "Simplification by cooperating decision procedures", *ACM Transactions on Programming Languages and Systems*, Vol. 1, No. 2, October 1979
[NeO 80]	Nelson, G., D. Oppen, "Fast decision procedures based on congruence closure", *J.ACM*, Vol. 27, No. 2., April 1980, pp. 356-354
[Sho 78]	Shostak, R. E., "An algorithm for reasoning about equality", *C.ACM*, Vol. 21, No. 7, July 1978, pp. 583-585

[Sho 79]   Shostak, R. E., "A practical decision procedure for arithmetic with function symbols", *J.ACM* Vol. 26, No. 2, April 1979, pp. 351-360

[ShS 82]   Shostak, R. E., Richard Schwartz, P.M. Melliar-Smith, "STP: A Mechanized Logic for Specification and Verification", *Proceedings of the Sixth Conference on Automated Deduction*, Springer-Verlag, June 1982

[Tar 75]   Tarjan, R.E., "Efficiency of a good but not linear set union algorithm", *J.ACM*, Vol. 22, No. 2, April 1975, pp. 215-225.

[Wen 78]   Wensley, J., et al., "SIFT: Design and Analysis of a Fault-tolerant Computer for Aircraft Control", *Proc. IEEE*, Vol. 66, No. 10, Oct. 1978

EXPONENTIAL IMPROVEMENT OF EFFICIENT BACKTRACKING A STRATEGY FOR PLAN-BASED DEDUCTION[*]

Tomasz Pietrzykowski
School of Computer Science
Acadia University
Wolfville, Nova Scotia, Canada

Stanislaw Matwin
Department of Computer Science
University of Ottawa
Ottawa, Ontario, Canada

ABSTRACT

The paper presents a method of mechanical deduction along the lines indicated in [3]. Attempts to find refutation(s) are recorded in the form of triples: **plan, constraints, conflicts**. A **plan** corresponds to a portion of AND/OR graph search space and represents purely deductive structure of derivation. **Constraints** form a graph recording the attempts of unification, while **conflicts** identify minimal subset of the plan, removal of which restores unifiability.

This method can be applied to any initial base of (non-necessarily Horn) clauses. Unlike the exhaustive (blind) backtracking which treats all the goals deduced in the course of proof as equally probable source of failure, this approach detects the exact source of failure.

In this method only a small fragment of solution space is kept on disk as as a collection of triples. The search strategy and the method of non-redundant processing of individual triples which leads to a solution (if it exists) is presented. This approach is compared - on a special case - with blind backtracking and an exponential improvement is demonstrated.

---

[*] This work was supported by National Sciences and Engineering Research Council of Canada grants A5111 and E4450.

I. Introduction

There are two major problems with the mechanical deduction systems based on depth-first search of the solution space. The first problem occurs when an infinite branch is generated during the proof. It is a well known fact that detection of such a situation is undecidable.

The second problem involves a procedure called "backtracking", which is invoked whenever an unsolvable subproblem is encountered. It seems that an efficient backtracking strategy is crucially important in a wider acceptance of otherwise attractive Logic Programming. Foundations of such an efficient algorithm - choosing an alternative action in the case when deduction fails - have been outlined in earlier papers by P.T. Cox and T. Pietrzykowski [3]. The system described is a natural continuation of their work; also some of the results presented here are stronger than the ones described in [4]. However, familiarity with [3] and [4] is not necessary: all the notions and definitions are presented here independently in a slightly different style.

The simplest backtracking strategy is embodied in the linear resolution method ([5], [6], [15]) implemented in well known PROLOG systems ([8], [13], [12]). In this method all subproblems, which have to be proven in order to find a deduction for a given problem, are linearly ordered. This means that if a failure occurs during solving of a subproblem, a PROLOG system will discard all the deductions between the point of failure and the last (in the sense of the linear order of subproblems) such subproblem for which there is an alternative solution. As there is no real reason to expect that such a procedure will lead the system directly to the source of failure of the proof, the name **blind backtracking** seems to suit well that method. A convincing example of unefficiency of this approach is contained in ([9], Ch. 7).

There exist two other approaches to the problem of locating more efficiently the source of failure of deduction. The first idea comes from Pereira ([8]). His approach is based on an algorithm which, by use of AND/OR trees, guides standard backtracking only to those alternative proofs where repetition of deduction failure may possibly

be prevented.
However, the choice of alternatives is restricted by an a priori order. Also, this approach does not separate the deductive part of the proof from its unification part, which during backtracking requires regeneration of the already computed unifications. Finally the above approach can be applied only to proofs based on Horn clauses.

The other approach to the backtracking problem has been first described in [3]. The idea is based on the separation of deductive structure of the proof, recorded in the special graph called **deduction plan,** from the unifications obtained during the proof and recorded in another graph, **graph of constraints.** In this paper we further develop ideas of [3] and [4] and rely only on their theoretical results. The approach of [3] does not address the key questions: how to organize a strategy of removing sources of non-unifiability from the plan, and how to keep record of such actions without loosing completeness of the search. A solution of these problems is presented in this paper. Finally, it should be mentioned that our approach is applicable not only to Horn clauses as Pereira's but to full predicate calculus.

The paper is organized as follows: in the next section we describe the basic concepts and present the overall idea of the system. In the third section we show by means of an example the underlying principles of operation of the system and compare them with other strategies of searching the solution space.

II. Basic Concepts of Plan Formation

Example 2.1. Consider the following set of clauses:
$$P(x) \quad Q(f(x,y)) \quad R(y)$$
$$-P(u) \quad V(g(u))$$
$$-Q(w)$$
$$-R(z) \quad S(z) -T(z)$$
$$-S(a)$$
$$-T(b)$$
$$-V(u) -P(g(u))$$

(For the standard definitions of literals, clauses, unifiers and other basic notions of mechanical deduction - see e.g. [2]).

From now on, the initial set of clauses will be called the **base.**

Please notice that the system described here is not restricted to Horn clauses.

Figure 1 shows a possible **deduction plan** for the base of Example

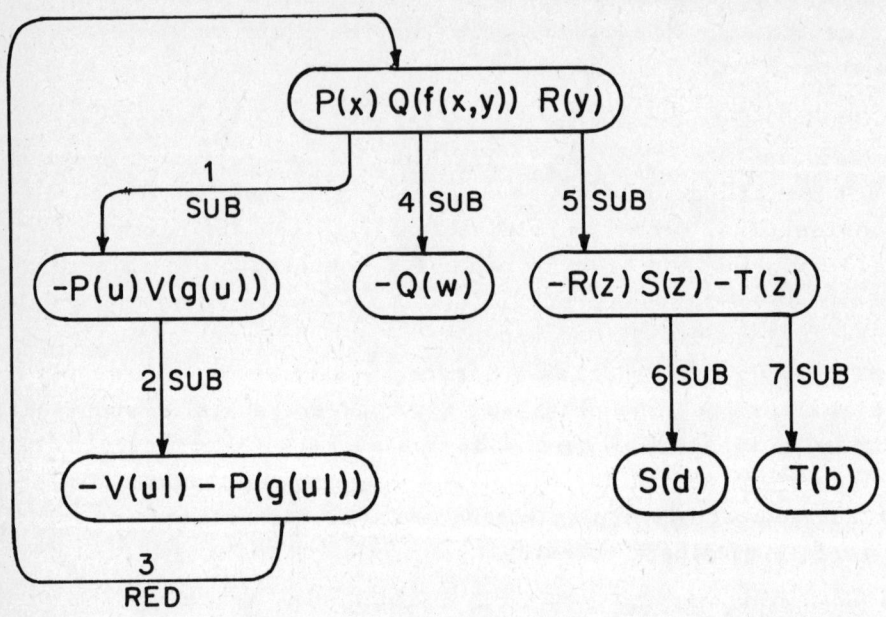

Fig. 1. A possible deduction plan for the set of clauses of Example 2.1.

A deduction plan is a directed graph the nodes of which are variants of clauses, so that the variable names in distinct nodes in the plan are distinct. The arcs of the plan are labelled by type which is SUB or RED, and by a pair of literals which are of the opposite sign and unifiable. Literals of such a pair are called **matching**. However, for graphical presentation, instead of explicitly listing the pair of literals labelling arcs, each arc starts under a specific literal of a node and points to a specific literal of another node. A literal of a node is called **key** if it is the second of literals which label the SUB arc ending in this node. Any other literal of this node is called the **goal**. In our graphical presentation the key is the literal which is pointed to by the arc.

If we ignore all the arcs labelled RED, the deduction plan is a tree. Let us call this tree a **SUB-tree.** Therefore, the notion of **ancestor** and **successor node** is well defined:the SUB arcs, called **substitutions,** go from ancestor to successor and the RED arcs, called **reductions,** go from successors to ancestors. For every pair of ancestor-successor nodes, the goal of the ancestor from which the path to the successor starts is called the **ancestor goal** of the successor node. In sequel we call the ancestor goal pointed by arc RED, and key literal pointed by arc SUB, the **target literal** of goal. Finally, no arcs start in the key and every RED arc ends in an ancestor goal. Each node of the plan, except the root of the SUB tree, has exactly one key. The root mentioned above has only goals.

As this descriptive definition appears to be quite different from the one in [3], we shall briefly list the discrepancies and establish correspondence between terms used in both papers. In [3] nodes were literals; here they are variants of clauses.The arcs of [3] were labelled by SUB,REPL,RED and FACT;in our approach, they are labelled by SUB,RED. Our SUB arcs combine SUB and REPL rules of [3] and correspond to the simple replacement case of the replacement rule of [3]. The nodes of [3] are literals: here the nodes are clauses. All the nodes of [3] connected by SUB arcs in the sense of [3] collapse here into one node. We also require that the literals which are connected in [3] by REPL and RED arcs are of opposite sign and unifiable.The RED arcs, as defined here, correspond to the reduction (RED) rule of [3]. The ancestory replacement and factorization rules of [3] are not used here. However, as proved in [3] the set of rules consisting of SUB and RED, as defined in our approach, is complete.

As stated above, the literals linked by plan arcs are unifiable. The unifiers are combined and recorded in the form of the **constraint graph,** where by a **constraint** we understand a set of variable and function names. Any two different constraints have no variables in common. The constraints are in turn linked into a directed graph. Existence of an arc in the graph means that a variable or function symbol from the tail of this arc is an argument of a function symbol from the head of this arc. Fig. 2 shows the constraints graph for the plan of Fig. 1. The directed arcs represent, as described above, the function-argument relationship between constraints. The undirected arcs with numbers show which plan arcs induced the sets of constraints; these arcs are not a part of the graph of constraints and have been introduced for illustration purposes only.

In general, we call a constraint graph **unifiable** (in [3] a plan which induced such graph was called correct) if no node of the graph contains more than one symbol and there are no circuits in the graph.

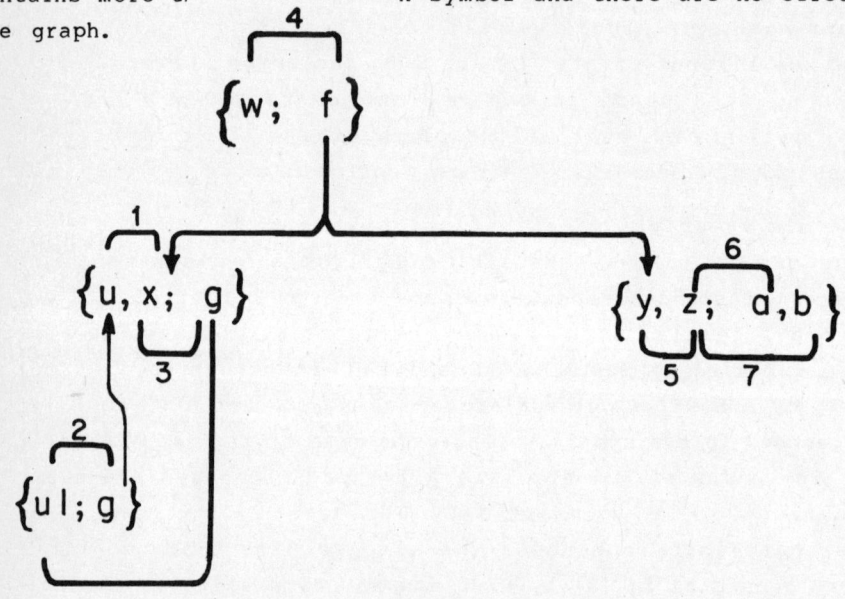

Fig. 2. The constraints graph for the plan of Fig. 1.

Constraint graph of Fig. 2 contains constraints reflecting two types of nonunifiability: one occurs when two different function symbols belong to the same constraint set (as in {y,z;a,b}). Nonunifiability of the other type occurs when a variable is unified with the formula within which it occurs (e.g. u is unified with g(u)). As our task is to obtain a (closed) plan with a unifiable constraint graph, we will have to modify the plan so that the induced constraint graph will be unifiable. Another data structure, the **conflict set**, will be used for that purpose. The elements of the conflict set, called **conflicts**, are conjunctions of plan arcs. Removal of **any** conflict (that is: all the arcs which belong to it) from the plan induces a new, unifiable constraint graph in which some previous constraint sets are split or disappear so that nonunifiability is removed. Conflicts are minimal sets of arcs satisfying the conditions described above. In general, there are many conflicts in the conflict set. In the example of Fig.2 we have the following conflicts: 1^5, 1^6, 1^7, 2^5, 2^6, 2^7, 3^5, 3^6, 3^7. Although removal of the conflicts involving arc 5 does not seem to isolate function symbol a from b, it is clear that in the absence of arc 5 constraint set {y,z;a,b} will never be generated. On the other hand, if we choose the conflict 1^6, the circuit (x,3,g->u1,2, g->u,1)

is broken by removal of 1 and function symbols a and b become isolated (as every connection of a and b involved arc 6).

It is easy to notice that each of the above conflicts is a minimal one. However, a further reduction is possible. For example, if we remove conflict 2^6 from the plan of Fig. 1, the resulting plan will be a proper subgraph of the plan obtained by deletion of 3^6. Since our objective is to cause as little destruction as possible, we consider the choice of 2^6 as redundant and also inferior to 3^6. With this in mind we can select from the conflict set the **reduced conflict set**, such that all the remaining conflicts are redundant. In our case the reduced conflict set consist of 3^6 and 3^7. A method of selection is described in section V of [7] and a proof of its completeness will be presented in [11]. However, in order to ensure the completeness of overall search we must provide a possibility of returning later to any of the initially rejected conflicts. That is achieved by appropriate recording of the choices and backtracking.

III. General Rules of Operation of the System

A. How Does the System Operate

Example 3.

1. Consider the following base:

$$
\begin{array}{ll}
P(x)\ Q(y)\ R(x,y)\ V(a,x) & A \\
-P(a) & B \\
-V(t,t) & C \\
-P(c) & D \\
-Q(w)\ V(w,w) & E \\
-R(z,z) & F \\
-R(u,v)\ S(u) & G
\end{array}
$$

```
-S(a)         H
-V(b,b)       I
-V(c,c)       J
```

Fig 3. shows a possible plan for this base. For graphical convenience the nodes are not fully presented, and are replaced by their names: A,B,...,J.

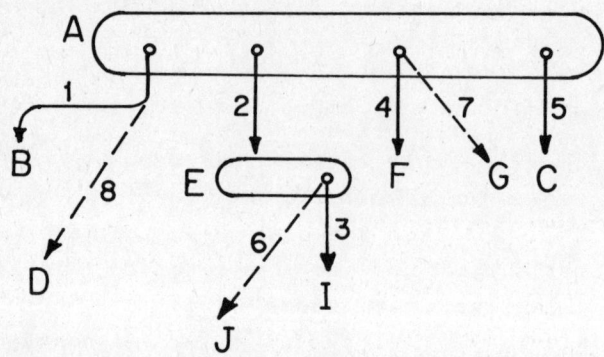

Fig. 3. A possible plan for the base of Example 3.1.
The resulting C plan is called ALPHA.

The reader, as an exercise, can identify remaining key and goal literals corresponding to the arcs of the plan. Arcs are accompanied by sets of possible arcs, denoted by broken lines, which could have been alternatively created instead of the actually generated ones. These sets of arcs are called **potentials**. Therefore D is an alternative for arc 1, G is an alternative for arc 4, J is alternative for 3 while other arcs have empty potentials. In order to build the potentials at every step of plan development, all we have to know are pairs of unifiable literals with opposite signs in the base. This information is static, that is independent of the plan development and is gathered during a special preprocessing phase.

Clearly a plan with potentials can be described in terms of AND-OR graphs. However, we will not use such notation because of its bulkiness in visual presentation. Further, if there is no danger of ambiguity, we shall refer to the plan with potentials briefly as plan.

The triple consisting of the deduction plan together with its constraint and conflict set is called a **C-plan**. In the case of plan presented on Fig.3 the constraints graph is a single node (See Fig. 4). Its reduced conflict set is {1^5,3,4}.

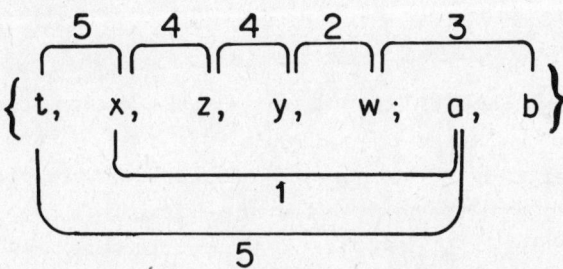

Fig. 4. The constraints graph for the plan of Fig.3.

Now we shall present an outline of the operation of our system. The detailed description is provided in [7]. First we introduce two concepts: **table plans** and **store plans**. Store plans form a collection of C-plans stored on the disk at given time. They provide alternatives in case the operation on the table plan lead to a dead end. Table plan or **table** is a C-plan residing currently in the memory and being processed. (The terms **table** and **store** are motivated by analogy with mass meat preparation. The ingredients (plans) on the table are cut and trimmed, and then kept in the cool storage to be used in the later phase of cooking). We start by developing an initial C-plan. To simplify this informal presentation suppose it is closed, although such an assumption is not necessary and our actual system is free from it. If its set of conflicts is empty, that is the constraints graph is unifiable, we have found a solution. Otherwise we place the inital C-plan in the store and enter the main loop. In our case the C-plan of Fig 3. is of the second category and is put into store. We shall refer to it as ALPHA. The loops consists of numerous actions. First we choose a C-plan from the store and move it to the table. In our case it has to be ALPHA. Since its conflict set is not empty we have to remove one of them. Let us choose a conflict consisting of 3. Now, two actions follow. First we modify the table plan ALPHA by removing 3 from the conflict set and if it is still non empty (which is our case: the conflict set is {1^5,4}) we send it to the store. Let us call this modified plan ALPHA'. Then we start processing the table plan. We remove arc 3 (since it was chosen as a conflict) and replace it by 6 from its potential.

Generally the system removes all arcs from the chosen conflict and replaces each of them with the ones chosen from the corresponding potentials, assuming that they are nonempty. In our example conflict involves a single arc 3, and there is only one potential for 3, i.e. 6. Replacing 3 by potential 6 results in a new constrained plan, shown on Fig. 5. and called BETA.

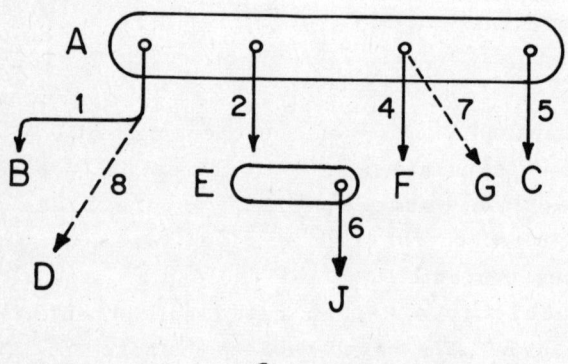

conflict set = {1 ∧ 5, 6, 4}

Fig. 5. C-plan BETA generated from ALPHA of Fig. 3.

Clearly, BETA is nonunifiable and as such is sent to the store. Again one of the plans residing on disk, i.e. ALPHA' and BETA, has to be chosen and placed on the table. Assume for the moment that BETA is fetched and that the arc 6 has been chosen as a conflict to be removed and BETA' is sent to the store. Clearly it was not a good decision, because 6 has no potential and moreover no arc on the path between 6 and A has a potential. Since replacement of 6 is impossible the system has to resort to backtracking. It prunes the path from J to A. Since A is the root, further pruning is impossible and the system comes to a local dead end. Consequently it abandons the table plan by not sending it to the store. It is easy to note that no plan developed from BETA would lead to a solution.

Assume now that ALPHA' is chosen as the next table plan (the conflict set of which is $\{1\char`\^5, 4\}$ because 3 has been removed from it when BETA has been built) and that $1\char`\^5$ is chosen as the conflict. We create ALPHA" with conflict set $\{4\}$ and send it to the store. However on the table we run in the same problem as in the last case. Although we can easily replace arc 1 with D, there is no way of replacing arc 5, because it has no potential. We shall again abandon the table plan residing in memory.

At this point the backup disk storage contains two C- plans, ALPHA" and BETA'. Assume we choose ALPHA", remove arc 4 from its conflict set and create ALPHA"'. Since its conflict set is empty we do not send it to the store: further processing is clearly useless. On the table, however, arc 4 is replaced by its only potential G and developed into C-plan GAMMA shown on Fig. 6.

Removal of 4 splits the constraint $\{t,x,z,y,w;a,b\}$ of ALPHA" so that constants a and b are isolated from each other and there are no more conflicts in the constrained graph. The plan is closed, therefore a refutation has been found.

IV. Comparison with Linear Resolution Method

In order to compare the approach being described here with linear resolution, we shall examine how both methods search a specific solution space.

Example 3.2. Consider the derivation plan of Fig. 7. As we are only concerned with the strategy of search in the solution space, we omit from the presentation the variants of clauses and assume in the remaining part of this section that whenever an arc is added to the plan, such an addition is possible, i.e. the respective literals, not

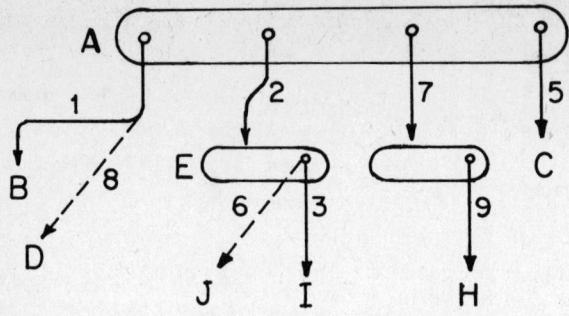

Constraints graph: $\{x, u, t; a\}; \{y, w; b\}$

conflict set: ϕ

Fig. 6. C-plan GAMMA generated from C-plan ALPHA" by removing the conflict 4.

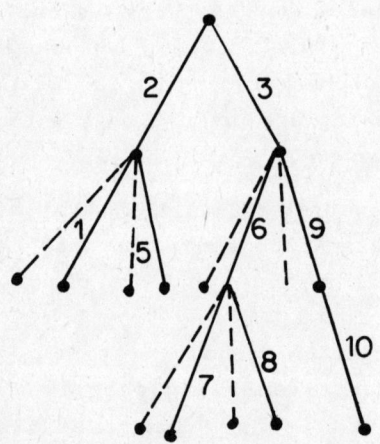

Fig. 7. A schematic plan illustrating the behaviour of linear resolution.

In the remainder of this section the potentials for the arcs of the plan will be denoted by the name of the actual arc with a "'" added, so that 9' denotes a potential for arc 9. Let us further assume that in the plan of Fig. 6 the conflict set consists of 10 and 4, but only replacement of 4 leads to a refutation.

Let us examine the behaviour of linear resolution on this example. The derivations will be described by the sequence of arcs with a left to right order. Each derivation is generated by the previous one through linear backtracking to the last arc in the sequence having an untried alternative. Arcs 2 and 3 are omitted as having no potentials:

    4 5 6 7 8 9 10
          9'10
        8'9 10
        8'9'10
    the last 4 lines repeated with 7 replaced by 7'
    the last 8 lines repeated with 6 replaced by 6'
    the last 16 lines repeated with 5 replaced by 5'
    and finally: 4'5 6 7 8 9 10 refutation!

Using the approach presented in this paper the history of deriving the solution will look as follows:

    4 5 6 7 8 9 10
         9'10
    4'5 6 7 8 9 10 refutation!

By comparison, with the blind backtracking approach presented above 168 deduction steps will be carried out while our method will involve only 9 deduction steps. Please note that in the example of application of intelligent backtracking we have chosen the **worst** case, i.e. starting with conflict 10 instead of 4.

In general, we may observe that linear resolution with blind backtracking will exhaust the space between the conflict and the real source of derivation failure, while the approach described here will only search these parts of the space which lay **above** the conflicts. Therefore our approach will never do more backtracking than linear resolution method; moreover for the class of derivations on which blind backtracking exhibits its worst performance, our approach constitutes an improvement of order exponential in number of backtracking steps.

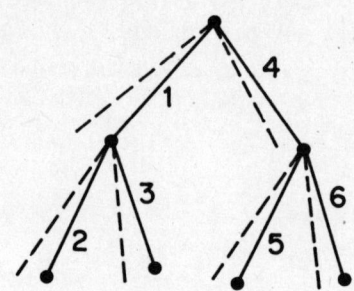

Fig. 8. A schematic plan, illustrating the worst case of linear backtracking.

Without going into details which will be fully developed in a separate paper [11], we shall justify improvement over the worst case of blind backtracking on the example of plan of Fig. 8. Using the notational conventions applied earlier in this paper, please examine the derivations produced by linear resolution with blind backtracking (conflict set = {2,6} and solution requires removal of 2):

derivation	number of deduction steps	
1 2 3 4 5 6	5	original deduction
6'	1	
5'6	3	$= (2^1 + 1)2^0$
6'		
4'5 6		
6'	10	$= (2^2 + 1)2^1$
5'6		
5'6		
3'4 5 6		
6'	36	$= (2^3 + 1)2^2$
4'5'6'		

        2' 3 4 5 6   (refutation!)

It is now clear that the number of backtracking steps of linear resolution may be expressed by the function of order:
$$O(2^{2i})$$
where i is the number of arcs between the conflict and the source of failure. In the class of derivations realizing the worst case of linear resolutions i=O(n), where n is the number of arcs in the plan, therefore the complexity is $\emptyset(2^{2n})$. Let us now examine the derivations generated by the approach presented in this paper, assuming the **worst** choices:

derivation	number of deduction steps
1 2 3 4 5 6	6 (original plan)
6'	
4'5 6	3
6'	1
2'3 4 5 6	1 (refutation)

It should be noted that the backtracking takes place only **above** conflict and therefore the number of backtracking steps in the worst case may now be characterized by the function:
$$O(2^{2k})$$
where k is the depth of the plan. Clearly $n = \emptyset(2^k)$ (n being the total number of arcs) what means **exponential** improvement over the worst case of linear resolution.

In a subsequent paper ([7]) we compare backtracking process involved in an execution of a PROLOG program with backtracking actions which our system will undertake for the same program. In the same paper we also outline the data structures consisting of the base of the system and discuss the method of implementation.

BIBLIOGRAPHY

[1] M. Bruynooghe, L. M. Pereira, "Reivision of Top-Down Logical Reasoning Through Intelligent Backtracking", Centre de Informatica da Universidade Nova de Lisboa, 1981.

[2] C. L. Chang, R. C. T. Lee, "Symbolic Logic and Mechanical Theorem Proving", New York, Academic Press, 1973.

[3] P. T. Cox, T. Pietrzykowski, "Deduction Plans: A Basis for Intelligent Backtracking", IEEE Trans. on Pattern Analysis and Machine Intelligence, Vol 3 N. 1, Jan. 1981.

[4] P. T. Cox, "Representational Economy in Mechanical Theorem-Prover", Proc. 4th Workshop Automat. Deduction, Austin TX Feb. 1979.

[5] D. W. Loveland, "A Linear Format for Resolution", Lect. Notes on Mathematics 125 (Symposium on Automatic Demonstration), Berlin, Spring Verlag 1970.

[6] D. Luckham, "Refinement Theorems in Resolution Theory", Lect. Notes on Mathematics 125 (Symposium on Automatic Demonstration), Berlin, Spring, Verlag 1970.

[7] S. Matwin, T. Pietrzykowski,"Plan based Deduction: Data structure and Implementation", this volume.

[8] L. M. Pereira, F. C. N. Pereira, H. D. Warren, "User's Guide to DEC System-10 Proglog", Laboratoria Nacional de Engenharia Civil, Lisbon.

[9] L. M. Pereira, A. Porto, "Intelligent backtracking and Sidetracking in Horn clause programs - the theory", Departmento de Informatica, Universidade Nova de Lisboa, 1979.

[10] L. M. Pereira, A. Porto, "Selective Backtracking for Logic Programs", Lect. Notes in Computer Science 81 (5th Conference on Automated Deduction), Springer-Verlag 1980.

[11] T. Pietrzykowski, S. Matwin, "Linear Complexity of Efficient Backtracking in Plan-Based Deduction", in preparation.

[12] G. Roberts, "Waterloo Prolog User's Manual", University of Waterloo, 1980.

[13] P. Roussel, "Prolog: manuel de reference et d'utilisation", Groupe d'Intelligence Artificielle, Marseille-Luminy, 1975.

[14] D. H. D. Warren, "Implementing Prolog", Department of Artificial Intelligence, Edinburgh University, 1976.

[15] N. K. Zamov, V. I. Sharonov, "On a class of strategies which can be used to establish decidability of the resolution principle", Issled po konstruktivnoye matematikye i matematicieskaye logikye III, vol 16, pp 54 - 64, 1969.

EXPONENTIAL IMPROVEMENT OF EXHAUSTIVE BACKTRACKING:
DATA STRUCTURE AND IMPLEMENTATION[*])

Stanislaw Matwin
Department of Computer Science
University of Ottawa
Ottawa, Ontario, Canada

Tomasz Pietrzykowski
School of Computer Science
Acadia University
Wolfville, Nova Scotia, Canada

ABSTRACT

The paper presents the data structure enabling an implementation of an efficient backtracking method for plan-based deduction. The structure consists of two parts. Static structure combines information about the initial set of clauses and unifiers. Dynamic structure records information about the history of the deduction process (plan graph), about the unifications built in this process (graph of constraints), and about the conflicts encountered. Dynamic structure consists purely of pointers and integers and is extremely economical.

The system operates on units. The units describe attempts to find the solution and reside on disk. A procedure controlling traffic of units between disk and memory and guaranteeing completeness of the search is presented. Some other procedures (development and pruning of plans, development of constraints graph and conflict detection) are also discussed.

The last section presents an example of a problem and follows its solution by a conventional PROLOG interpreter and by our system.

---

[*]) This work was supported by National Sciences and Engineering Research Council of Canada grants NO A5111 and E4450.

I. Introduction

The paper describes underlying data structures and outlines the general approach to implementation of a mechanical deduction system with a backtracking mechanism exponentially better than the traditional, exhaustive and redundant backtracking. This work is a natural follow-up of [9], where basic terms and philosophy of the system are introduced. The example in sec. VI will provide the reader with insight of system's principles and rules of operation.

Problems of implementation of an efficient backtracking algorithm have been discussed earlier in [3], [4] and [6]. However, the method presented in this paper solves the key problem of search methodology, not addressed in [3], and [6], and provides several important improvements over [4] and [6], e.g. order of magnitude reduction of the size of the conflict set, economy and efficiency of variables' representation, and an innovative data structure for dynamic constraints. In addition, our system addresses also the problems of implementation in the presence of storage limitation, by explicitly assigning parts of the underlying data structure to backup storage.

In the following sections we shall describe in more detail the data structure and algorithms implementing the four main parts of the system:
- (i) basic structure
- (ii) plan developer and pruner
- (iii) constraint generator
- (iv) conflict checker

Such a design reflects the idea of separation of different components of the derivation process: clauses and their variants, deduction structure, results of the unification and the conflict set[*]).

The basic criterion in system's design was storage economy and the possibility of parallel processing. The data structures are almost entirely composed of multi-level, multi-linked lists of pointers. In these parts of the system which undergo dynamic growth, i.e. (ii), (iii) and (iv) above, care has been taken to avoid duplication of data using the references to a single copy of this data. This being the prevailing concept, the system achieves almost ultimate storage economy.

II. Basic Structure

Basic structure represents the clauses from the base[1]) and pairs of unifiable literals in the form best suiting fast access to this data by the other parts of the system [2]).

---

[*]) For the descriptive definition of conflict set, see sec. II of [9].

1) For the definitions of base, matching literals, and constraint see Sec. II of [9].

2) The algorithm notation used throughout this paper is Pascal-like and should be self-explanatory, with the exception of two consistently applied abbreviations:
 - if **p** is a pointer variable, pointing to a record type with a field **x**, then - if no ambiguity arises - instead p .x we shall write **p.x**
 - if is a name of the type, then   denotes the type list (d). e.g. **arc_ins** = list(**arc_in**)

 - ∅ denotes both an empty list and a null pointer.

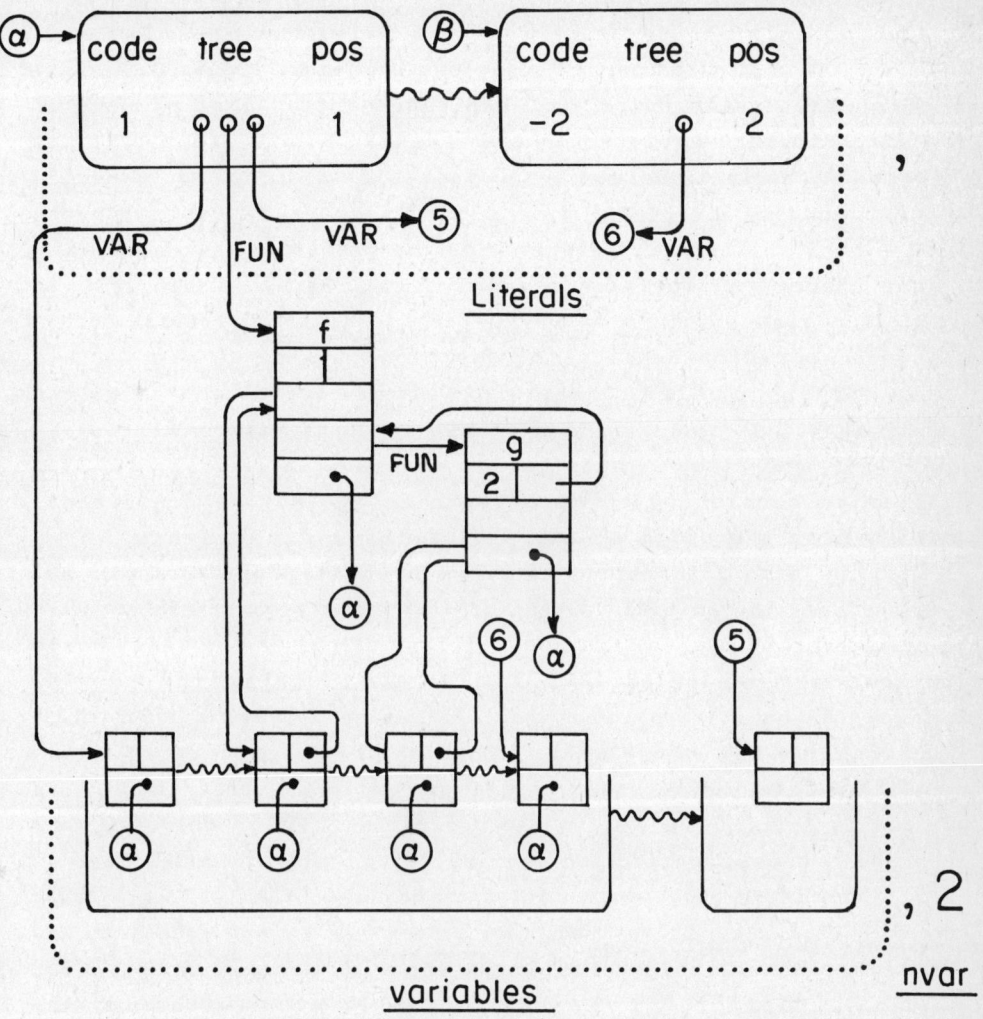

Fig. 1. Data structure, representing the <u>clause</u> for C.

```
basic_structure = clauses
clause = (literals, variables)
literal = (code, trees, position, p-clause, unifs)
```

We shall outline the design of the basic structure on an example of a base consisting of a single clause $C = P(x,f(x,g(x)),y) -Q(x)$. Fig. 1 shows this clause as represented in our system. The list **literals** of this clause has obviously two elements, **P** and **Q**.

Component **variables** of a **clause** represents a list of sets of occurences of every variable in the clause. In our example it has two elements, as in C there are two variables x and y. In general, component **code** of a literal is a signed integer representing the literal symbol and **tree** represents the subexpressions of the literal. Fields **p_clause** and **position** identify the literal within its own clause, as **p_clause** points to the clause in which the literal occurs and integer **position** denotes the position of literal within this clause. Component **unifs** denotes the list of literals matching the given literal (on Fig. 1 field **p_clause** and **unifs** are ommitted, as there is only one clause in the base). **unifs** of a literal is a list of all pairs consisting of matching literal and the constraints resulting from the unification, called **basic constraints**. They represent the same information as **dynamic constraints** (see sec. IV) and in fact are the initial values for the dynamic constraints.

It is worth noticing that the **unifs**, as well as the whole basic structure, are generated once and forever during the special preprocessing run, before the actual derivation process starts. As a consequence, unlike in the well known PROLOG implementations ([7], [11], [12]), only the necessarily dynamic action - constraints' sets merging - is carried out during the derivation. As to the structure of component **trees**, motivation for its design shall become clear in sec. V. Here it is only worth noticing that when the leaf of a tree, which is a variable, is reached, the system has a fast access to any other occurrence of this variable in the clause (through the list **variables**) and to the literal in which it occurs (through the pointers α and β on Fig. 1).

III. Plan Developer and Pruner

Plan[1]) is a directed graph with some special properties,

---

1) For the definition of plan, goal, key, ancestor goal, SUB and RED arcs, potential and dynamic constraint graph, see sec. II of [9].

representing deductive structure of the derivation process. Let us first define the structure and meaning of information, contained in a node:

node = (modifier, p_clause, key, goals)

The fields of this record represent the following information:
- **modifier** is an integer, defined at the time the node is generated and used later, (during unification), to ensure that variables occurring in different nodes are distinct.
- **p_clause** is a pointer to the clause, variant of which is associated with the node. The only information identifying this variant is **modifier** and **p_clause**, the other information in **node** represents the graphical structure of the plan. It is worth noticing that such a design achieves significant storage economy over other implementations of linear resolution.
- **key** represents the key literal in the clause, pointed to by **p_clause**.

key = (k_pos, arc_in)

where k_pos is an integer, denoting the position of the key literal in the clause pointed to by **p_clause**. The immediate ancestor goal of the key literal is identified by the component **arc_in** of the key (for detailed description of **arc_in** see below).

To explain the meaning of the **goal** requires some clarification. Record, representing a **goal** has the following components:

goal = (arc_out, arc_ins, potential)

We also assume that the list **goals** in the node is ordered in the same way as the list of literals in the corresponding clause. The **arc_out** component of a given goal g points to a node n containing target literal for g:

arc_out =± p_node

where '+' denotes a SUB arc, '-' denotes a RED arc. The target literal itself is identified differently for SUB and RED arcs. In case when the arc out is a SUB arc, the target literal in the node **n** is determined by the field **k_pos** of the key of **n**. In case when the arc_out is a RED arc, we have to inspect the **arc_ins** of every goal of **n**.

arc_in = (p_node, g_position)

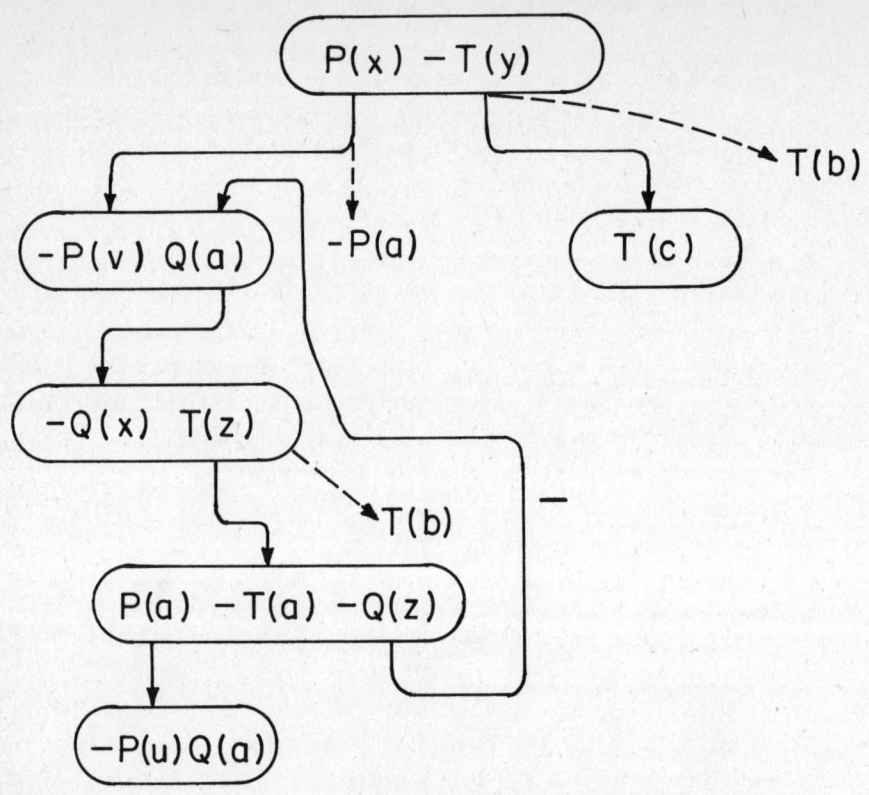

Fig. 2a. Example of a plan as defined in [9] sec. II. Dashed arcs denote potentials.

Besides the above mentioned application, **arc_in** is used for other purposes in plan developer and conflict checker.

Clearly, if there are no RED arcs coming into the goal, the list **arc_ins** is empty.

The last field of a goal, **potential**, is a pair of lists of pointers indicating respectively matching literals in the clauses of the base and literals corresponding to ancestor goals:

        potential = (L_pots, D_pots)
        L_pot   = p_literal
        D_pot   = p_goal

Fig. 2.a shows an example of the plan and Fig. 2.b shows the

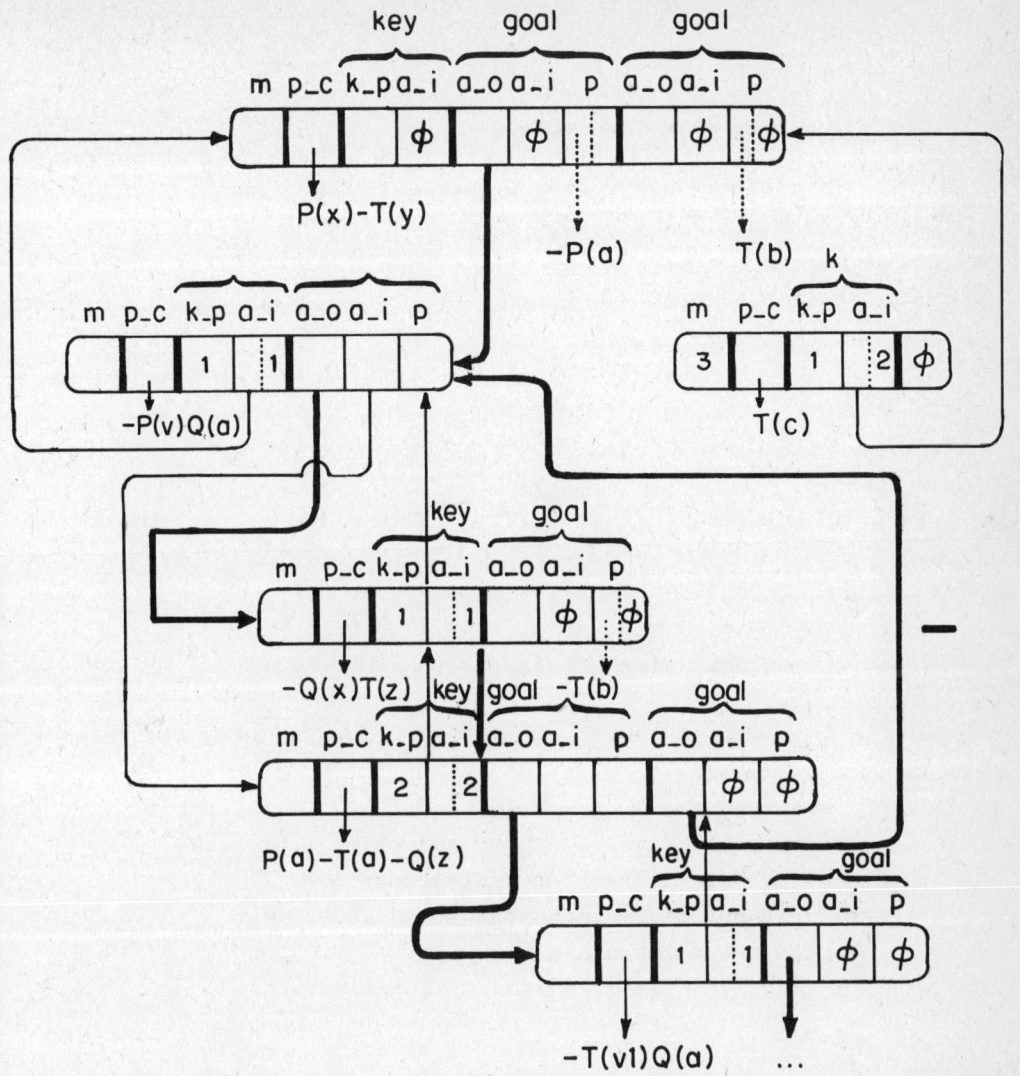

Fig. 2 b. Representation of the plan of Fig. 2a. Arrows in bold face correspond to plan arcs of Fig. 2a. m stands for modifier, p_c for p_clause k_p for k_pos, a_i for arc_in, a_o for arc_out.

representation of the same plan in the sense of the above definition.

As we pointed out earlier, the key idea of the system is separation of the deductive structure from the results of unification, as well as the conflict sets. This basic assumption is reflected in the global data structure, called **unit** (it corresponds to the C-plan in the sense of [9]):

where
$$\textbf{unit} = (\textbf{plan, dcg, conflicts, opens, mod})$$
$$\textbf{plan} = \text{list}(\textbf{node})$$

represents the deductive structure of the derivation process Component **dcg** (dynamic constraint graph) denotes the graph of constraints and is discussed in the next subsection. The **conflict** is a set of arcs which have to be removed simultaneously (conjunction) in order to make the dynamic constraint graph (**dcg**) unifiable. On the other hand, **conflicts** is the list of **all** such conflicts (disjunction).

The field **opens** represents all the open goals, i.e. such goals for which arc_out = ∅.

Finally, integer **mod** is the global variable storing the last unique variable name, represented as an integer.

Now, we can present the main module, acting as the monitor of the whole system. (In accordance with the terminology of [9], the collection of units, which reside on disk in a given time and provide alternatives in case the operations on the table plan lead to a dead end, is called the **store**. The **table plan**, or simply **table**, is the unit residing currently in the memory and being processed.)

```
procedure main (in BS: basic structure;BCG:
   {basic-constraint-graph; out result:BOOLEAN);
   basic-constraint-graph is the result of the
   prepocessing run of the unification module.
   It is used in the constraint generator (module
   add-constraint)}
   local stop:  BOOLEAN;
      SU: = set-of-units; {SU represent the store}
      U: = unit
   begin
      drop:= result: = FALSE;
```

```
SU:= initial unit
   {mod = 0, plan = top node, dcg = BCG, conflict = 0,
    opens = all the literals of the top node}
while SU ≠ 0 and not result do
   {choose a unit in SU, remove it from SU an assign to U.
    Different choice criteria are possible - some will be
    discussed later.  SU resides on disk}
   delete (SU,U);
   if U. conflicts ≠ 0 then remove conflicts (U, drop)
   if not drop then
      develop (U);
      if U.opens = 0 and U.conflicts = 0
      then
         results: = true
      else
         SU:= SU v {U}
   if SU = 0 and not result then
   "no refutation"
end
```

The system stops when either all the units have been processed and rejected or a refutation has been generated. Of course, removal of the condition not result from the outermost loop will generate all possible refutations, which is important for Logic Programming applications. A refutation is obtained if there are no conflicts and no open goals remain in the plan.

In case, however, if there are conflicts in the unit fetched from disk, the procedure **remove conflicts** chooses one of these conflicts and places the unit with the modified conflict set (if not empty) back in store. From the plan remaining on the table all the arcs belonging to the conflict are removed and a new plan is generated. After having removed the arcs of the conflict, the system prunes all the ancestor nodes until it encounters an ancestor goal **g** (of a node) with non-empty potential. During this untypical backtracking, limited to the path between our node and **g** the plan is pruned appropriately. Removal of plan nodes has to be accompanied by changing dynamic constraint graph accordingly. It is worth noticing that **dcg** may be updated by modifying only this part of it which has been affected by removal from the plan of the pruned arcs, instead of recomputing the **dcg**.[*])

The next major action of the main module consists in expanding the plan (procedure **develop**). Development of the plan shall stop in one of the two situations: there are no more open goals or a user-defined interrupt intervenes. An example of a possible definition of such an interrupt is to stop plan development when the last plan arc created adds a constraint containing a function symbol, which in turn may lead to a conflict. Plan development process creates new nodes and arcs, induced by closing of the open goals. For each such goal, one element of its potential is chosen in order to generate the new **arc_out**. One of the possible choice strategies may be to try one of the RED_arcs of the potential first (or last). Investigation of good strategies for plan development, choice of potential and choice of unit from the store will be subject of further research.

IV. Constraint Generator

The task of constraint generator is to integrate constraints, resulting from unification of two literals, with the already existing graph of constraints. As mentioned in sec. II constraints obtained by unifying any two literals are known since the preprocessing phase, as the component **unifs** of a literal. These statically determined graphs of constraints will be referred to as **basic constraints,** or **bc's**.

When a new plan arc connecting literals **L1** and **L2** is generated in the procedure **develop,** the basic constraint **bc1**, belonging to **L1.unif** and representing the unifier of **L1** and **L2,** is converted into **dc1**. **dc1** is a dynamic constraint; at this stage it contains exactly the same data at **bc1** but structured differently and names of variables from the base are modified to agree with the names in the nodes. The most typical operation involved in the integration of a new graph of constraints **dc1** with the existing graph of constraints **dcg** is checking whether there is in **dcg** a constraint **dc2** such that **dc1** and **dc2** have a variable in common. If yes, **dc1** is merged with **dc2** (see [1]).

As this process is involved repeatedly during plan development, much emphasis has been put on its efficiency and storage economy.

---

[*]) In the current version, this feature is not provided and will be the subject of further research.

This is accommodated through a particular design of data structure.

Each variable belonging to a constraint **dc** has a corresponding element in the component **dummies** of **dc**; **dummies** is a circular list of pointers. The dummies, besides pointing to the next element, have also a sign. Exactly one of them is different (has a negative sign). The purpose of this structure is to check whether any two variables belong to the same constraint. In order to accomplish that, we set up a register of all the variables belonging to all the constraints in a form of an array of pointers, each pointing to exactly one dummy. The array is indexed by variable names, represented as integer numbers.

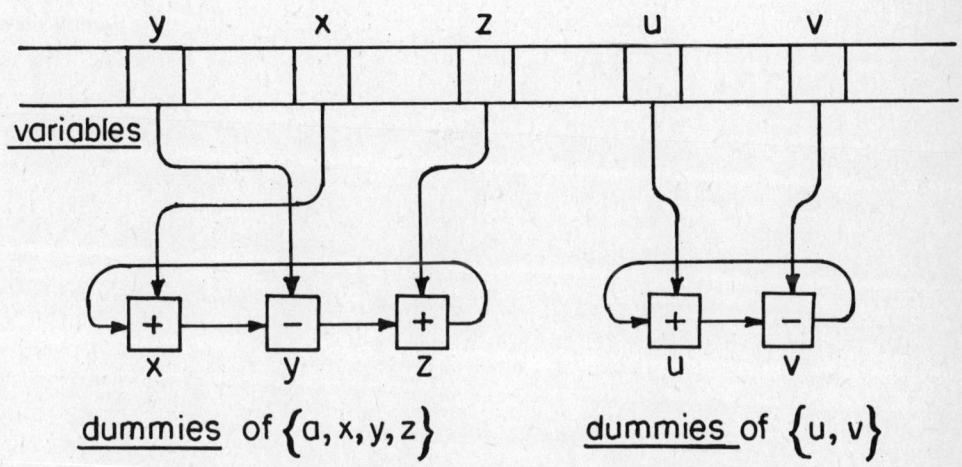

Fig. 3. Graphic representation of coustraints {a, x, y, z} and {u, v} in the gdc.

We shall illustrate this structure on the following example: let us consider two constraints {**a,x,y,z**} and {**u,v**} (a is a constant). Fig. 3 shows mutual relationship of variables and dummies. Suppose that we want to establish whether a pair **z,u** is in the same constraint. Starting at the dummy pointed to by **z** we traverse the list to which it belongs until we reach a negative pointer, i.e. the one cor responding to **y**. Then we do the same for

u, and reach **u**. As **y** and **u** are not the same, **z** and **u** belong to different constraints (since there is exactly one negative element in the list of dummies of a **dc**). The reader may wonder why we have the list of dummies instead of having direct pointers from the array **variables** to appropriate **dc's**. The reason is that merging two dc's would involve search of the full array **variables** in order to redirect corresponding pointers, whereas our approach we will merge two dc's in computationally neglectable time.

The formal description of the whole dc is:

$$dc = (\textbf{dummies}, \textbf{arc\_up}, \textbf{superhead}, \textbf{plan\_links}).$$

Component **arc-up** points to the function symbol in another dc (its parent) and represents the argument-function relationship, analogically as in the basic structure. Component **superhead** represents function symbol (or symbols in the case of non-unifiability) occurring in the constraint set. Component **plan-links**[*]) identifies these plan arcs, which induced the constraint represented by **dc**.

All **dc's**, together with the array **variables**, enter the dynamic constraint graph dcg:

$$dcg = (\textbf{variables}, \text{list}(\textbf{dc}), \textbf{clashes})$$

List **clashes** represents pointers to such dc's superheads of which contain two distinct function symbols, therefore causing non-unifiability.

V. Conflict Checker

The objective of the conflict checker module is to generate the conflict set. Each conflict is a collection of such plan arcs, removal of which from the plan induces a new plan with a unifiable constraint set. In order to define a conflict set, let us define the set **P** of all the paths in the plan, linking any two nodes containing any two functions symbols indicated by the clashes.

The conflict contains two kinds of arcs:
(i) any set of arcs, removal of which breaks all the paths in **P**

---

[*]) The **plan-links** are necessary in order to modify the dcg as the result of pruning. In the current version of the system, they are not yet used.

(ii) any set of arcs, removal of which breaks all the
circuits in **dcg**.

(The conflicts of type (ii), as it is known, are not detected in PROLOG systems). Conflict set is a set of all conflicts. Reduced conflict set contains conflicts, consisting of all the RED-arcs of a conflict and those SUB-arcs which are minimal in the sense of partial ordering in the SUB-tree (see [9] sec. II).

The two following procedures, **walk** and **eliminate**, carry out the task of computing the reduced conflict set.

Procedure **walk** (designed along the ideas of P. T. Cox's dissertation[3]) starts at a given goal g and builds the stack of all the entries corresponding to the goals of the path. Each entry has the following structure:

entry = (element, step_stack)
step_stack = list (**step**)
step = (p_goal, p_node)

where **element** describes the **goal h** of the path which is currently visited and **step_stack** consists of the target goal of **h** (see [9] sec. II), the ancestor goal of **h** if **h** is the key and those goals which point to **h** via a RED arc. We shall refer to these goals as reachable from **h**.

element = (p_arc_down, p_node, traverse, p_step)

Component **p_arc_down** points to the node of the tree in the basic structure, **p_node** indicates the plan node currently reached by the path, **p-step** points to the element of path representing the goal being visited. The role of the last component, **traverse**, will be explained using Fig. 4. Suppose that **p_arc_down** of **element** points to the function symbol **a**. The tree of the subexpression of **P** to which **a** belongs is traversed in the upward direction, and the information about symbols encountered and their occurences as arguments is recorded. Let us notice that the necessary information is available in the basic structure, and in our example we obtain the **traverse** =(a,1,g,2,f). The top of the subexpression tree of **P** is encountered and all the goals reachable from P through arc_out and all the arc_ins of P are stacked. One of them (suppose it is the goal -P determined by the arc_out) is chosen and included in the path. The system then traverses the tree of -P,

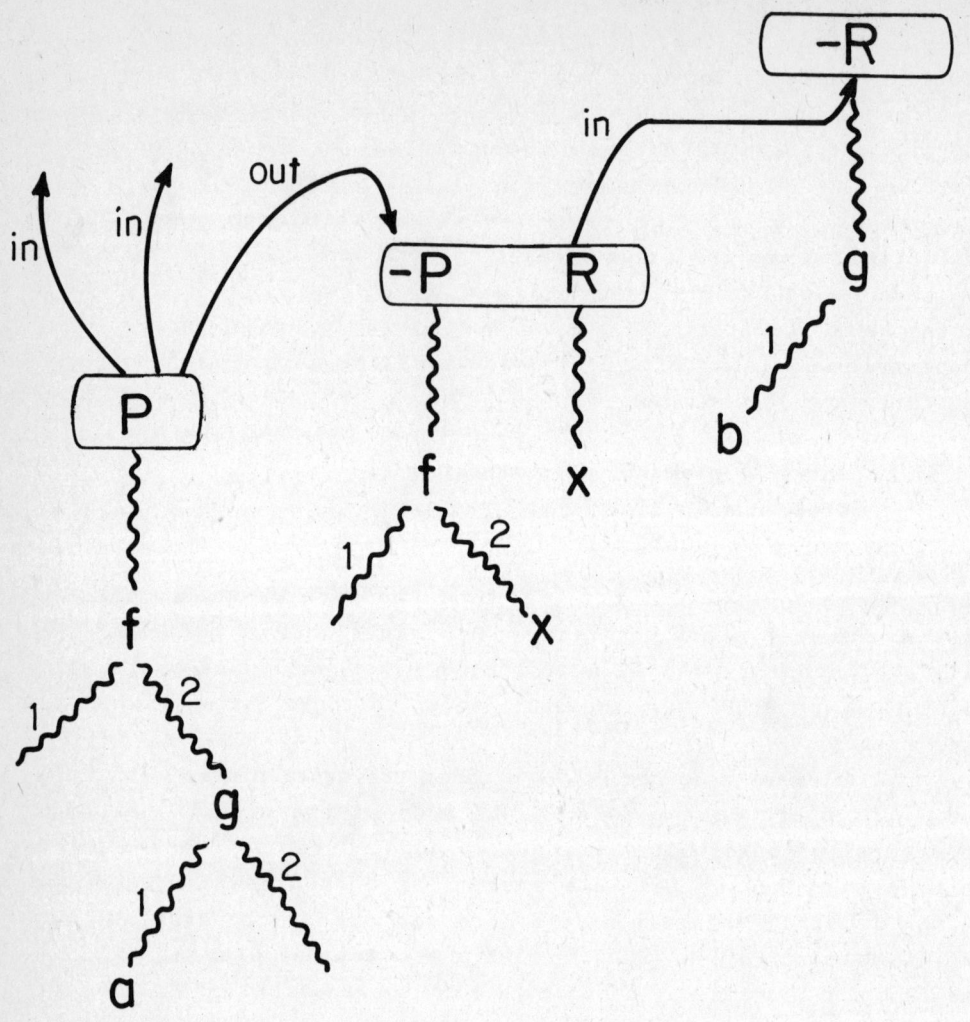

Fig. 4. Fragment of the plan, on which acts the procedure <u>walk</u>. Wavy lines represent the arcs_down of the basic structure.

in the same time erasing them from **traverse**, so that when the variable x is encountered,

traverse = (a,l,g).

All the other occurrences of **x** in the node are then stacked, and one of them (suppose **x** occurs in R) is chosen as the subsequent **element**. The subexpression of R is traversed downwards without any effect (a variable is encountered immediately), all the goals reachable from R are stacked and one them (-R) is chosen. The tree of -R is traversed downwards accordingly to **traverse**, and **b** is encountered instead of **a** which is at the bottom of **traverse**.

Therefore, a conflict has been detected and a path between **a** and **b** is returned by **walk**.

Procedure **eliminates** selects from the path, produced by **walk**, the arcs which subsequently are used in forming the reduced conflict set. On Fig. 5 those arcs of the path which are chosen by **eliminate** are marked by bold arrows. The algorithm used in **eliminate** can be briefly described as follows: suppose that **eliminate** operates on a path Q; it will then return a subset of Q, consisting of only these arcs, which point to plan nodes initiating no arcs from Q.

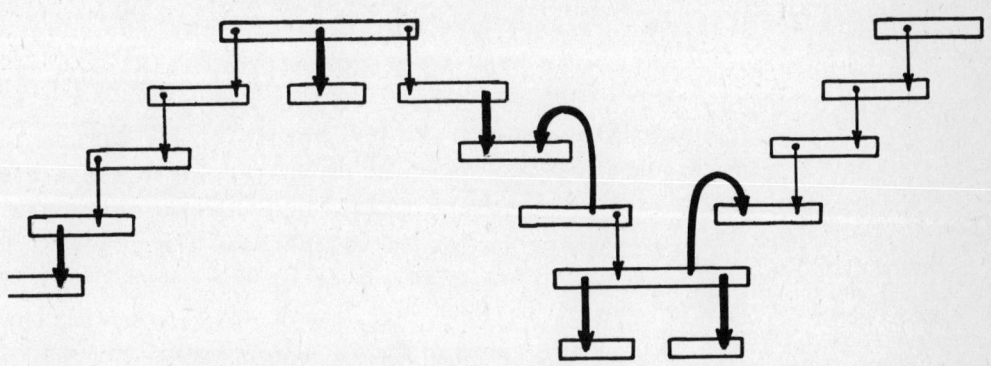

Fig. 5. Only the arcs in bold lines are selected by the procedure <u>eliminate</u>.

VI. Comparison with exhaustive backtracking

VI. Comparison with exhaustive backtracking

In this section we shall show, on an example of a PROLOG program, the differences between exhaustive backtracking, Pereira's intelligent backtracking and the method presented in this paper.

Let us consider a simple relational data base describing, for several European countries, their membership in military alliances, as well as the existence of a common border between some pairs of those countries: [1])

    nato (uk).   **nato (italy).**
    nato (france).   **nato (w germany).**
    wpact (hungary).   **wpact (poland).**
    wpact (ussr).
    na (austria).   **na (sweden).**
    na (switzerland).
    bord (italy, austria).
    bord (italy, france).
    bord (italy, wgermany).
    bord (france, w germany).
    bord (austria, switzerland).
    bord (hungary, austria).
    bord (uk, ireland).

Our objective is to check the existence of the non-aligned country having a border with both a NATO country and a Warsaw Pact country:

  betw(X,Y,Z) :- nato(X), wpact(Y), na(Z), bord(X,Z), bord(Y, Z).

Execution trace of this PROLOG program shows that with exhaustive backtracking 32 deduction steps are necessary to achieve success. Such an amount of backtracking is caused by the fact that, given the order of unit clauses in the data base, UK is picked up as the first NATO country. As UK has no border with a non-NATO country, this choice ultimately leads to deduction failure. However, before the next clause matching **nato(X)** is chosen, all alternatives for the goals following **nato(X)** will be exhaustively tried.

In order to examine the way our system will handle this problem, we shall refer to the constrained plan (C-plan) build during deduction process. Fig. 6 shows the plan at the time when the derivation process has reached for the first time the goal **bord(X,Z)**.

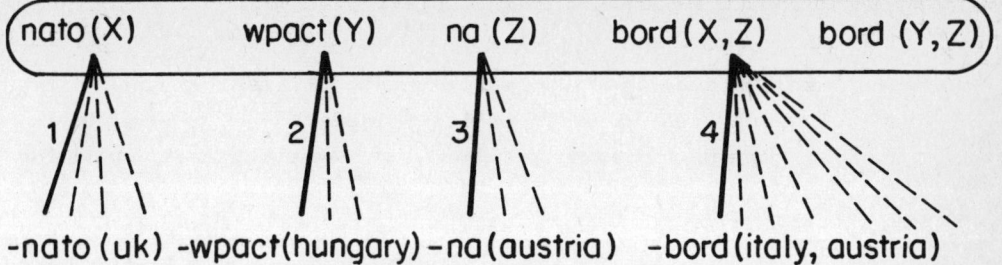

Fig. 6. The plan during derivation process. Broken arcs represent potentials.

Clearly, the constraint graph is then {X;uk,it},{Y,hung},{Z,ans}. The conflict set is therefore {1,4}. As our system does not impose any ordering on conflicts within the conflict set, it is the question of strategy which conflict will be chosen for removal. Let us stress that, at this point, our system differs essentially from Pereira's concept. If our understanding of the paper [8] is correct, in his approach the conflicts are ordered and arc 4 has to be removed first. In our case, if the particular conflict choice strategy used to solve this problem will remove arc 1 first, we shall reach the solution in 9 deduction steps.

Suppose however that our strategy is to consistently choose for removal this among the arcs in the conflict set, which has the most "bushy" potential set, i.e. its potential set is the largest. This means that arc 4 will be the first deleted from the conflict set of the current C-plan (referred to as ALPHA) and replaced by one of its potentials, arc 4'. This will produce another constrained plan ALPHA'. Let us notice that in ALPHA' arc 4' has only five potentials, because original arc 4 has been tried and failed. As ALPHA' has a non-empty conflict set, it will be sent to the store and one of the plans already in the store will be put on the table. Assume ALPHA' is chosen and - following our strategy - 4' is replaced by one of its potentials 4". Again a constrained plan with a conflict is obtained, conflict 4" is chosen for removal etc. Let us notice, however, that after four potentials of the original arc 4 of ALPHA have been subsequently tried and all have failed, arc 1 becomes the most "bushy" element of the conflict set. It is then chosen for replacement, which - as noticed earlier - leads to a success. With the "bushiness" strategy described above, it takes thirteen deduction steps to succeed.

Again, the difference with Pereira's approach is that his method as described in [8] will try all six potentials of arc 4, and take fifteen steps to succeed. The difference of costs between his method and our does not seem to be drastic. However, it has to be noticed that the tree height of the plan in the example above is one. With deeper plans one can expect a larger difference in costs between the two methods.

BIBLIOGRAPHY

[1] L. D. Baxter, "A practically linear unification algorithm", University of Waterloo, CS-76-13, 1976.

[2] C. L. Chang, R. C. T. Lee, Symbolic Logic and Mechanical Theorem Proving", NY, Academic Press, 1973.

[3] P. T. Cox, "Representational Economy in Mechanical Theorem-Prover", Proc. 4th Workshop in Automat. Deduction, Austin, TX, Feb. 1979.

[4] P. T. Cox, "Deduction Plans: A Graphical Proof Procedure for the First-order Predicate Calculus", Ph.D. dissertation, Dept. Computer Sci., University of Waterloo, 1977.

[5] P. T. Cox, T. Pietrzykowski, "Deduction Plans: A Basis for Intelligent Backtracking", IEEE Trans. on Pattern Analysis and Machine Intelligence, Vol.3, Jan. 1981.

[6] D. Forster, "GTP: A Graph-based Theorem Prover", Master Thesis, University of Waterloo, 1980.

[7] L. M. Pereira, F. C. N. Pereira, D. H. D. Warren, "User's Guide to DEC system-10 PROLOG", LNEC, Lisbon, 1978.

[8] L. M. Pereira, "Selective Backtracking for Logic Programs", proceedings of CADE-5, Les Arcs, 1980.

[9] T. Pietrzykowski, S. Matwin, "Exponential Improvement of Efficient Backtracking: A Strategy for Plan-based Deduction", this volume.

[10] G. Roberts, "Waterloo Prolog User's Manual", University of Waterloo, 1980.

[11] P. Roussel, "PROLOG: manuel de reference et d'utilization", GIA, Marseille - Luminy, 1975.

[12] D. H. D. Warren, Implementing PROLOG, Department of Artificial Intelligence, Edinburgh University, 1976.

[13] P. T. Cox, "Deduction plans: A graphical proof procedure for the first-order predicate calculus "Ph.D.dissertation, Dep. Comput. Sci., University of Waterloo, 1977.

INTUITIONISTIC BASIS FOR NON-MONOTONIC LOGIC

by

Dov M. Gabbay
Project Rohrer
University of Stuttgart

and

Bar-Ilan University

Abstract

McDermott and Doyle [4] suggested a system, denoted by $\mathrel{|\!\sim}$, of non-monotonic logic. This notion was intended to formalise non-monotonic reasoning as involved in real situations and in artificial intelligence. McDermott and Doyle also list in their paper several difficulties and problems in their approach. Their semantics seems to be inadequate and there are several counterintuitive results obtained in their system.

McDermott and Doyle base their provability notion $\mathrel{|\!\sim}$ on the provability notion $\vdash$ of classical logic. We introduce in this note two logical systems based on the provability notion $\Vdash$ of intuitionistic logic. We show that in the resulting non-monotonic logic most of the problems disappear. We further show that intuitionistic $\Vdash$ is indeed the reasoning adopted implicitly by available TMS (Truth Maintainance Systems).

§1. Introduction

Classical provability, which we denote by $\vdash$, as well as intuitionistic provability, which we denote by $\Vdash$, are monotonic notions. This means they satisfy the rule:

(1.0) $\dfrac{A \vdash Y}{A, B \vdash Y}$

In other words whatever can be obtained from assumption A can still be obtained if we add to A more assumptions.

In many cases, reasoning is non-monotonic. Consider the story below of the World-Travel-Agency (WTA). The story shall serve as the basis for introducing a system of non-monotonic logic.

WTA offers package tours from New York to Paris and Stuttgart. Every Sunday, a group of New Yorkers board a WTA plane in New York, some get off at Paris and the rest continue to Stuttgart. The Stuttgart office arranges for hotels and tours in Europe. Our non-monotonic reasoning will be done by the Stuttgart office.

At $12^{\underline{00}}$ Stuttgart time, 3 hours after the scheduled New York take-off of the WTA group, the Stuttgart office learns that a terrorist attack was launched at Paris Airport, 2 planes were hijacked and that all flights nearing France are rerouted.

Stuttgart has to decide what to do with the WTA-group. All telephone lines to New York are busy. The only established fact is that the WTA-plane did take off (Some passenger's wife happened to leave a message).

Stuttgart reasons that in this case the WTA-plane will be rerouted to London, the Paris passengers will stay in London and the Stuttgart passengers will arrive tonight. This reasoning is performed on the basis of knowledge of WTA procedures and connections.

(1.1) Thus the reasoning at 12^{00} Stuttgart time is as follows:
- (a) <u>Established as true at 12^{00} hours</u>:
    p = Group flight took off.
    q = Terrorist attack on Paris airport.
- (b) <u>Assumed true at 12^{00} hours on basis of company procedures, and no evidence to the contrary</u>:
    r = Flight rerouted through London.
- (c) <u>Conclusion at 12^{00}</u>:
    C = The Stuttgart passengers are arriving tonight from London.

Two hours later, at 14^{00}, Stuttgart manages to establish connection with New York. They learn that the plane was two and a half hours late in take-off, mainly because a new large group of Paris passengers was added. The New York office did not know what the captain would decide to do in midair, when hearing of the terrorist attack. Again, based on company policy and without evidence to the contrary, that the Stuttgart office assumes the captain would return the short distance to New York and let the passengers spend the night at home.

(1.2) Thus at 14^{00} we have:
- (a) <u>Established as true at 14^{00}</u>:
    p,q and
    p' = Take-off delayed in New York. A large number of Paris passengers is on flight.
- (b) <u>Assumed true on the basis of company procedures, and no evidence to the contrary</u>:
    r' = Flight returning to New York.
- (c) <u>Conclusion at 14^{00}</u>:
    -C: Stuttgart passengers not arriving from London tonight.

To formalise the above Stuttgart reasoning, we need three notions, besides the usual connectives $-, \wedge, \vee, \rightarrow$.

(1.3a) Notion $\vdash^*$ of deducibility, the nature of which to be determined.

(1.3b) A notion of <u>Having established A, we further assume B, on the basis of our knowledge of the way our world runs under normal conditions and no evidence to the contrary</u>.

(1.3c) A notion $\models^*$ for the non-monotonic reasoning involved, which is somehow defined using (a) and (b) above and possibly other considerations.

The chain of reasoning in this particular story can be represented as follows:
(1.4) At 12^{00}:
(a) We know $p \wedge q$, we assume or expect r.
(b) $(p \wedge q \wedge r) \models^* C$.
(c) $p \wedge q \models^* C$.

(1.5) At 14^{00}:
(a) We know $(p \wedge q \wedge p')$, we assume or expect r'.
(b) $p \wedge q \wedge p' \wedge r' \models^* -C$.
(c) $p \wedge q \wedge p' \models^* -C$.

It is clear from (1.4c) and (1.5c) that $\models^*$ is non-monotonic.

McDermott and Doyle [4] introduce the extra connective M into the language and read MC as *it is consistent to assume C*.

The main formal equation for M, besides properties following from its intuitive meaning is:
(1.6) $A \models^* MC$ iff $A \not\models^* -C$

McDermott and Doyle take $\models^*$ to be classical logic provability $\vdash$. They define $\models^*$ from $\vdash$ using M in a certain way.

Thus 1.4a above is formalised in their system as $p \wedge q \vdash Mr$ where $\vdash$ is classical provability.

Our approach, to be introduced in detail in section 8, is to introduce a binary relation A>B reading __B is expected on the basis of A__, and some intuitive axioms for >. We take $\models^*$ to be intuitionistic provability $\Vdash$ and define $\Vdash^*$ using $\Vdash$ and >. Thus for example 1.4a is represented as $(p \wedge q) > r$ in our system with >.

The exact details and the justification for our approach are given in section 8.

The essential difference between us and McDermott and Doyle is in basing the non-monotonic system on intuitionistic logic. Whether one uses M or > is not so important, as differences can be compensated in the manner of defining $\models^*$ from $\models^*$ and M or from $\models^*$ and >.

We are going to introduce two systems of non-monotonic logic. One, called μ, which is essentially McDermott and Doyle type (using M) based on intuitionistic logic and another, called γ, is also based on intuitionistic logic but with > as primitive.

We proceed as follows. In §2 we study the difficulties in McDermott and Doyle's system. §3 is a general discussion; §§4,5,6,7 study μ. §8 introduces the system γ with >; and §9 discusses the connection with McCarthy's circumscription, and other general remarks.

§2. **The need for an intuitionistic basis.**

McDermott and Doyle formally add to the language of classical logic the additional operator M. They use ⊢ to define the notion |∼ of non-monotonic provability in a certain way.

(2.0) Here are some examples of their deductions, given in their paper. PC stands for the classical predicate calculus.

(a) The theory T1 = PC ∪ {MC→-D, MD→-C} non-monotonically proves MC∨MD.

(b) The theory T2 = PC ∪ {MC→-C} is non-monotonically inconsistent.

(c) The deduction theorem does not hold for non-monotonic logic.

(d) The theory T3 = PC ∪ {MC→C} proves non-monotonically exactly C (which is not classically provable).

(e) The theory T4 = PC ∪ {∀x[MP(x) → [P(x)∨(∀y ≠ x)-P(y)]]} proves non-monotonically ∃xMP(x).

(f) The theory T6 = PC ∪ {MC → -C,-C} is non-monotonically consistent and proves -MC, but its subtheory {MC → -C} is not consistent.

McDermott and Doyle in [4], page 69, list the following difficulties with their logic as evidence that it "fails to capture a coherent notion of consistency".

Difficulty A: The theory T13 = PC ∪ {MC → D,-D} is inconsistent in their logic, mainly because -C does not follow from T13, thus allowing MC to be assumed. They say in [4] that "this can be remedied by extending the theory to include -C, the approach taken by the TMS, but this extension seems arbitrary to the casual observer".

We therefore want that our system of non-monotonic logic will prove:
   -D, MC → D |∼ -C.

Difficulty B: They continue to say: "Another incoherence of our (i.e., McDermott and Doyle) logic is that consistency is not distributive; MC does not follow from M(C∧D) ".

Difficulty C: Another difficulty is that the logic "tolerates axioms which force an incoherent notion of consistency, as in T14 = PC ∪ {MC, -C}.

Let us see what is required to remedy the above difficulties:

From the elimination of Difficulty A we want the validity in the non-montonic logic of the following inference:

$$\frac{MC \to D, \; -D}{-C}$$

or equivalently:

$$\frac{-D, -D \to -MC}{-C} \quad .$$

Since D is arbitrary, the rule really means:

$$\frac{-MC}{-C}$$

(One can obtain this also by taking D = MC.)

263

The second difficulty leads to the rule
$$\frac{M(A \wedge B)}{MA}.$$
The third difficulty requires that the set {MC,-C} be inconsistent. We can thus take the rule:
$$\frac{-C}{-MC}.$$
This rule seems intuitive. If we are given -C it is indeed not consistent to add C. In fact, one of the goals of TMS is to prevent both -C and MC being "in".

The above already shows that classical reasoning is not involved with the intuitions behind M. MC and C are equivalent, according to the above rules, if understood classically, and it is not surprising that they are not valid in [4].

Furthermore, let us see what TMS does in case of contradiction. We quote page 67 of [4].

"For example, the existing theory may be {MC→E} in which both MC and E are believed. Adding the axiom MD → -E leads to an inconsistent thoery, as MD is assumed (there being no proof of -D), which leads to proving -E. The dependency-directed backtracking process would trace the proofs of E and -E, find that two assumptions, MC and MD were responsible. Just concluding -MCv-MD does no good, since this does not rule out any assumptions, so the TMS adds the new axiom E→-D, which invalidates the assumption MD and so restores consistency".   End of quote.

We thus have the rule:
$$\frac{MC \to E,\ MD \to -E}{E \to -D}.$$
This deduction is intuitionistic.

We think that at this stage it is safe to conclude that there is enough evidence for trying to base non-monotonic logic on intuitionistic logic and see whether we can get a satisfactory system!

3. Caution!

The construction of McDermott and Doyle's system, as indeed the construction of any logical system stems from two basic considerations.

(3.1) The intuitive interpretation and intended meaning for the notions to be formalised.

(3.2) A partial list of theorems and non-theorems that the expected logic must have.

We have the above for non-monotonic logic.

We have argued that it is worthwhile checking whether non-monotonic logic can be based on intuitionistic logic. We have to be careful in our examination of the new candidate.

It is not enough to say: Here is a logic that can derive what we want to have as theorems and keeps out what we don't want to get (i.e., a system for M compatible with (3.2)). It must also be the case that a natural interpretation for the new system is also available, which is compatible with (3.1).

To give an example from tense logics (see [1]), suppose we introduce the operator Nq for the English progressive reading: "q is true now and has been true a little bit before now and will continue to be true a little bit in the future of now".

This is the main property of the progressive in English:
For example:

Nq = "He is writing now",

has this reading. If we write down the formal properties of the progressive N, we obtain:

⊢Nq→q

⊢Nq→NNq

⊢N(p∧q)↔Np∧Nq.

We can come to you and say: We know of such a system. It is one of Lewis' modal systems called S4 (for necessity $\Box$). Its interpretation is:

$\Box$q is true at a certain situation iff q is true in all resembling situations. So $\Box$ may have all the properties of N and hence satisfy (2) but it is wrong to say that the progressive is really a sort of a modal necessity operator and its logic is the logic of necessity. The flaw is that it may not satisfy (1). To show its weakness, suppose we want to talk about N'q, which is not "He is writing" but "He has just started writing". N' is very similar to N. Can you find a $\Box$' very similar to $\Box$ to match N'? The notion of "beginning" does not enter at all the interpretation of $\Box$.

So when we present our system μ we must watch that both (3.1) and (3.2) above are well matched. For this reason we start by presenting μ through its semantical interpretation.

§4.. Semantical presentation of μ.

We are looking for a logical system in the language of - (negation), ∧(conjunction), ∨(disjunction), →(if - then), ∀(universal quantifier), ∃(existential quantifier) and the special additional unary connective M.

We want the following to hold for ⊩ , the provability of our system.

(4.0) If A ⊮-p then A⊩Mp.

Mp reads "p is consistent".

Consider a flow of time $(T, \leq)$, where T is the set of moments of time and $\leq$ is the earlier - later relation; it is transitive and reflexive. We write $t \leq s$ to mean, s is in the future of t or s is t itself. Atomic propositions represent unit statements. At each moment of time there are those statements which are known to hold true. As time goes on we learn more and more about the world and more and more statements become true. <u>So the advance of time does not bring new events but more knowledge!</u>

Mp reads: it is consistent to assume at this stage that p is true. So we know it is possible that sometimes in the future we establish p. But we are not sure; it may turn out that -p is established. Thus we have the following properties:

(4.1) If A is established now then A will always be established.

(4.2) If -A is established now then -A will always hold.

(4.3) A→B is established now iff it is already clear now that if we establish A then we must establish B. We may not know yet that A is established but whenever it is, B will also follow.

(4.4) Mp holds now if it is consistent with what we know now that p is true.

(4.5) ∃xA(x) holds now iff for some element a, A(a) holds now.

(4.6) ∀x A(x) is established now iff it is clear now that for any later moment and any new or old element a, A(a) will be established.

Notice that (4.0) above holds in this interpretation.

In fact

A⊩ -pvMp is valid.

(4.7) Formal description of the model μ in the propositonal case:

The language contains -,∧,v,→,M. The structures are of the form (T,≤,h). The function h is the assignment. For atom q and each t∈T, h(t,q) gives a truth value. If h(t,q) = truth and t≤s then h(s,q) = truth.

The function h can be extended on all Wffs as follows:

(a) h(t,A∧B) = truth iff h(t,A) = h(t,B) = truth.

(b) h(t,AvB) = truth iff h(t,A) = truth or h(t,B) = truth.

(c) h(t,-A) = truth iff ∀s≥t (h(s,A) = false.

(d) h(t,A→B) = truth iff ∀s≥t (if h(s,A) = truth then h(s,B) = truth).

(e) h(t,MA) = truth iff ∃s≥t (h(s,A) = truth).

(f) We say A proves B (intuitionistically) (notation A⊩B)) iff for any t,h h(t,A) = truth implies h(t,B) = truth.

The non-monotonic provability notion |⊢~ will be based on ⊩. It will be defined in the end of this section in 4.10.

Example 4.8:

Let us continue to get the feel of the model. The following are valid rules. Remember that the logic is based on intuitionistic logic and so many classical equivalences may not hold!

(a) ⊩ MAV-A

(b) -MA is equivalent to -A.

(c) MA→B is equivalent to -AvB.

(That is, either -A if not then MA can be assumed and hence B.)

(d) A particular case of (c) is: MC→C is the same as -CvC.

(e) CV-C is not a theorem of the logic because it is equivalent to MC→C and it is not a theorem. It says that if C is not proved now then it is inconsistent to assume it!

(f) MC→-C is equivalent to -C.

(g) If A∧Mq ⊩B, we can still find out that -Mq (-q) and so it is quite possible that also A∧-q ⊮B.

Remark 4.9:

We have to be careful how to translate from McDermott and Doyle logic into our logic. Since their logic is based on classical logic and ours on intuitionistic logic, different formulations equivalent over classical logic may not be equivalent in our logic. We must therefore translate the meaning into our system, which amounts to choosing one of the equivalent classical versions.

A simple example is A∨-A. It is equivalent to A→A in classical logic but not in intuitionistic logic.

The logic μ is not really weaker than classical logic. It just affords more opportunities for formulation and therefore is much richer. For example for any A,B of classical propositional logic we have

$\quad$ A⊢ -B $\quad$ iff $\quad$ A⊪ -B
$\quad$ classically $\qquad$ intuitionistically

This shows that intuitionistic logic is really richer, allowing for more play.

The reader can continue to check that many desirable properties of non-monotonic reasoning are available. Note further that the rule $\frac{A}{MA}$ is valid.

Note that the interpretation which was used to introudce the system is quite compatible with the intuitive meaning behind M.

Let us see whether difficulties A,B,C are resolved.

Difficulty A. The rule

$$\frac{-MC}{-C}$$

is indeed valid.

Difficulty B. The rule

$$\frac{M(A\wedge B)}{MA}$$

is also valid.

Difficulty C. {MC,-C} is indeed inconsistent!

Let us see now what the TMS would do in case MC→E and MD→-E. According to our logic μ, this means we cannot have both MC and MD and hence we can add either -C or -D immediately. If we add -D, then we can still hold MC and hence obtain E (compare with the description in section 2, where TMS takes E→-D). Or we can add -E, in which case we can still hold MD and hence get -E. Notice that in our logic MC→E is the same as -C∨E and MD→-E is the same as -D∨-E and so we can also let TMS take on from here.

The logic μ is not exactly the same as McDermott and Doyle's logic. We pay the price for resolving difficulties A,B,C.

For example the theory $\{MC \to C\}$ does not prove C as in McDermott and Doyle. It proves $-C \lor C$ which says either C is determined now or $-C$ is determined now. It does not say which one. McDermott and Doyle's logic would say, since neither is determined, assume MC, therefore C. This is what TMS would do. But this is unjustified since $\{MC \to C, -C\}$ is also consistent and is equivalent to $-C$.

Further Remarks.

Observe that in proposing the logic μ we have only proposed a new basis for non-monotonic logic. We can further strengthen the logic by allowing for a procedure for finding fixed points as proposed by McDermott and Doyle. If we do that, we would get $\{MC \to C\} \Vdash C$ as they do. The difficulties A,B,C will still be removed since they are removed by certain rules. These rules will still be available after the fixed points procedure is added.

Another difference between the two logics is in case $\{MC \to -C\}$. This theory is consistent in our case and proves $-C$. According to McDermott and Doyle, this theory is non-monotonically inconsistent. They also get in their logic that $\{MC \to -C\}$ is inconsistent but if they add $-C$, i.e. have $\{MC \to -C, -C\}$, then what they get is consistent!

We regard this as an undesirable feature and in our logic both theories are consistent and are equivalent to $-C$.

The theory $T4$ of (2.0f), i.e. the theory $\forall x [MP(x) \to (P(x) \land (\forall y \neq x) - P(y))]$ does not prove $\exists x P(x)$ in our logic, but rather $\forall x (P(x) \lor -P(x))$. As we shall see in section 9, this is a McCarthy's circumscription on P as represented in our system.

We are now in a position to define our notion of non-monotonic provability, based on the intuitionistic system N. We denote this provability by $\Vdash$. It is different from McDermott and Doyle's and more like Reiter's [5] default reasoning.

Definition 4.10:

We say that $A \Vdash B$ iff there exists formulas $C_0 = A, C_1, \ldots, C_n = B$, called intermediate stages of the non-monotonic deduction from A to B and set of formulas

$$MX_1^1, \ldots, MX_{k(1)}^1$$
$$\vdots \qquad \vdots$$
$$MX_1^n, \ldots, MX_{k(n)}^n$$

called the extra assumptions (similar to defaults), such that for all $1 \leq i \leq n$, $1 \leq j \leq k(i)$ we have:

$$C_i \land \bigwedge_{j=1}^{k(i)} MX_j^i \Vdash C_{i+1}$$

Example 4.11:

(a) $(MC \to C) \land MC \Vdash C$

Thus $(MC \to C) \Vdash C$ using MC as default.

(b) If $A \wedge Mq \Vdash C_1$
$C_1 \wedge Mr \Vdash B$.
Then by definition
$A \Vdash\!\!\sim B$.
We reason as follows:
With q consistent, we get C_1 from A. Assume Mq, get C_1. Having C_1, r is plausible. Believe r. So since $C_1 \wedge Mr \Vdash B$ we got B. Hence $A \Vdash\!\!\sim B$ non-monotonically.

(c) $(MC \to C) \wedge M \text{-}C \Vdash \text{-}C$
Hence $(MC \to C) \Vdash\!\!\sim \text{-}C$.
We thus get $(MC \to C)$ can non-monotonically prove either -C or C. This means we can take either alternative and go on from there.

Remark 4.12:

We can take McDermott and Doyle's construction of fixed points and obtain, by basing their construction on μ, another non-monotonic provability compatible with theirs. We can denote it by $\Vdash\!\!\sim_{MD}$.

Personally we prefer the default type $\Vdash\!\!\sim$ of 4.10.

§5. Further properties of the system μ.

(I) Let α be a wff built up from $-, \vee, \wedge, \to, M$. Then there exists β_i built up without M and a α^* such that α^* is built up from β_i using $\wedge, \vee, M$ only, with the property $\Vdash \alpha \leftrightarrow \alpha^*$. α^* is said to be in normal form.

To prove this claim we observe that M can be pulled out from under the scope of $\to, -$. The following are valid rules rules to be used:

(a) $\Vdash (x \vee y \to z) = (x \to z) \wedge (y \to z)$.
$\Vdash (z \to x \wedge y) = (z \to x) \wedge (z \to y)$.
$\Vdash -(x \vee y) = (-x \wedge -y)$.

We can thus consider only nested occurrences of the following forms:
(1) $x \wedge My \to z \vee Mv$
(2) $-(x \wedge MA)$.

Notice that the following are further valid rules to be used:
(b) $\Vdash (x \wedge My \to z \vee Mv) \leftrightarrow [(x \wedge -v) \to (-y \vee z)]$. Also
(c) $\Vdash -(x \wedge MA) \leftrightarrow (x \to -A)$.

(II) Let α be in normal form. Then there exists a wff γ without M satisfying
$\Vdash -\alpha \leftrightarrow \Vdash \gamma$.
The proof is long. We do not give it here.

(III) For any α, β without M
$\alpha \Vdash -\beta$ iff $\alpha \vdash -\beta$
(where $\vdash$ is classical provability).

(III) is an important rule, connecting the μ logic with classical logic. It says that without M, we still have classical logic if we want.

§6. **An axiom system for quantified μ with ∀ and ∃.**

The language contains M,-,∧,→,∀,∃,∨.

I. Axioms for wffs without M:
1. A→(B→A)
2. [A→(B→C)] → [(A→B) → (A→C)]
3. A → (-A→B)
4. (A→-B) → (B→-A)
5. A→(B → A∧B)
6. A∧B→A
7. A∧B→B
8. (A→C)→((B→C)→(A B → C))
9. A → A∨B
10. B → A∨B
11. ∀xA(x) → A(y); A(y) → ∃xA(x).

II. Axioms for M:
12. A → MA
13. MMA → MA
14. -(x∧MA) ↔ (x→-A)
15. (x∧My → z∨M⊭) ↔ (x∧-⊬ → -y∨z)
16. ∀x[A∨MB(x)] ↔ ∀x[-B(x)→A]
17. ⁻∃y[A∧MB(y)] ↔ ∀y(A→-B(y))

III. Rules:

$$\frac{\Vdash A, \; \Vdash A \to B}{\Vdash B} \qquad \frac{\Vdash A \to B(x)}{\Vdash A \to \forall x B(x)} \, , \quad \frac{\Vdash B(x) \to A}{\Vdash \exists x B(x) \to A}$$

x not free in A.   x not free in A.

The completeness theorem can be proved for this axiom system and the semantics given earlier.

7. **Further evidence for μ:**

We can give further evidence that μ is a good basis for non-monotonic logic by showing how the rule

$$A \not\Vdash -B \; \text{iff} \; A \Vdash MB$$

is represented in our system μ. We have seen that -B∨MB is valid but that is not exactly the same as the rule above.

Consider the fragment with -,∧,∀ only. Let T be the set of wffs of this fragment. Thus the moments or states of knowledge, in this case, are the formulas themselves. Let A≤B mean ⊩B→A. So if B proves A, it represents a state of knowledge greater and stronger than A.

For any atom q let:
    h(A,q) = truth iff A⊢q.
I.e an atom q is true at a state of knowledge A iff A proves q.
  It can be shown that for any B
    h(A,B) = truth iff A⊩B.
This means for any B, B is true at state of knowledge A iff A proves B.
  The above is exactly the interpretation we want for M. Since
      A⊮-B iff A∧B is consistent
        iff the state of knowledge represented by A∧B is consistent and of
course A≤A∧B. So A⊩MB holds, since, as we have seen, there is a future state
(namely A∧B) in which B is established.
  Under this very natural interpretation,
      A⊮ -B iff A⊩MB
holds.

§8. The system γ with > (preliminary version)
  The language of our system contains the connectives -,∧,∨,→,> and quantifiers
∀ and ∃. The system is based on intuitionistic provability ⊩ and has all the axioms
of intuitionistic logic and the additional defining axioms for >.
  The following is an axiom system for γ:
(8.1) Axioms and rules for intuitionistic logic, namely group axioms I of §6 and
group rules III of §6.
(8.2) Axioms for >.
(a) ⊩(A>B)∧(A>C) ↔ (A>B∧C)
(b) $\dfrac{\vdash A \to B}{\vdash A > B}$; $\dfrac{\vdash A \leftrightarrow A'}{\vdash (A>B) \leftrightarrow (A'>B)}$
(c) $\dfrac{\vdash (A > \text{falsity})}{\vdash -A}$
(d) ⊩ (A>B)∧(A∧B>C)→A>C
(e) ⊩(A>B)∨-(A>B)

(8.3) The following features are <u>not</u> taken as axioms because they are <u>not</u> wanted.
(a) ⊮ (A>B)∧(B>C)→(A>C)
(b) ⊮ (A>B)→(-B>-A)
(c) ⊮ (A>B)→(A∧A'>B)
(d) ⊮ B→(A>B)

  To discuss the axioms for >, note that 8.2a and 8.2b come intuitively from
the meaning of A>B as "B is expected on the basis of A and world knowledge".
8.2c says one cannot expect falsity. This axiom makes > more restrictive than
M because A⊩MB says B is consistent, so it is possible that both B and -B are
consistent with A. A>B chooses only one of the two.

  Axiom 8.2d is a form of transitivity. We reject full transitivity (8.3a)
because it is not intuitive. For example, on the basis of total stockmarket

collapse (=A), we may expect Jones to lose his savings (=B) and on the basis of Jones losing his savings we may expect his wife to give him hell (=C) but we may not expect Jones' wife to give him hell on the basis of total stockmarket collapse. This is a counterexample to 8.3a. It is not a counterexample to 8.2d because we may not expect Jones' wife to give him hell on the basis of total stockmarket collapse and his losing his savings.

The meaning of 8.2e is that $A>B$ is independent of whether we know A is true or not; we can expect a drunken driver to cause an accident independently of whether there are such drivers today. For a similar reason 8.3d is rejected. The fact that B is true does not imply it was expected. The form in which 8.2e is written has to do with the fact that our base is intuitionistic logic. It says that right at the beginning it is established whether $A>B$ or $-(A>B)$.

The rejection of 8.3c has to do with non-monotonicity. 8.3c is an axiom for monotonicity. 8.3b is rejected for the same reason. 8.3b does not allow for our expectation to go wrong.

We now define non-monotonic provability $\Vdash_\gamma$. It should be compared with the non-monotonic $\Vdash_\mu$ of 4.10.

Definition 8.4:

Let $A \Vdash_\gamma B$ iff for some X such that $A>X$ we have $A \wedge X \Vdash B$.

Lemma 8.5:

If $A \Vdash_\gamma B$ and $A \wedge B \Vdash_\gamma C$ then $A \Vdash_\gamma C$.

Proof:

If $A \Vdash_\gamma B$ then for some X, $(A>X) \wedge (A \wedge X \Vdash B)$ by axiom 8.2b we get $(A \wedge X) > B$ and from 8.2d we get $A>B$.

From $A \wedge B \Vdash_\gamma C$ we get that for some Y, $A \wedge B > Y$ and $A \wedge B \wedge Y \Vdash C$. We use 8.2d again to get $A>Y$. But since also $A>X$, we get by axiom 8.2a $A>(X \wedge Y)$.

Thus we have that for some $X \wedge Y$, $A>(X \wedge Y)$ and $A \wedge X \wedge Y \Vdash C$. (since $A \wedge X \Vdash B$ and $A \wedge Y \wedge B \Vdash C$) and hence $A \Vdash_\gamma C$.

Remark 8.6:

The previous lemma allows us to define $\Vdash_\gamma$ without resorting to a sequence of intermediaries $X_1, X_2, \ldots$ as we did in definition 4.10 for $\Vdash_\mu$.

§9. Further remarks

(a) Propositional μ is decidable. The result follows from the decidability of the intuitionistic propositional calculus (see [2]).

We don't know whether propositional γ is decidable; probabliy it is.

(b) μ, as well as McDermott and Doyle's system cannot be based on classical logic. M will collapse (MC = C) if we want to have reasonable rules (no difficulties). The system γ can be based on classical logic. It would not be satisfactory, however, but $>$ will not collapse.

(c) We can represent some cases of McCarthy's circumscription in our system. Take a wff $A(\underline{P})$, where the notation signifies that A "talks" about the property $\underline{P}$. Then circumscription on $\underline{P}$ will yield that exactly those elements x that are said to satisfy $\underline{P}$ do satisfy $\underline{P}$ and no more.

Thus if we are at a stage of knowledge t (which can be taken to be A itself, according to section 7), then if we know $\Vdash \not\vdash \underline{P}(a)$ then we assume by circumscription that $-\underline{P}(a)$. So what we are saying in this case is that

$$\underline{P}(a) \vee -\underline{P}(a)$$

holds.

Remember that in our logic $z \vee -z$ is not a theorem. We have an intuitionistic basis!

Thus

A + McCarthy's circumscription $\vdash$ B

is the same as

A + $\{\forall x [\underline{P}(x) \vee -\underline{P}(x)]$ for all $\underline{P}\} \Vdash$ B.

We cannot deal with other forms of circumscription, e.g. those giving induction.

References:

1. D.M. Gabbay, <u>Investigations in Modal and Tense Logic with Applications</u>, D. Reidel, 1976.

2. D.M. Gabbay, <u>Semantical Investigations in Heytings' Intuitionistic Logic</u>, D. Reidel, 1981.

3. John McCarthy, Circumscription: A form of non-monotonic reasoning, in <u>Artificial Intelligence</u> 13 (1980), pp. 27-39.

4. D. McDermott and J. Doyle, Non-monotonic logic I, in <u>Artificial Integlligence</u> 13 (1980), pp. 41-72.

5. R. Reiter, A logic for default reasoning, in <u>Artificial Intelligence</u> 13 (1980), pp. 81-132.

Knowledge Retrieval as Limited Inference

Alan M. Frisch

James F. Allen

Computer Science Department

The University of Rochester

Rochester, NY 14627

Abstract

Artificial intelligence reasoning systems commonly employ a knowledge base module that stores a set of facts expressed in a representation language and provides facilities to retrieve these facts. A retriever could range from a simple pattern matcher to a complete logical inference system. In practice, most fall in between these extremes, providing some forms of inference but not others. Unfortunately, most of these retrievers are not precisely defined.

We view knowledge retrieval as a limited form of inference operating on the stored facts. This paper is concerned with our method of using first-order predicate calculus to formally specify a limited inference mechanism and to a lesser extent with the techniques for producing an efficient program that meets the specification. Our ideas are illustrated by developing a simplified version of a retriever used in the knowledge base of the Rochester Dialog System. The interesting property of this retriever is that it performs typical semantic network inferences such as inheritance but not arbitrary logical inferences such as modus ponens.

1. INTRODUCTION

Artificial intelligence reasoning systems commonly employ a knowledge base module (KB) that stores a set of facts expressed in a representation language and provides facilities to retrieve these facts. Such a module is then used by the reasoner in performing its task-specific inference. In designing a knowledge base, it is important to maintain a fine balance between efficiency and usefulness. The module should perform only those inferences for which it has adequate control knowledge to perform efficiently. For instance, we present a KB that performs inheritance as found in semantic networks but not general logical inference such as modus ponens.

Thus, the fundamental issue in designing a KB retriever is how to limit inference. One possibility is to limit the expressive power of the representation language so that an efficient search space is produced. This is a bad idea as the reasoner may depend on being able to represent and retrieve arbitrary facts. The other alternative is to leave the representation language completely general and to limit the inferences that could occur during retrieval. This paper describes our methodology for specifying such a limited inference process.

The methodology uses first-order predicate calculus (FOPC) as the specification formalism. The use of FOPC notation in a representation language is not uncommon (Nilsson, 1980). And, as the study of logic programming (Kowalski, 1979) tells us, a retrieval algorithm can be specified in logic. Though the representation logic and the retrieval logic are different, they are not independent! In particular, the representation logic must have a well-defined semantics in order to be useful to the reasoner. The retrieval logic must respect this semantics by specifying only valid inferences. An example of a relationship between the two logics that meets this criteria is expressed by the statement, "The sentence '$\alpha \wedge \beta$' can be retrieved if and only if the sentence 'α' and the sentence 'β' each can be retrieved."

We continue this paper by refining our notion of retrieval and then present our method of using FOPC to give a functional specification of a retriever. We illustrate these methodological points by a two-part presentation of a simplified version of the knowledge base module of the Rochester Dialog System (Allen, 1980). The first part of the presentation considers a retriever based on logical inference and the second part extends the system to handle semantic-network inference. We conclude with a brief discussion of techniques for implementing an efficient program that meets the specification.

This paper uses conventional notation for FOPC. The symbol used for logical implication or entailment is $\Rightarrow$ and the statement it forms is called a *sequent*. The symbol used for logical equivalence or bi-directional entailment is $\equiv$ and the statement it forms is called an *equivalence*. As is traditional, the order of precedence of the logical symbols from highest to lowest is; $\sim$, $\wedge$, $\vee$, $\rightarrow$, $\leftrightarrow$, $\forall$, $\exists$. In logical formulae, variables appear in lower case while constant symbols (predicates and functions) appear in upper case. Greek letters are used in schemata as variables that range over formulae or constant symbols.

2. WHAT IS RETRIEVAL?

At any point the KB will have a set of facts called its base facts, on the basis of which it responds to queries made by the reasoner. The issues addressed by this section are what form a query should take and what the retriever should do to respond to the query. We will characterize two extreme positions that can be taken and then present our position as lying somewhere between them.

At one extreme the set of base facts can be treated as an abstract data structure and queries as calls to utilities for examining it. This approach neither prevents a reasoner from performing any kind of operations it wants nor commits a reasoner to any kind of operation it doesn't want. Yet such a retriever doesn't provide much assistance. Presumably, there are operations that can be factored out of the reasoner to be done automatically and efficiently by the retriever on each query.

At the other extreme the set of base facts can be treated as a representation and a query as a request to determine if a given sentence logically follows. This approach has several difficulties that are caricatured by the fact that there is no assurance that the retrieval process will ever terminate. Allocating the retriever such power is putting the muscle of the system in the wrong place. It is the reasoner that is specialized for appropriately controlling inference in its task domain.

Our viewpoint is that retrieval must respect the semantics of the representation language and is therefore inference. A query is a request for the retriever to attempt to infer a specified sentence of the representation language. (For purposes of this paper we will assume that the sentence is closed. It is a trivial extension (Green, 1969; Luckham and Nilsson, 1971) to consider a query as a request to prove an instance, or even all instances, of an open sentence and to return the instance.) If this attempt to infer the queried sentence must be guaranteed to terminate then either the representation language must be severely restricted or the inference engine must be severely restricted and thus incomplete. Restricting the representation language would be a serious mistake. The role of a representation is to define the set of valid inferences that *could* be made, not those that *are* made. Even if the retriever only makes a small portion of all valid inferences the remaining possibilities must be available for the reasoner to consider. Our refusal to restrict the representation language leaves us with the problem of designing a limited inference engine.

One common approach to limiting inference is to restrict the amount of resources used in the computation (Norman and Bobrow, 1975; Bobrow and Winograd, 1977; Robinson and Sibert, 1980). This can be done by restricting computation time, the total number of inference steps taken, or the depth of the inference. These approaches are unsuitable for knowledge retrieval because they limit all forms of inference uniformly. For example, if inference is limited to a depth of 5, then properties cannot be inherited down 6 levels of a type hierarchy. In general, there may be some kinds of inference that we want to be to be computed completely and others that we want to be ignored completely. A methodology for limiting inference should provide the knowledge base designer with enough control to pick and choose the inferences that he wants done.

Another class of limited inference systems are the incomplete theorem provers that are fairly common in the literature - for example, (Brown, 1978). Typically, these systems are not guaranteed to terminate. They often have the undesirable property that the prover has a base fact yet cannot respond to the query that consists of that very fact. It is not surprising that these inference systems are ill-suited for knowledge retrieval - they were not designed for that purpose.

For a while we attempted to design a specialized resolution theorem prover suited for knowledge retrieval. We tried to conjure up a scheme for limiting the resolutions in the search space. The attempt failed, again because it was so difficult to limit inference and yet have all base facts retrievable.

It is instructive to consider why it is so difficult to restrict a traditional proof system without infringing on the retrievability of the base facts. Posit the base fact P→Q and also the query P→Q. A common way of proving the query is to assume P and derive Q by modus ponens. If modus ponens is not restricted then the retriever may not terminate but if it is restricted then P→Q may not be retrievable. Situations similar to this are pervasive since theorem provers typically prove a complex sentence by using inference to recursively constuct the sentence from its constituent parts. We seek a proof technique that can respond to a query simply by noticing that it is a base fact.

3. SPECIFYING LIMITED INFERENCE

Our method of specifying limited inference requires a shift in viewpoint from an algorithmic description of the inference engine to a higher level functional description. This is done by focussing on the retriever's query set, Q(B), which is the set of queries that succeed given a particular set of base axioms, B. A specification of Q(B) is more appropriate than a specification of its decision procedure for certain puposes, such as proving that the retrieval system has properties like the ones we are about to discuss.

The previous section argued informally that a retriever must have certain properties. We now examine these more closely from the new viewpoint. The first requirement is derived from the stipulation that all retrievals must terminate. A set is recursive if and only if there is an algorithm that in finite time determines whether any given object is a member of the set. This places the following requirement on the query set:

(Req$_1$) Q(B) is recursive.

It is this requirement which necessitates limitations on a retriever's inference. If the retriever is a complete theorem prover, then Q(B) is not necessarily recursive.

The second of the requirements is that if a query is in the query set then it logically follows from the base axioms.

(Req$_2$) Q(B) $\subseteq$ {q | B $\Rightarrow$ q}

This essentially states that the KB retrieval mechanism is sound. The third and final requirement is that all base facts can be retrieved. That is, Q(B) contains B.

(Req$_3$) Q(B) $\supseteq$ B

Our method for specifying representation language retrievability (i.e., Q(B)) is somewhat analogous to the way in which object language provability is traditionally expressed in a meta-language. Because we do not introduce a meta-language term to name each base fact, we use the expressions "representation language" and "retrieval language" rather than "object language" and "meta-language."

The specification of a retriever has two components: a mapping, M, of representation language sentences to retrieval language sentences, and a set of retrieval axioms, R. (On occasion, M will be applied to a set, in which case it designates the set derived by applying the mapping to each member.) A query, q, succeeds if and only if its retrieval language representation, M(q), logically follows from the retrieval language representation of the base axioms, M(B), and the retrieval axioms, R. The retriever does not try to decide the truth of B $\Rightarrow$ q but rather the much simpler decision of the truth of M(B) $\cup$ R $\Rightarrow$ M(q). Retrieval is *limited* inference since the latter sequent implies the first, but not vice-versa.

It will be seen that a quite elaborate retriever can be specified clearly and succinctly with this method. This is possible because the mapping allows representation language constructs to be embedded in the retrieval language, rather than interpreted by it. For example, it would suffice to require that representation language sentences are mapped to retrieval language sentences by quotation. However, doing so would require that a universally quantified variable be given its meaning in the retrieval language by defining operations such as substitution. This is what we have referred to as *interpretation*. On the other hand, the meaning of a universally quantified variable could be *embedded* in the retrieval language simply by mapping representation language variables to retrieval language variables. Interpretation is a more general technique but, when possible, embedding will be used for its simplicity.

It is crucial to observe that in our system everything is mapped to the retrieval language and it is there that retrievability is decided. For all intents and purposes, the representation language has been discarded. This is in contrast to FOL (Weyhrauch, 1980) which uses reflection rules to map back and forth to its META representation when desired and the Bowen and Kowalski (1981) proposal which allows meta-language and object language to be mixed freely.

4. A LOGICAL RETRIEVAL SYSTEM

In this section we develop a logical retrieval system by specifying M and R. We call it a *logical* system because the mappings and retrieval axioms deal specifically with each logical symbol but treat all non-logical symbols uniformly. We have *deliberately* made this system quite weak because it has no control knowledge of how to handle any specific predicate or function symbols. We strengthen the system in section 5 by dealing explicitly with a small set of predicates used to structure semantic networks.

The mapping, M, which takes representation language sentences to retrieval language sentences, will be explained by demonstrating how it handles several sample formulae. A literal in the representation language maps to a term in the retrieval language by mapping predicate symbols and function symbols to corresponding function symbols and the logical operator ~ to the function symbol NOT. Thus the representation language literal ~DRINKS(JB,BEER) maps to the retrieval language term NOT(DRINKS(JB,BEER)). Since representation language and retrieval language are never mixed, we adopt the convention of mapping function symbols and predicate symbols to retrieval language function symbols of the same name. It should always be clear from context which language a given formula is in.

A representation language disjunction of literals is mapped to the retrieval language as a set of terms. For example, ~DRINKS(JB,MILK) V DRINKS(JB,BEER) maps to the term {NOT(DRINKS(JB,MILK)),DRINKS(JB,BEER)}. By virtue of sets being *unordered* collections, the retrieval language accounts for the fact that representation language V is commutative and associative.

Now consider how the base fact

(1)   DRINKS(JB,MILK) → DRINKS(JB,BEER)

is mapped to the retrieval language. First the base fact is put into prenex conjunctive normal form, which in this case yields the single conjunct:

(2)   ~DRINKS(JB,MILK) ∨ DRINKS(JB,BEER)

This disjunction of literals is then mapped to a retrieval language term that is then asserted to be retrievable with the predicate RET. Thus (2) is mapped to the atomic sentence

(3)   RET({NOT(DRINKS(JB,MILK)),DRINKS(JB,BEER)})

If the sentence to be mapped has more than one conjunct, such as

(4)   DRINKS(ALAN,MILK) ∧ DRINKS(ALAN,BEER)

then each conjunct is mapped as before. The resulting atomic sentences are then made into a conjunction. Thus, mapping (4) would result in the retrieval language sentence

(5)   RET({DRINKS(ALAN,MILK)}) ∧ RET({DRINKS(ALAN,BEER)})

It can be seen that object language conjunction is mapped to retrieval language conjunction. This method of defining the meaning of a representation language construct by mapping it directly to its corresponding retrieval language construct is what we have referred to as embedding.

Object language quantification is also embedded in the retrieval language. Consider the sentence

(6)   ∀x (∃y DRINKS(x,y)) ∧ (∃y ~DRINKS(x,y))

which in prenex conjunctive normal form becomes

(7)   ∀x ∃y ∃z DRINKS(x,y) ∧ ~DRINKS(x,z)

The matrix is mapped as before with the additional consideration that variables are mapped to corresponding variables. The prefix remains essentially unchanged by the mapping except for the mapping of variables to corresponding variables. Thus, mapping (7) to the retrieval language results in

(8)   ∀x ∃y ∃z RET({DRINKS(x,y)}) ∧ RET({NOT(DRINKS(x,z))})

Observe that since the representation language connectives ∨ and ~ have been mapped to retrieval language function symbols, their meaning will be interpreted by the retrieval language. There will be axioms which enable deductions to be made with the terms constructed from NOT and { }. These will mimic sound representation language inferences. However, the representation language proof system must not be totally mimicked by the retrieval language axioms if (Req_1) is to be met. OR and NOT will be given much weaker interpretations than ∨ and ~. On the other hand, representation language ∧ and quantification has been mapped to retrieval language ∧ and quantification. This embedding gives these representation language constructs their full-blown logical meaning.

Before considering retrieval axioms, let us examine this retriever with no retrieval axioms. It can immediately be seen that this retriever meets (Req_3). It also meets the first two requirements, but showing so involves more subtlety than can be dealt with in this paper.

But the query set contains much more than just the base axioms! For example, the query

   (9)   ~∃x (∀y DRINKS(x,y)) ∨ (∀y ~DRINKS(x,y))

would succeed in a KB that had (6) as a base fact. Though not immediately obvious, (6) and (9) both map to (8) in the retrieval language and are therefore logically equivalent. The normalization process that accounts for this by equating many logical paraphrases is a powerful technique. Mapping disjunctions to sets is also a form of normalization because it makes the retrieval language insensitive to the ordering of representation language disjunction. Thus, if A ∨ B were added to the KB, the query B ∨ A would succeed.

In addition, the embedding of representation language variables and conjunction means that these will be handled with their full logical meaning. For example, the query

   (10)   ∃x ∃y DRINKS(x,y)

can be answered positively by this KB if either (4) or (7) had been added. The retrieval language representation of this query,

   (11)   ∃x ∃y RET({DRINKS(x,y)})

logically follows from (5), the retrieval language representation of (4), or from (8), the retrieval language representation of (7).

Because there are no retrieval axioms that chain base facts together (i.e., use more than one base fact in an inference) this system has the property that

(Prop$_1$) A query succeeds *only if* each conjunct of the query follows from one conjunct of one base fact.

We will now look at ways of strengthening this system by adding retrieval axioms. These retrieval axioms strengthen the system but not so much as to destroy (Prop$_1$). The first retrieval axiom is based on the fact that disjunction is monotonic. That is, if a disjunct is true, then adding additional disjuncts to it cannot change its truth value. This inspires the logical retrieval axiom:

(R$_1$) $\forall y\, (\exists x\, \text{RET}(x) \wedge x \subseteq y) \rightarrow \text{RET}(y)$

The above retrieval axiom allows a set to be manipulated in a manner consistent with the meaning of the disjunction it encodes. The next retrieval axiom does the same for NOT and the negation it encodes. It says that an atomic sentence and its negation cannot both be retrievable.

(R$_2$) $\forall x\, \sim\text{RET}(\{x\}) \vee \sim\text{RET}(\{\text{NOT}(x)\})$

(R$_2$) does not change the query set of the current retriever but it is needed in the semantic network retrieval system.

Because (Prop$_1$) is a desirable design objective, (R$_1$) and (R$_2$) are the only logical retrieval axioms used. There will be other retrieval axioms, such as those derived from the study of semantic networks undertaken in the next section. Although we consider the three requirements which we presented to be constraints on all KB designs, the design objective was chosen based on our particular task. We shall now discuss the rationale for this design objective.

As discussed before, and as can be seen clearly now, the KB has no special knowledge of any representation language predicates. It therefore treats them uniformly and this is what we mean when we say that the KB is domain independent. Without such knowledge, a strong inferencial component cannot perform effectively and thus we have taken a conservative approach to inference. In this sense the system presented here can be viewed as a KB kernel on top of which a more powerful, and possibly domain-dependent KB could be built. A reasoner can compensate for a KB that is efficient but deductively weak by making deductions of its own. However, it is hard for a reasoner to compensate for a KB whose excessive deductive power leads to inefficiency.

5. A SEMANTIC NETWORK RETRIEVAL SYSTEM

In this section, we investigate a very simple semantic network system by formalizing it in FOPC. This investigation yields a primitive set of predicates that can be used to structure knowledge and a set of axioms relating these predicates. This formalization takes place without consideration of the role that the axioms should play in a KB. We then take up the problem of how to integrate these axioms into the KB so that the retriever can use them to make the kind of inferences typically performed by semantic network interpreters.

The fact that our KB uses the notation of FOPC for the representation language is methodologically important yet it says little about how a domain should be represented. Modern semantic networks - those since (Woods, 1975) - have made steps in suggesting knowledge-structuring primitives, though a great deal of freedom still remains in choosing how to represent a domain. Others have referred to these knowledge-structuring primitives as epistemological primitives (Brachman, 1979), structural relations (Shapiro, 1979) and system relations (Shapiro, 1971).

Elsewhere (Frisch, 1981; Allen and Frisch, 1982), we have shown how semantic networks can motivate a logic with a fixed set of predicates and how the relationship between these predicates can be axiomatized. Here we use a much simpler scheme that has 4 predicates, TYPE, SUBTYPE, ROLE, and =, and 3 axioms, labelled (Asn-1), (Asn-2), and (Asn-3). We then integrate these axioms into the KB so that they can drive the retriever.

The use of type hierarchies is characteristic of semantic networks. The domain of individuals is divided into sets called types. TYPE(I,T) asserts that an individual I belongs to a type T and SUBTYPE(T,T') asserts that the type T is a subset of the type T'. There are two axioms involving these predicates:

(Asn$_1$) $\forall x,t,t'$ TYPE(x,t') $\wedge$ SUBTYPE(t',t) $\rightarrow$ TYPE(x,t)

(Asn$_2$) $\forall x,t',t$ SUBTYPE(t',x) $\wedge$ SUBTYPE(x,t) $\rightarrow$ SUBTYPE(t',t)

If events such as 'Jellybean drank milk' are represented as DRANK(JB,MILK) then, as pointed out by Davidson (1967), there is no way to quantify over events and their components. This prevents the representation of assertions such as 'The actor of an action causes that action.' For this reason and for the purpose of making all relations binary, semantic networks traditionally represent 'Jellybean drank milk' as

(12)   TYPE(DRANK01,DRANKEVENT) $\wedge$ ACTOR(DRANK01,JB) $\wedge$ OBJECT(DRANK01,MILK)

Thus, 'The actor of an event causes that event' can be expressed as

(13) $\forall x,y \; TYPE(x,ACTION) \land ACTOR(x,y) \rightarrow CAUSE(y,x)$

However, in this representation, there is no way to state 'Role fillers are unique' or to query 'Is there an event involving Jellybean and milk.' Because we do not restrict ourselves to binary relations, we can generalize the above representation by making roles into objects in their own right. Thus, (12) becomes

(14) $TYPE(DRANK01,DRANKEVENT) \land ROLE(ACTOR,DRANK01,JB) \land$
        $ROLE(OBJECT,DRANK01,MILK)$

and the query 'Is there an event involving Jellybean and milk' is represented as:

(15) $\exists e,r,r' \; TYPE(e,EVENT) \land ROLE(r,e,JB) \land ROLE(r',e,MILK)$

Notice that (15) logically follows from (14).

The third and final semantic network axiom states that role fillers are unique and can now be expressed as:

(Asn_3) $\forall r,e,f,f' \; ROLE(r,e,f) \land ROLE(r,e,f') \rightarrow f = f'$

How can (Asn_1), (Asn_2) and (Asn_3) be integrated into the KB so that they can be used by the retriever in its deductions? We would like to obtain the property that the query set is closed with respect to the derivation of true sentences from these axioms.

First of all, these axioms should be added to the set of base facts in the KB. Mapping them to the retrieval language yields:

(16) $\forall x,t,t' \; RET(\{NOT(TYPE(x,t')),NOT(SUBTYPE(t',t)),TYPE(x,t)\})$

(17) $\forall x,t',t \; RET(\{NOT(SUBTYPE(t',x)),NOT(SUBTYPE(x,t)),SUBTYPE(t',t)\})$

(18) $\forall r,e,f,f' \; RET(\{NOT(ROLE(r,e,f)),NOT(ROLE(r,e,f')),f = f'\})$

But this is not enough. Let us consider the situation in which TYPE(JB,DOG) and SUBTYPE (DOG,MAMMAL) are also base facts and the query that we wish to succeed is TYPE(JB,MAMMAL). Thus, in the retrieval language there are two base facts,

(19) $RET(\{TYPE(JB,DOG)\})$

(20) $RET(\{SUBTYPE(DOG,MAMMAL)\})$

and the query

(21) RET({TYPE(JB,MAMMAL)})

But notice that (21) does not follow from the base facts. The query fails because disjunction in (Asn_1) is weakened when mapped to (16) in the retrieval logic. We can get the query to succeed by using the following retrieval axiom based on (Asn_1):

(R_3) $\forall x,t,t'$ RET({NOT(TYPE(x,t'))}) $\lor$ RET({NOT(SUBTYPE(t',t))}) $\lor$ RET({TYPE(x,t)})

(21) now logically follows from (19), (20), (R_2) and (R_3). The increased power of the retriever is due to the fact that the disjunction in (Asn_1) has been mapped to a disjunction in (R_3) rather than a set as was done in (16).

Likewise, the two other semantic network axioms are made into retrieval axioms:

(R_4) $\forall x,t',t$ RET({NOT(SUBTYPE(t',x))}) $\lor$ RET({NOT(SUBTYPE(x,t))}) $\lor$
         RET({SUBTYPE(t',t)})

(R_5) $\forall r,e,f,f'$ RET({NOT(ROLE(r,e,f))}) $\lor$ RET({NOT(ROLE(r,e,f'))}) $\lor$ RET({f = f'})

However, a deficiency still remains. Consider the situation in which 'All mammals drink a liquid' is added to the above KB and the query 'Jellybean drinks a liquid' is made. Adding

(22) $\forall m$ TYPE(m,MAMMAL) $\rightarrow$ $\exists l$ TYPE(l,LIQUID) $\land$ DRINKS(m,l)

yields the retrieval language base fact

(23) $\forall m \exists l$ RET({NOT(TYPE(m,MAMMAL)),TYPE(l,LIQUID)}) $\land$
      RET{NOT(TYPE(m,MAMMAL)),DRINKS(m,l)})

Though (21) logically follows, (23) is not strong enough to logically imply the queried sentence

(24) $\exists l$ RET({TYPE(l,LIQUID)}) $\land$ RET({DRINKS(JB,l)})

More generally, the KB as it now stands cannot infer a property of an individual based on a property asserted of all members of one of its types. The problem is that M combines TYPE atoms with other atoms thus subjecting types to the system's inference limitations. The problem could be solved by altering M to factor out appropriate representation language TYPE atoms into their own retrieval language RET atoms. For example, the query, (24), would succeed if (22) were mapped to the retrieval language as

(25) $\forall m \, \exists l \, \text{RET}(\{\text{TYPE}(m,\text{MAMMAL})\}) \rightarrow \text{RET}(\{\text{TYPE}(l,\text{LIQUID})\}) \wedge \text{RET}(\{\text{DRINKS}(m,l)\})$

Rather than complicating the mappings to recognize and factor out the appropriate type information, the representation language is extended so that certain type information can be written in a factored-out manner initially. The extended language, called typed first-order predicate calculus or TFOPC, includes formulae of the form $\forall x{:}\tau \; \varphi$ and $\exists x{:}\tau \; \varphi$, where φ is a formula and τ is the name of a type. The meaning of this notation is defined by two equivalence schemata:

(26) $\quad \forall x{:}\tau \; \varphi \equiv \forall x \, \text{TYPE}(x,\tau) \rightarrow \varphi$

(27) $\quad \exists x{:}\tau \; \varphi \equiv \exists x \, \text{TYPE}(x,\tau) \wedge \varphi$

Thus, (22) could be written in TFOPC as

(28) $\forall m{:}\text{MAMMAL} \; \exists l{:}\text{LIQUID} \; \text{DRINKS}(m,l)$

M is appropriately modified to handle the extended representation language. As before, the representation language base fact or query is put in prenex conjunctive normal form. The same process that converts a FOPC sentence to prenex conjunctive normal form (Robinson, 1979) also converts a TFOPC sentence. Here's why: The conversion process is based on many equivalence schemata, six of which deal with quantification. There are six corresponding TFOPC schemata that can easily be proved and therefore, quantifiers can be moved about as if the types weren't even there.

Once the query or base fact is in prenex conjunctive normal form, it is mapped as before but with the additional consideration that formulae of the form $\forall x{:}\tau \; \varphi$ map to

(29) $\quad \forall x \, \text{RET}(\{\text{TYPE}(x,\tau)\}) \rightarrow M(\varphi)$

and those of the form $\exists x{:}\tau \; \varphi$ map to

(30) $\quad \exists x \, \text{RET}(\{\text{TYPE}(x,\tau)\}) \wedge M(\varphi)$

Therefore, applying the complete M mapping to the base fact (28) yields

(31) $\quad \forall m \, \text{RET}(\{\text{TYPE}(m,\text{MAMMAL})\}) \rightarrow \exists l \, \text{RET}(\{\text{TYPE}(l,\text{LIQUID})\}) \wedge \text{RET}(\{\text{DRINKS}(m,l)\})$

The query 'Jellybean drinks a liquid,' which can be stated in TFOPC as

(32) $\quad \exists l{:}\text{LIQUID} \; \text{DRINKS}(\text{JB},l)$

gets mapped to the retrieval language as

(33)   ∃l RET({TYPE(l,LIQUID)}) ∧ RET({DRINKS(JB,l)})

And finally, the deficiency stated at the outset has been overcome. The example situation now behaves as desired since (33) logically follows from (19), (20), (31), and the retrieval axioms.

As a final observation, notice that typing existential variables has not extended the power of the system as has typing universal variables. This is due to the fact that a type on an existential variable is an abbreviation for a conjunction and the limited inference handles conjunction fully. Typed existential variables have been added to the language for the sake of uniformity.

The semantic-network retriever presented in this section meets (Req_1), (Req_2) and (Req_3). Though the retriever can now chain base facts together in deriving TYPE and SUBTYPE relations, Q(B) is still recursive because the type hierarchy is finite.

6. IMPLEMENTING RETRIEVAL

The KB module of the Rochester Dialog System has a retriever that is an extended version of the semantic-network retriever presented here. The representation language has been extended with a set of abbreviations tailored to our domain and the semantic network retrieval axioms have been extended to handle a larger set of knowledge-structuring primitives (Allen and Frisch, 1982; Frisch, 1981). All communication with the KB is in the representation language - the retrieval language is totally invisible. We will only briefly discuss the techniques used in the implementation since we have not yet proved their correctness.

Our method of producing a program that meets the specification developed in this paper is to map queries and base facts to the retrieval language and treat the specification as the logic component of an algorithm (Kowalski, 1979). This logic component can then be transformed to an equivalent specification - one for which we can produce an efficient control component. The transform we employ makes all retrieval language sentences into Horn clauses. Notice that all of the semantic-network retrieval axioms are disjunctions of positive RET literals of singleton sets. For concreteness consider

(R_3) $\forall x,t,t'$ RET({NOT(TYPE(x,t'))}) ∨ RET({NOT(SUBTYPE(t',t))}) ∨ RET({TYPE(x,t)})

There are three sentences that have a single positive literal which logically follow from (ASN$_1$) and

(R$_2$) $\quad \forall x \; \sim\text{RET}(\{x\}) \vee \sim\text{RET}(\{\text{NOT}(x)\})$

They are:

(34) $\quad \forall x,t,t' \; \text{RET}(\{\text{NOT}(\text{TYPE}(x,t'))\}) \vee \sim\text{RET}(\{(\text{SUBTYPE}(t',t))\}) \vee \sim\text{RET}(\{\text{NOT}(\text{TYPE}(x,t))\})$

(35) $\quad \forall x,t,t' \; \sim\text{RET}(\{\text{TYPE}(x,t')\}) \vee \text{RET}(\{\text{NOT}(\text{SUBTYPE}(t',t))\}) \vee \sim\text{RET}(\{\text{NOT}(\text{TYPE}(x,t))\})$

(36) $\quad \forall x,t,t' \; \sim\text{RET}(\{(\text{TYPE}(x,t'))\}) \vee \sim\text{RET}(\{\text{SUBTYPE}(t',t)\}) \vee \text{RET}(\{\text{TYPE}(x,t)\})$

Thus all sentences of *n* literals are rewritten into *n* sentences, each of which has one positive literal and *n-1* negative literals. Once the rewritting is completed (R$_2$) is no longer needed. This rewriting is clearly sound and, in general, incomplete. The important question remains open: Is this rewriting technique complete for our class of theories? The new set of sentences need not be equivalent to the original set of sentences - it need only preserve the truth of the sequent M(B) ∪ R ⇒ M(q). We point out that this rewriting is related to Meltzer's proposal (1966) and appears to yield a system equivalent to linear input resolution (Loveland, 1978).

It can be seen that these rewritten sentences can then trivially be made into Horn clauses. We then use a PROLOG-like theorem prover called HORNE (Allen and Frisch, 1981) to interpret the clauses. Failure to prove a query is construed as meaning that the query is not in Q(B) (Clark, 1978). We now discuss the selection and search strategies (Van Emden, 1977) that we use to control the theorem-prover.

Since all clauses that result from retrieval axioms are fixed, we manually order the literals within the clauses to take advantage of HORNE's left-to-right selection strategy. As discussed by Clark and McCabe (1979), several orderings are specified for each clause depending on which variables are bound. Cuts are also manually added to these clauses based on the bindings. Ordering of literals within base facts is not crucial since they are atomic with the exception of the TYPE literals. Ordering of literals within a query is crucial and must be done dynamically. Nothing has been done in this regard but some thought has been given to using a method similar to Chat-80's query optimization (Warren and Pereira, 1981).

Search strategy plays a relatively minor role in the retriever's performance. When the retriever is looking for all answers to a query or when there are no answers to a query, the entire search space must be examined. These situations are not unusual and clearly search strategy is irrelevant in them. Therefore, HORNE's depth-first search is a good choice because it lends itself to efficient implementation through the use of stack allocation. The order that branches are chosen is arbitrary.

However, if our search space were dynamically pruned, search strategy would be a factor. Not only could dynamic pruning eliminate deductive paths guaranteed to fail (Perreira and Porto, 1980) but it could eliminate deductive paths guaranteed to yield redundant solutions.

We have realized great improvements in efficiency by using TFOPC as the retrieval language and accordingly extending HORNE to handle TFOPC Horn clauses. Type checking is done on unification by recursively calling the retriever to test the appropriate TYPE and SUBTYPE relations. This organization reorders the goals in a proof and appears to result in a much smaller search space. The rationale for this is that, when possible, the retriever reasons about classes of individuals rather than about the individuals themselves. This is a minimum commitment strategy similar to that obtained in MOLGEN (Stefik, 1981) by the use of constraints.

ACKNOWLEDGEMENTS

We thank Pat Hayes for his insightful comments on this paper and the Artificial Intelligence Study Group for patiently hearing our arguments and providing valuable counterarguments. This research has been supported in part by NSF grant IST-801248 and ONR grant N00014-80-C-0197.

REFERENCES

Allen, J. F., The Rochester natural language understanding project. *1980-81 Computer Science and Computer Engineering Research Review.* Computer Science Department, University of Rochester, 1980.

Allen, J. F. and Frisch, A. M., What's in a semantic network? Submitted to 20th Annual Meeting of the Association for Computational Linguistics, 1982.

Allen, J. F. and Frisch, A. M., HORNE user's manual. Internal report, Computer Science Department, University of Rochester, 1981.

Bobrow, D. G. and Collins, A. M. (Eds.), *Representation and understanding.* New York: Academic Press, 1975.

Bobrow, D. G. and Winograd, T., An overview of KRL, a knowledge representation language. *Cognitive Science,* 1977, *1,* 3-46.

Bowen, K. A. and Kowalski, R. A., *Amalgamating language and metalanguage in logic programming.* Technical Report, School of Information and Computer Science, Syracuse University, 1981.

Brachman, R. J., On the epistemological status of semantic networks. In Findler, 1979.

Brown, F. M., Towards the automation of set theory and its logic. *Artificial Intelligence,* 1978, 10, 281-316.

Clark, K. L., Negation as failure. In Gallaire and Minker, 1978.

Clark, K. L., and McCabe, F. G., The control facilities of IC-PROLOG. In Michie, 1979.

Davidson, D., The logical form of action sentences. In Rescher, 1967.

Elcock, E. W. and Michie, D. (Eds.), *Machine intelligence 8.* Chichester, England: Ellis Horwood, 1977.

Findler, N. V. (Ed.), *Associative networks: Representation and use of knowledge by computers.* New York: Academic Press, 1979.

Frisch, A. M., A formal study of knowledge representation and retrieval. Ph.D. thesis proposal, Computer Science Department, University of Rochester, 1981.

Gallaire, H. and Minker, J. (Eds.), *Logic and data bases.* New York: Plenum Press, 1978.

Green, C., Theorem-proving by resolution as a basis for question-answering systems. In Meltzer and Michie, 1969.

Kowalski, R. A., *Logic for problem solving.* New York: North Holland, 1979.

Loveland, D. W., *Automated theorem proving: A logical basis.* Amsterdam: North-Holland, 1978.

Luckham, D. C., and Nilsson, N. J., Extracting information from resolution proof trees. *Artificial Intelligence,* 1971, 2, 27-54.

Meltzer, B., Theorem-proving for computers: Some results on resolution and renaming. *Computer Journal,* 1966, 8, 341-343.

Meltzer, B., and Michie, D. (Eds.), *Machine intelligence 4.* Ediburgh: Edinburgh University Press, 1969.

Michie, D. (Ed.), *Expert systems in the micro electronic age.* Ediburgh: Edinburgh University Press, 1979.

Nilsson, N. J., *Principles of artificial intelligence,* Palo Alto, CA.: Tioga, 1980.

Norman, D. A., and Bobrow, D. G., On data-limited and resource-limited processes. *Cognitive Psychology,* 1975, 7, 44-64.

Pereira, L. M., and Porto, A., Selective backtracking for logic programs. 5th *Conference on Automated Deduction Proceedings,* Springer-Verlag, 1980.

Rescher, N. (Ed.), *The logic of decision and action.* Pittsburgh: U. of Pittsburgh Press, 1967.

Robinson, J. A., *Logic: Form and function.* New York: North Holland, 1979.

Robinson, J. A., and Sibert, E. E., LOGLISP - An alternative to PROLOG. Technical Report,

School of Computer and Information Science, Syracuse University, December, 1980.

Shapiro, S. C., The SNePS semantic network processing system. In Findler, 1979.

Shapiro, S. C., A net structure for semantic information storage, deduction and retrieval. *Proceedings of the 2nd International Joint Conference on Artificial Intelligence*, 1971.

Stefik, M. J., Planning with constraints. *Artificial Intelligence*, 1981, *16*, 111-140.

Van Emden, M. H., *Programming with resolution logic.* In Elcock and Michie, 1977.

Warren, D. H. D., and Pereira, F. C. N., An efficient easily adaptable system for interpreting natural language queries. DAI Research Paper No. 155, Department of Artificial Intelligence, University of Ediburgh, 1981.

Weyhrauch, R. W., Prolegomena to a theory of mechanized formal reasoning. *Artificial Intelligence*, 1980, *13*, 133-170.

Woods, W. A., What's in a link: Foundations for semantic networks. In Bobrow and Collins, 1975.

On Indefinite Databases and the Closed World Assumption

Jack Minker
University of Maryland
College Park, Maryland 20742

Abstract

A database is said to be indefinite if there is an answer to a query of the form Pa ∨ Pb where neither Pa nor Pb can be derived from the database. Indefinite databases arise where, in general, the data consists of non-Horn clauses. A clause is non-Horn if it is a disjunction of literals in which more than one literal in the clause is positive.

Horn databases, which comprise most databases in existence, do not admit answers of the form Pa ∨ Pb where neither Pa nor Pb are derivable from the database. It has been shown by Reiter that in such databases one can make an assumption, termed the Closed World Assumption (CWA), that to prove that $\overline{Pa}$ is true, one can try to prove Pa, and if the proof for Pa fails, one can assume $\overline{Pa}$ is true.

When a database consists of Horn and non-Horn clauses, Reiter has shown that it is not possible to make the CWA. In this paper we investigate databases that consist of Horn and non-Horn clauses. We extend the definition of CWA to apply to such databases. The assumption needed for such databases is termed the Generalized Closed World Assumption (GCWA). Syntactic and semantic definitions of generalized closed worlds are given. It is shown that the two definitions are equivalent. In addition, given a class of null values it is shown that the GCWA gives a correct interpretation for null values.

1. Introduction

We consider a relational database to consist of a set of *extensional data*, and a set of *intensional data*. The extensional data consists of data as stored in any relational database. The intensional data consists of a set of axioms, or views, from which new relations can be derived using both the axioms and the extensional data.

Databases which are exclusively extensional, or whose axioms are all Horn are of greatest interest as they arise frequently. An axiom is said to be Horn if it consists of a conjunction which implies a single atomic formula. That is, if it is of the form $R_1 \wedge \ldots \wedge R_n \rightarrow R_{n+1}$, where R_i, $i=1,\ldots,n+1$ are positive literals. Alternatively, it may be considered to be a disjunctive form, in which there is at most one positive literal, i.e., $\overline{R}_1 \vee \ldots \vee \overline{R}_n \vee R_{n+1}$. Each R_i is an n_i-ary expression of the form $R_i(t_1,\ldots,t_{n_i})$, where the t_i are terms. A term is a constant or a variable. We do not permit function symbols or Skolem constants (zero-ary function symbols). An axiom is assumed to be universally quantified and hence the universal quantifier symbol over the variables in an axiom are omitted.

Reiter [1978a,1978b] has considered databases in which one has both an extensional and an intensional database and has shown that Horn data bases always have definite answers. That is, there can be no answers to questions where the answers are of the form Pa ∨ Pb and neither Pa nor Pb can be derived from the database.

Clearly there are databases (DB) which have indefinite data. The trivial DB given by DB : {Pa ∨ Pb}, and consists of no other data either extensional or intensional has no definite answers. The indefiniteness arises because the only data in the database is non-Horn. Not all databases that have non-Horn axioms result in indefinite answers. Consider the following : DB : {Pa ∨ Pb, $\overline{Pa}$}. The answer to the question of whether Pb is true is "yes". That is, although Pa ∨ Pb is in the data base, Pb can be derived since $\overline{Pa}$ is true and Pa ∨ Pb is true, then Pb must be true. Hence, there are no indefinite answers in this database. This database has, in fact, an equivalent Horn database, Horn DB : {$\overline{Pa}$, Pb}.

Our focus in this paper is that of indefinite databases. Before addressing indefinite databases some background is necessary.

1.1 Open and Closed World Data Bases

The terms open and closed world databases were introduced by Reiter [1978a,1978b]. An *open world* is associated with a first-order theory. That is, first-order predicate logic. Data is represented by clauses, and negative data is listed explicitly in the database. Answers to queries may be either looked-up or derived from the data and the axioms. In an open world, negative data must be listed explicitly. When the database complies with this assumption concerning negative data, the database is said to satisfy the *open world assumption* (OWA). A problem arises in that negative data may overwhelm a system.

An alternative to an open world is a closed world database that arises from an assumption concerning negative data. That is, the data base contains no negative data and, to determine whether one can derive a negative fact from the data base, one attempts to prove the positive fact true. If one fails to prove the positive fact, then the negative data is assumed to be true. That is, under the *closed world assumption* (CWA) certain answers are admitted as a result of *failure* to find a proof. The need for a closed world assumption in deductive databases was described first, perhaps, by Nicolas and Syre [1974]. It is also related to the negative operators of PLANNER (Hewitt [1972]) and PROLOG (Roussel [1975]) where, in these languages, negative means "not provable". For a description of negation as failure and its truth function semantics see Clark [1978].

Given a DB, one can associate an extended data base $\overline{EDB}$ which consists of negative data whose positive counterparts cannot be proved from DB. More formally, $\overline{EDB}$ = {$\overline{P\vec{c}}$ | P is a predicate letter, $\vec{c}$ a tuple of constants and $P\vec{c} \notin$ DB or $P\vec{c}$ cannot be proved from DB, (written DB $\not\vdash P\vec{c}$). This statement can be simplified to : $\overline{EDB}$ = {$\overline{P\vec{c}}$ | P is a predicate letter, $\vec{c}$ a tuple of constants, and DB $\not\vdash P\vec{c}$}.

By a DB we assume a finite set of constants which belong to a single universe. We do this for simplicity and note that Reiter assumes a universe that is subdivided into types, and variables belong to types. The work described in this paper also encompasses types.

An answer to a query under the CWA, as defined by Reiter, is given as follows. Let EDB be given as above, then $\vec{c}^{(1)} + \ldots + \vec{c}^{(r)}$ is a CWA answer to $<\vec{x}|(\exists \vec{y})W(\vec{x},\vec{y})>$ (with respect to DB), and W is a well-formed formula if and only if $DB \cup \overline{EDB} \vdash \bigvee_{i \leq r} (\exists \vec{y}) W(\vec{c}^{(i)}, \vec{y})$. Reiter has shown that in Horn databases it is not necessary to have explicit knowledge of $\overline{EDB}$ to find answers to queries. Furthermore, since either $P\vec{c}$ or $\overline{P}\vec{c}$ can be derived as an answer, there is no indefiniteness possible in answers under the CWA. Additionally, Reiter has shown that queries of the form given above can be transformed into the union and intersection of queries on atomic formulae. Horn data bases are shown to comply with the CWA.

The concept of a CWA has been addressed from a semantic point of view by van Emden [1977] who considers the issue of database consistency under the CWA. A model of a set of clauses is a set of ground atomic formulae (atomic formulae that only contain constants) that make all formuale true. A "minimal model" of a data base as defined by van Emden is the intersection of all its models. If this minimal model is itself a model of the database, then the database is consistent with the CWA. It has been shown by van Emden and Kowalski [1976] that Horn databases have a minimal model.

We note that DB : Pa $\vee$ Pb does not have a minimal model. Consider all of the models : M_1 : {Pa}
M_2 : {Pb}
M_3 : {Pa, Pb}
$M = M_1 \cap M_2 \cap M_3 = \{\emptyset\}$.

This can be seen as follows. M_1 consists of only Pa since Pa "true" is a model of DB. Similarly M_2 consists of Pb and M_3 consists of both Pa and Pb. The intersection of the three sets is empty. This denotes that $\overline{P}a$ and $\overline{P}b$ must be "true". But Pa and Pb "true" cause the DB to be false. Hence it has no model. The DB is clearly not consistent with the CWA since we cannot prove Pa then $\overline{P}a$ is assumed "true" and since we cannot prove Pb, $\overline{P}b$ is assumed "true". Hence, $\overline{EDB} = \{\overline{P}a, \overline{P}b\}$. But, $DB \cup \overline{EDB} = \{Pa \vee Pb, \overline{P}a, \overline{P}b\}$ is always false and the database is not consistent with the CWA.

A database DB is termed *mixed world* if part of the data complies with the closed world assumption, and part with the open world assumption.

1.2 Approaches To Handling Indefinite Data

We describe three approaches to handle DBs with non-Horn clauses that have been proposed in the literature. These are : (1) Viewing the perceived world as a first order theory as discussed by Nicolas and Gallaire [1978]. (2) Mixing the OWA and CWA by the use of if-and-only-if definitions as described by Clark [1978] and Kowalski [1978]. (3) Splitting the DB into a disjunction of DBs each of which contains only

Horn clauses, as proposed by Loveland [1978]. In Section 2 we provide a definition
for a GCWA which applies to mixed worlds and serves to extend the concept of the CWA
applicable only to Horn databases.

1.2.1   The Perceived World As A First-Order Theory

In the perceived world as a first-order theory as discussed by Nicolas and
Gallaire [1978], a DB is formalized in terms of first-order logic where the elementary
facts (relational database) and the general axioms are considered as the proper axioms
of a first-order theory with equality. The theory is defined as follows:

- its set of constants (respectively predicate names) is the set of elements (relations) appearing within information;
- its set of proper axioms is the set of well-formed formulae (wff) associated with information in the following way : if the statement is known to be true, then the wff is an axiom, and if it is known to be false then the negation of the wff is an axiom.

Viewing the world as a set of wffs extends the concept of relational databases
in several ways. First, it permits negative data to be represented explicitly in the
database. Second it allows indefinite data to be represented, which is not possible
in a relational data base. Third, it allows general axioms as part of the database,
which corresponds to the concept of a "view" in database terminology. Whereas "views"
generally are not recursive, general axioms have no such restriction.

The above advantages of a first-order theory for databases must be tempered by
its disadvantages. First, whereas in a relational database negative data is implicitly stored and corresponds to the CWA, all negative data must be stored explicitly in
the DB. The amount of negative data that needs to be stored may be overwhelming as
it is generally an order of magnitude larger than the positive data. Consider, for
example, a large university with 30,000 students and a course in which 30 students
are enrolled. To determine who is not in the course, either 29,970 facts must exist
explicitly listing those not in the course or if and only if definitions are used as
explained in the following section.

A DB which contains indefinite data becomes more complex than one which does not
contain such information. In general, one is forced to include negative data to
answer questions, and cannot rely upon the CWA. Furthermore, a general purpose
inference mechanism must be used. If one did not have indefinite data, then a
straightforward inference mechanism termed LUSH resolution (Hill [1974]), complete
and sound only for the Horn clauses can be used. Indefinite data requires a general
purpose inference mechanism that is complete and sound for arbitrary clauses. Useful
inference mechanisms are linear resolution with selection function (SL) as developed
independently by Kowalski and Kuehner [1971], Loveland [1968], and Reiter [1971] or
a variant of these approaches that allows somewhat greater flexibility, linear resolution with selection function based on trees (LUST), developed by Minker and Zanon

[1979]. A discussion of why a general purpose inference mechanism is needed can be found in Loveland and Stickel [1973]. In contrast to LUSH resolution, the other inference systems mentioned above require two added operations - ancestry resolution and factoring.

1.2.2 **The Use of If-And-Only-If Definitions**

It has been noted by Clark [1978] and Kowalski [1978] that Horn clauses express only the if-half of definitions which are otherwise expressed completely by using the full intended if-and-only-if definition. For example, the clauses (or facts, or assertions)

(1) TEACHES (A,PROGRAMMING)
(2) TEACHES (B,PROGRAMMING)

state that A and B teach programming. They could have been written equivalently as "if" statements :

(1') $E(x,A) \rightarrow$ TEACHES (x,PROGRAMMING)
(2') $E(x,B) \rightarrow$ TEACHES (x,PROGRAMMING),

where E denotes the equality predicate. That is, x teaches programming if x equals A. If the intent is to mean that "only A and B teach programming", then the "if" statements do not capture the intent. Under the CWA, the meaning is assumed to be that A and B are the only ones. In first order logic, by including the intended "only-if" part of the definition, both Clark and Kowalski note that one makes the statement explicit. Thus, one would write

(3) $E(x,A) \lor E(x,B) \leftrightarrow$ TEACHES (x,PROGRAMMING).

Whereas (1) and (2) (or (1') and (2')) are sufficient to find all instances of the TEACHES relation, they are inadequate to conclude that some individual "C" does not teach programming (unless some assumption is made) when "C" is not found among the instances of the relation. However, (3) does capture this relationship.

While Reiter captures negative data implicitly through the CWA, Clark and Kowalski capture the same information explicitly through if-and-only-if definitions. Using such definitions precludes the necessity for listing all negative information. Thus, (3) is sufficient to prove that C does not teach programming and does not require that $\overline{\text{TEACHES}}$ (C,PROGRAMMING) be stored explicitly.

While not having to store negative data, the introduction of if-and-only-if definitions precludes the use of LUSH resolution for the clauses are now non-Horn. Derivations now become more complex than when Horn clauses exist to evaluate queries.

1.2.3 **Splitting a Non-Horn DB into a Disjunction of Horn DBs.**

We have noted the desirability of working with a Horn DB above. It would be of interest to determine if a DB could be factored in some way so as to become a Horn DB. Loveland [1978] (cf. pp. 100-102) shows how this may be accomplished. As an example of his method consider a DB which consists of only Horn clauses to which we add the non-Horn clause $D = (a_1 \lor a_2 \lor C)$, where C is a disjunction of negative atomic formulae. Then $DB \land D \equiv DB \land (a_1 \lor a_2 \lor C) \equiv (DB \land (a_1 \lor C)) \lor (DB \land (a_2 \lor C))$ is a tautology.

Thus, a negated query, $\overline{Q}$, together with DB∧D is unsatisfiable if-and-only-if $\overline{Q}$ taken individually with DB∧$(a_1 \vee C)$, and with DB∧$(a_2 \vee C)$ are both unsatisfiable. The test for unsatisfiability in each of the two DBs is not independent. Specifically, variables from a split clause in each DB must be identical. Notice what happens if they are not independent. Consider the DB as follows :

DB: Ra∧ Rb∧ (Px∨ Qx∨ $\overline{R}$x).
Split DB (DBS) : (Ra∧ Rb∧ (Px∨ $\overline{R}$x))∨ (Ra∧ Rb∧ (Qx∨ $\overline{R}$x)).
Let the query be
 Q : Pa∨ Qb.
Now,
 DB∧ $\overline{Q}$ = Ra∧ Rb∧ (Px∨ Qx∨ $\overline{R}$x)∧ $\overline{P}$a∧ $\overline{Q}$b
is satisfiable since {$\overline{P}$a,Pb,Qa,$\overline{Q}$b,Ra,Rb} all true constitutes a model. However,
 DBS∧$\overline{Q}$ yields a split into
 DBS$_1$∧ $\overline{Q}$ = Ra∧ Rb∧ (Px∨ $\overline{R}$x)∧ $\overline{P}$a∧ $\overline{Q}$b and
 DBS$_2$∧$\overline{Q}$ = Ra∧ Rb∧ (Qx∨ $\overline{R}$x)∧ $\overline{P}$a∧ $\overline{Q}$b
both of which are unsatisfiable when viewed independently. One must first prove DBS$_1$ $\overline{Q}$ unsatisfiable. Once this is done, bindings to variables in DBS$_2$ that arise from clauses that were split must be made identical. Thus, in DBS$_2$∧$\overline{Q}$ a derivation can be found with x = a. If DBS$_2$∧$\overline{Q}$ has x bound to a before the proof, we find DBS$_2$∧$\overline{Q}$ to be satisfiable. If we back up to find an alternative proof in DBS$_1$∧$\overline{Q}$ to be the case when x = b, binding x to b in DBS$_2$∧$\overline{Q}$ now leads to a derivation of the null clause.

The above illustrates the care that must be used in performing splits. The splitting process is easily generalized to having n non-Horn clauses with n_i positive literals in the ith non-Horn clause. For a single non-Horn clause with n positive literals D = $a_1 \vee \ldots \vee a_n \vee C$, a_i are positive literals and C_i a clause with all negative literals DB∧ D ≡ (DB ∧ ($a_1 \vee C_1$))∨ ... ∨(DB ∧ ($a_n \vee C_1$)). For n non-Horn clauses we have a split into $\prod_{i=1}^{n} n_i$ different databases. It should be clear from the above that splitting becomes extremely cumbersome even for small values of n and n_i. For example if n_i = 2 for i = 1,...,n, there are 2^n different spaces in which a query must be solved.

1.2.4 The Non Monotonic Character of Closed World Definitions

An important property of first order logic is that it is *monotonic*, i.e., if P and Q are sets of first order formulae and P ⊢w then P∪Q ⊢w. Formulae valid with respect to the formulae in P remain valid when additional information is added. The closed world assumption is *non-monotonic*. This may be seen as follows :

Let DB ={ p∨$\overline{q}$}. Then we may assume $\overline{p}$ and $\overline{q}$ by the closed world assumption. Thus, DB ⊢$\overline{p}$. Now, if to DB we add q, then DB∪{q} ={ p∨$\overline{q}$,q} and DB∪{q}⊢p. Hence when dealing with a closed world assumption one is dealing with a non-monotonic logic. The generalization of the closed world assumption as described in subsequent sections of this paper is also a non-monotonic logic. We shall show this later.

Some interesting descriptions of non-monotonic logics are given by McCarthy [1980],

Mc Dermott and Doyle [1980], Davis [1980], Reiter [1980], and Weyhrauch [1980] in a special issue of the *Artificial Intelligence Journal* of April 1980 devoted to this subject.

2. Indefinite Databases and the Closed World Assumption

2.1 Need for an Extended Closed World Assumption Definition

Consider the simple database

(2.1 - 1) $\quad \left\{ \begin{array}{l} Pa \lor Pb \\ Pc \lor \overline{P}d \end{array} \right\}$.

We have seen previously that the closed world assumption does not apply to this database and leads to an inconsistent database if applied. However, (2.1 - 1) does admit a definition of atoms whose negations can be assumed to be true. In particular, there is no loss if we assume $\overline{P}c$ and $\overline{P}d$ to be true. Notice that if we add $\overline{P}c$ and $\overline{P}d$ to (2.1 - 1) to obtain

(2.1 - 2) $\quad \left\{ \begin{array}{l} Pa \lor Pb \\ Pc \lor \overline{P}d \\ \overline{P}c \\ \overline{P}d \end{array} \right\}$,

that this set of clauses is consistent. In a sense this is a maximally consistent set of clauses since if $\overline{P}a$ is added we can prove Pb from (2.1 - 2) and Pb cannot be proven from (2.1 - 1). Similarly if we add $\overline{P}b$ (but not $\overline{P}a$) we can prove Pa. If both $\overline{P}a$ and $\overline{P}b$ are added to (2.1 - 2), then the set is inconsistent.

We shall, in the following sections, make precise our definitions of a closed world assumption that encompasses both Horn and non-Horn databases.

2.2 Semantic Definition for a Closed World Definition of Databases

We shall first give a definition of a closed world assumption which applies to all databases. We shall refer to this as the GCWA in contrast to the CWA which applies only to Horn databases. The semantic definition will be a suitable generalization of the definition given by van Emden and Kowalski. We first need some preliminaries.

The database to which we shall refer is function-free and therefore contains no Skolem constants. An expression (term, literal, set of clauses) is *ground* if it contains no variables. We refer to the database as DB, where DB is a set of clauses. The set of all *ground atomic formulas* $P(t_1,...,t_n)$ where P occurs in the set of clauses DB and $t_1,...,t_n$ belong to the Herbrand universe H of DB is called the Herbrand base, $\hat{H}$, of DB. A Herbrand interpretation simultaneously associates, with every n-ary predicate symbol in DB a unique n-ary relation over H. The relation $\{(t_1,...,t_n) : P(t_1,...,t_n) \in I\}$ is *associated* by I with the predicate symbol P in DB.

      (1) A ground atomic formula A is *true* in a Herbrand interpretation I iff $A \in I$.

      (2) A ground negative literal $\overline{A}$ is *true* in I iff $A \notin I$.

      (3) A ground clause $L_1 \ldots L_n$ is *true* in I iff at least

one literal L_i is true in I.
(4) In general a clause C is *true* in I iff every ground instance $C\sigma$ is true in I. ($C\sigma$ is obtained by replacing every occurrence of a variable in C by a term in H. Different occurrences of the same variable are replaced by the same term).
(5) A set of clauses is *true* in I if and only if each clause in DB is true in I.

Given a Horn database van Emden and Kowalski have shown that the intersection of all models is a model. This is referred to as the model intersection property. As shown in Section 1.1, non-Horn data bases do not have a corresponding intersection property. The *model intersection* property of Horn clauses results in a unique minimal model. We generalize this concept for non-Horn databases.

A *model* of a set of clauses DB is an interpretation I such that every clause in DB is true in I. The set of Herbrand interpretations that are models of DB are denoted by $M(DB)$. A Herbrand interpretation I that forms a model for DB, i.e., $I \in M(DB)$ such that no smaller subset of I is a model is termed a *minimal model Herbrand interpretation* or simply a *minimal model*. The set of *minimal models of* DB is denoted by $MM(DB)$.

We illustrate the minimal model concept by the following examples.

Example 1.

Let DB = $\begin{Bmatrix} Pa \vee Pb \\ Pc \vee \overline{Pd} \end{Bmatrix}$.

the following constitute $M(DB)$.

$$\begin{Bmatrix} \{Pa\}, \\ \{Pb\}, \\ \{Pa, Pb\}, \\ \{Pa, Pc\}, \\ \{Pb, Pc\}, \\ \{Pa, Pb, Pc\}, \\ \{Pa, Pc, Pd\}, \\ \{Pb, Pc, Pd\}, \\ \{Pa, Pb, Pc, Pd\} \end{Bmatrix}.$$

The set of minimal models is given by :

$$MM(DB) = \begin{Bmatrix} \{Pa\}, \\ \{Pb\} \end{Bmatrix}.$$

Note that there are two minimal models. Note that $\overline{Pc}$ and $\overline{Pd}$ are in both minimal models since the positive atoms Pc and Pd do not appear in the minimal models.

Example 2.

Consider the following database :

$$DB = \begin{Bmatrix} Qa, \\ Pa \lor Pb, \\ Pc \lor Pd, \\ Rx \lor Qx \lor \overline{P}x \end{Bmatrix}.$$

It is clear that DB can be replaced by an equivalent set of clauses :

$$DB_1 = \begin{Bmatrix} Qa, \\ Pa \lor Pb, \\ Pc \lor Pd, \\ Ra \lor Qa \lor \overline{P}a, \\ Rb \lor Qb \lor \overline{P}b, \\ Rc \lor Qc \lor \overline{P}c, \\ Rd \lor Qd \lor \overline{P}d \end{Bmatrix} = \begin{Bmatrix} Qa, \\ Pa \lor Pb, \\ Pc \lor Pd, \\ Rb \lor Qb \lor \overline{P}b, \\ Rc \lor Qc \lor \overline{P}c, \\ Rd \lor Qd \lor \overline{P}d \end{Bmatrix}$$

where the clause $Qa \lor Ra \lor \overline{P}a$ is subsumed by the clause Qa and is therefore deleted.

It can be seen that there are twelve minimal models of DB.

$\Gamma_1 = \{Qa, Pa, Pc, Rc\}$,
$\Gamma_2 = \{Qa, Pa, Pc, Qc\}$,
$\Gamma_3 = \{Qa, Pa, Pd, Rd\}$,
$\Gamma_4 = \{Qa, Pa, Pd, Qd\}$,
$\Gamma_5 = \{Qa, Pb, Pc, Rb, Rc\}$,
$\Gamma_6 = \{Qa, Pb, Pc, Rb, Qc\}$,
$\Gamma_7 = \{Qa, Pb, Pc, Qb, Qc\}$,
$\Gamma_8 = \{Qa, Pb, Pc, Qb, Rc\}$,
$\Gamma_9 = \{Qa, Pb, Pd, Rb, Rd\}$,
$\Gamma_{10} = \{Qa, Pb, Pd, Qb, Rd\}$,
$\Gamma_{11} = \{Qa, Pb, Pd, Rb, Qd\}$,
$\Gamma_{12} = \{Qa, Pb, Pd, Qb, Pd\}$.

We note that Qa is in every minimal model, and $\overline{R}a$ is in every minimal model.

Let DB be a database whose Herbrand base is $\widehat{H}_{DB}$. Let $MM(DB) = \{\Gamma_i\}$, $i=1,\ldots,n$. be the set of minimal models of DB. Let $\Gamma = \left\{ \bigwedge_{i=1}^{n} \Gamma_i \right\}$, and let $\widetilde{\Gamma}$ be the set of ground atomic formulae not in any minimal model of DB. Thus, $\widetilde{\Gamma} = \left\{ \bigwedge_{i=1}^{n} (\widehat{H}_{DB} - \Gamma_i) \right\}$. Let $\widetilde{\Gamma D} = \Gamma \cup \widetilde{\Gamma}$. Every atomic formula in $\widetilde{\Gamma D}$ is said to be definite. Let ΓI be the set of atomic formulae not in $\widetilde{\Gamma D}$. Thus, $\Gamma I = \widehat{H}_{DB} - \widetilde{\Gamma D}$. Then every atomic formula in ΓI is said to be indefinite. An atomic formula in $\widetilde{\Gamma}$ is interpreted to be false and hence its negation is interpreted to be true. The atomic formulae in $\widetilde{\Gamma}$ are said to form the *generalized closed world assumption for databases* (GCWA). If we let $\overline{\Gamma} = \{\overrightarrow{Pc} \mid \overrightarrow{Pc} \in \widetilde{\Gamma}\}$, then $\overline{\Gamma}$ represents the negative formulae true in the database. Similarly, if we let $\Gamma D = \Gamma \cup \overline{\Gamma}$, then ΓD represents the atomic formulae and negation of atomic formulae that can be assumed true in the DB.

It should be clear that the above definition is consistent with the CWA for Horn databases. In this case Γ represents the set of all atomic formulae that can be proven from DB since there is a unique minimal model. The set $\widetilde{\Gamma}$ then contains all atomic formulae that cannot be proven from DB. Thus, the definition for CWA is contained within the definition for GCWA.

Example 3.
$DB = \{Pa \lor Pb\}$
$\Gamma_1 = \{Pa\}$
$\Gamma_2 = \{Pb\}$ } only minimal model interpretations of DB.
$\Gamma = \tilde{\Gamma} = \{\emptyset\}, \Gamma D = \{\emptyset\}, \Gamma I = \{Pa, Pb\}$

Example 4.
$DB = \{Qa, Pa \lor Pb\}$
$\Gamma_1 = \{Qa, Pa\}$
$\Gamma_2 = \{Qa, Pb\}$ } only minimal model interpretations of DB.
$\Gamma = \{Qa\}, \tilde{\Gamma} = \{Qb\}, \widetilde{\Gamma D} = \{Qa, Qb\}, \Gamma I = \{Pa, Pb\}$.

There are two definite atomic formulae Qa and Qb, and two indefinite formulae. We can assume $\overline{Qb}$ since it is a member of both minimal models.

Given a consistent set of Horn clauses in a DB, then by van Emden and Kowalski there is a unique minimal model, the intersection of all models. If an atomic formula does not appear in the intersection, its negation is true. Thus, Γ corresponds to the set of positive atomic formulae that are true in every model, while $\tilde{\Gamma}$ corresponds to the set of positive atomic formulae whose negations are true. $\Gamma \cup \tilde{\Gamma}$ then corresponds to the set of all atomic formulae in the Herbrand base. In Horn databases every atomic formula is definite.

When one has indefinite data, $\Gamma \cup \tilde{\Gamma}$ contains the atomic formulae that are definite while the remaining atomic formulae are indefinite. If we have two indefinite formulae, say Pa and Qb, and we want to determine whether or not Pa∨Qb is true, it suffices to find in every minimal model if either Pa or Qb is true. If this is the case, then Pa∨Qb is true. The above definition of a GCWA thereby generalizes the concept of a CWA to be applicable to Horn databases, non-Horn databases, or mixed databases.

2.3 Syntactic Definition for a Closed World Definition of Databases

In Horn databases if from DB $\not\vdash$ Pc, then $\overline{Pc}$ can be assumed to be true. We seek a suitable definition that corresponds here.

Let DB be a consistent set of clauses. Let $E = \{\vec{Pc} \mid DB \vdash \vec{Pc} \lor K$, and K is either the null clause, or K is a positive clause such that $DB \not\vdash K\}$. That is, E corresponds to the set of atoms that appear in positive clauses derivable from DB. Let $\widetilde{EDB} = \hat{H}_{DB} - E$, where $\hat{H}_{DB}$ is the Herbrand base of DB. We claim that $\widetilde{EDB}$ corresponds to the set of atoms whose negation may be assumed to be true in DB. Hence $\widetilde{EDB}$ corresponds to the syntactic definition of the GCWA assumption. Let $\overline{\overline{EDB}} = \{\overline{\vec{Pc}} \mid \vec{Pc} \in \widetilde{EDB}\}$, then $\overline{\overline{EDB}}$ consists of the negated atoms assumed to be true in the DB. We must show that the syntactic and the semantic definitions are equivalent.

For Horn databases Reiter has defined $\overline{EDB} = \{\overline{\vec{Pc}} \mid DB \not\vdash \vec{Pc}\}$ to be the definition of the closed world assumption (CWA). As shown by Reiter, if DB is a consistent database consisting only of Horn clauses, then we can derive only definite answers. Hence, E reduces to $E = \{\vec{Pc} \mid DB \vdash \vec{Pc}\}$, and since $\widetilde{\overline{EDB}} = \hat{H}_{DB} - E = \{\vec{Pc} \mid DB \not\vdash \vec{Pc}\}$, and $\overline{EDB} = \{\vec{Pc} \mid \vec{Pc} \in \widetilde{EDB}\}$, $\overline{EDB} = \widetilde{\overline{EDB}}$. Hence, for Horn databases $\overline{EDB}$ encompasses $\widetilde{EDB}$.

We can show the following.

Theorem 1

Let DB be a consistent database, and K a clause.

If $DB \vdash K$, then every minimal model of DB contains a literal of K.

Proof : If $DB \vdash K$, then K is true in every model of DB. Hence, K is true in every minimal model of DB.

3. Equivalence of Generalized Closed World Assumption (GCWA) Syntactic and Semantic Definitions for Databases

3.1 Proof of Equivalence

We now show that the semantic and syntactic definitions of the GCWA are equivalent. The proof will be based on the following lemma.

Lemma 1

Let DB be a consistent set of clauses C, and $C = CP \cup CNP$ where CP consists of all positive clauses provable from DB, and CNP is the set of all clauses provable from DB that contain at least one negated atom. Then, every minimal model of C is a minimal model of CP.

Proof :

Let M be a minimal model of C. Let M_{CP} be a minimal model of M restricted to the clauses in CP. If $M = M_{CP}$, we are done. Hence, assume $M_{CP} \subset M$. There must be a non empty minimal mode M_{CP} of M restricted to the positive clauses in C since M is a model of CP.

There must be a clause in CNP such that every positive atom in the clause is not in M_{CP} and every negated atom in the clause is in M_{CP}. If this were not the case, then M_{CP} would be a model of every clause in CNP. But then M_{CP} would be a model of C. Since $M_{CP} \subset M$, it contradicts the assumption that M is minimal. Let a clause having this property be :

(1) $L \vee \overline{B}_1 \vee \ldots \vee \overline{B}_n$,

where L is a positive clause all of whose atoms are not in M_{CP}, and the B_i, $i=1,\ldots,n$ are positive atoms such that $B_i \in M_{CP}$.

Let the clauses in CP containing the B_i be of the following forms :

(2) $\{\bigwedge_{i_1=1}^{K_1} D_{1i_1} \vee B_1, \bigwedge_{i_2=1}^{K_2} D_{2i_2} \vee B_2, \ldots, \bigwedge_{i_n=1}^{K_n} D_{ni_n} \vee B_n\}$

where the D_{ji_j} are positive clauses that contain no B_t for $t \neq j$, and

(3) {clauses containing two or more B_i, $i = 1,\ldots,n$}.

Assume that (2) does not contain any clauses of the form $\bigwedge_{i_j=1}^{K_j} D_{ji_j} \vee B_j$ for some j. Then, the only clauses in which B_j appear in CP are in (3). But B_j always appears with some other atom $B_i \in M_{CP}$. Hence, $M_{CP} - \{B_j\}$ is a model of CP. But M_{CP} was assumed to be minimal and we have a contradiction.

Now assume that B_j must appear in a clause in (1) for $i = 1,\ldots,n$. That is, there must be at least one clause in CP of the form $\mathcal{D}_{ji_j} \in B_j$ for $j = 1,\ldots,n$. Resolving clauses in (1) with clauses in (2) yields the following set of clauses in CP:

(4) $\{ \bigcap_{i_1=1}^{K_1} \ldots \bigcap_{i_n=1}^{K_n} L \vee \mathcal{D}_{1i_1} \vee \ldots \vee \mathcal{D}_{ni_n} \}.$

Since L is false in M_{CP}, and every clause in (4) is in CP, an atom in each of the clauses (4), coming from the $\mathcal{D}_{ti_t}$ must be in M_{CP}. Now, there must be at least one B_j such that there is an atom for each $\mathcal{D}_{ji_j}, i_j = 1,\ldots,k_j$ that is true in M_{CP}. If this were not the case, i.e. each B_j has a $\mathcal{D}_{ji_j}$ all of whose atoms are falsified in M_{CP}, we would have constructed a clause (4) in CP (or a clause in CP that subsumes a clause in (4)) that is falsified in M_{CP}, which is clearly not possible since M_{CP} is a minimal model of CP. Since there is an atom in every $\mathcal{D}_{ji_j}, i_j=1,\ldots,k_j$ that is true in M_{CP}, then every clause in (2) of the form $\mathcal{D}_{ji_j} \vee B_j$ is true in $M_{CP}-\{B_j\}$ and furthermore, every clause in (3) is true in $M_{CP}-\{B_j\}$, and these are the only clauses in CP that contain B_j, $M_{CP}-\{B_j\}$ is a model of CP and M_{CP} cannot be a minimal model of CP. Hence, we have a contradiction. The above argument applies for every clause in CNP. Hence, we must have $M_{CP}=M$ which was to be proven.

Theorem 2
Let DB be a consistent database. An atom L is in $\widetilde{\Gamma}$ iff L is in $\widetilde{EDB}$. That is, $\widetilde{EDB} = \widetilde{\Gamma}$.

Proof: We show first that $\widetilde{\Gamma} \subseteq \widetilde{EDB}$ and then that $\widetilde{EDB} \subseteq \widetilde{\Gamma}$.

Proof that $\widetilde{\Gamma} \subseteq \widetilde{EDB}$. To show that if $L \in \widetilde{\Gamma}$, then $L \in \widetilde{EDB}$ is equivalent to showing that if $L \in \widetilde{\Gamma}$, then $L \notin E$. Assume that $L \in \widetilde{\Gamma}$ and $L \in E$. Since we assume $L \in E$, by definition $DB \vdash L \vee K$ where K is either the null clause or a positive clause such that $DB \not\vdash K$. If $DB \vdash L$, then L is in every minimal model and hence, L cannot be in $\widetilde{\Gamma}$ which is a contradiction. In the case $DB \vdash L \vee K$, every minimal model must contain an atom of $L \vee K$. But, since $DB \not\vdash K$, there is a minimal model that contains no atom of K (for otherwise $DB \vdash K$). This minimal model must contain L since $DB \vdash L \vee K$. Hence L appears in some minimal model and hence L cannot be in $\widetilde{\Gamma}$ which is a contradiction.

Proof that $\widetilde{EDB} \subseteq \widetilde{\Gamma}$. Let $L \in \widetilde{EDB}$. Then, $DB \not\vdash L \vee K$, where K is either null or a positive clause such that $DB \not\vdash K$. Now, L cannot appear in any positive clause in CP (the set of all positive clauses derivable from DB). For, if it did then it would not be in $\widetilde{EDB}$. Hence, $L \notin CP$ and since every minimal model of DB is a minimal model of CP as shown by the above lemma, L is in no minimal model of DB. But then

$L \in \widetilde{\Gamma}$. Hence, $\widetilde{\widetilde{EDB}} \subseteq \widetilde{\Gamma}$.

Since we have shown that $\widetilde{\widetilde{EDB}} \subseteq \widetilde{\Gamma}$ and $\widetilde{\Gamma} \subseteq \widetilde{\widetilde{EDB}}$, we have $\widetilde{\Gamma} = \widetilde{\widetilde{EDB}}$.

Corollary 2.1 :

$\overline{\overline{EDB}} = \overline{\Gamma}$.

Proof : The corollary is obvious from the definitions of $\overline{\overline{EDB}}$, $\overline{\Gamma}$ and the equivalence of $\widetilde{\widetilde{EDB}}$ and $\widetilde{\Gamma}$.

3.2 Properties of GCWA

We derive some consequences of the above. Let $\overline{\overline{EDB}}$ be the negated atoms of $\widetilde{\widetilde{EDB}}$ defined previously.

Theorem 3

Let DB be a consistent database, then $DB \cup \overline{\overline{EDB}}$ is consistent.

Proof : Every minimal model of DB is free of atoms of $\widetilde{\widetilde{EDB}}$ since $\widetilde{\Gamma} = \widetilde{\widetilde{EDB}}$. Hence, every minimal model of DB is a minimal model of $DB \cup \overline{\overline{EDB}}$. Hence $DB \cup \overline{\overline{EDB}}$ is consistent.

Theorem 4

Let DB be a consistent database and let K be a positive clause. If $DB \not\vdash K$, then $DB \cup \overline{\overline{EDB}} \not\vdash K$.

Proof : The previous theorem states that not only is $DB \cup \overline{\overline{EDB}}$ consistent, but every minimal model of DB is a minimal model of $DB \cup \overline{\overline{EDB}}$. If $DB \cup \overline{\overline{EDB}} \vdash K$ an atom of K must appear in every minimal model of DB. However, this means that $DB \vdash K$ which is a contradiction.

The above theorem states that we cannot prove any more positive clauses from $DB \cup \overline{\overline{EDB}}$ than we can prove from DB. However, there are some non-positive clauses that can be proven from $DB \cup \overline{\overline{EDB}}$ that cannot be proven from DB aside from the unit clauses in $\overline{\overline{EDB}}$.

Example 5.

Let $DB = \{P \vee Q \vee \overline{R}\}$.

Then, $\Gamma_1 = \{\emptyset\}$ is the only minimal model. Hence $\Gamma = \{\emptyset\}$ and $\widetilde{\Gamma} = \{P,Q,R\}$. Hence $\widetilde{\widetilde{EDB}} = \{P,Q,R\}$, and $\overline{\overline{EDB}} = \{\overline{P},\overline{Q},\overline{R}\}$. Now, $DB \cup \overline{\overline{EDB}} = \{P \vee Q \vee \overline{R}, \overline{P}, \overline{Q}, \overline{R}\}$. $DB \cup \overline{\overline{EDB}} \vdash \overline{P} \vee \overline{R}$ and $DB \not\vdash \overline{P} \vee \overline{R}$.

The set of clauses $\overline{\overline{EDB}}$ when added to DB forms a maximal set of clauses in the following sense. A set of clauses DB *is maximally consistent with the closed world assumption for databas*es if the addition of a negative atom to DB, where the atom, say Pa, is taken from the Herbrand base of DB and is not subsumed by DB satisfies the following conditions.

(1) $DB \cup \{\overline{Pa}\}$ is inconsistent or

(2) $DB \cup \{\overline{Pa}\} \vdash C$, where C is a positive clause and $DB \not\vdash C$.

Theorem 5

Let DB be a consistent database, then $DB \cup \overline{\overline{EDB}}$ is maximally consistent.

Proof : Assume that $DB \cup \overline{\overline{EDB}}$ is not maximally consistent. From Theorem 3, $DB \cup \overline{\overline{EDB}}$ is consistent. If $Pa \in \widetilde{\widetilde{EDB}}$, then $\overline{Pa} \in \overline{\overline{EDB}}$ and the addition of $\overline{Pa}$ does not change the status of $DB \cup \overline{\overline{EDB}}$. Now, let Pa be an atom in a positive clause in CP. Consider

$DB \cup \overline{EDB} \cup \{\overline{Pa}\}$. Since $DB \vdash Pa \vee K$, where K is null or positive and $DB \not\vdash Pa$, $DB \not\vdash K$, then every minimal model for $DB \cup \overline{EDB}$ that contains Pa cannot be a model of $DB \cup \overline{EDB} \cup \{\overline{Pa}\}$. If every model of $DB \cup \overline{EDB}$ contains Pa, then the addition of $\overline{Pa}$ causes the set of clauses to be inconsistent. The remaining minimal models of $DB \cup \overline{EDB}$ must be minimal models of $DB \cup \overline{EDB} \cup \{\overline{Pa}\}$. Hence, since every minimal model of $DB \cup \overline{EDB}$ contains either an atom of Pa or an atom of K, the minimal models of $DB \cup \overline{EDB} \cup \{\overline{Pa}\}$ must contain atoms of K. Hence $DB \cup \overline{EDB} \cup \{\overline{Pa}\} \vdash K$ and therefore $DB \cup \overline{EDB}$ is maximally consistent with the GCWA.

The concept of maximally consistent is useful. However, it may be possible to make a weaker assumption about databases that might prove computationally efficient. For example let the database be $DB \cup \{Pa \vee Pb, Pc\}$, where DB is Horn and contains no predicate letter of the form P. We may define every atom of P to be indefinite. Using a Horn theorem prover, failure to prove Pa would not permit us to assume $\overline{Pa}$, which is correct. We could also prove Pc which is again correct. However, if we could not find Pd we could not assume $\overline{Pd}$ which is not correct according to the GCWA. We could only conclude that Pd could be proven with another atom which would not be correct. However, this might be satisfactory for some database applications. With the GCWA we obtain a maximally complete set so that there would be no ambiguity concerning the status of Pd. That is, $\overline{Pd}$ could be assumed to be true.

In Section 1.2.4 we noted that the GCWA had the non-monotonicity property. This can be seen easily as follows. Let, $DB = \{p \vee q \vee r\}$. Then, assuming the GCWA, $DB \vdash \overline{r}$. Consider now $DB \cup \{r\}$. Then, $DB \cup \{r\} \not\vdash \overline{r}$. Hence the GCWA has the non-monotonicity property. Thus, when adding new information to DB, we must *revise our concept of $\overline{EDB}$*. That is our belief about what negative data holds true must be modified.

3.3 Null Values and the GCWA

The concept of a null value in database systems has many connotations. It may denote, "value at present unknown, but one of some finite set of known possible values", or "value at present unknown, but it is not necessarily one of some finite set of known possible values".

We shall consider the case where the unknown value must be among the finite set of constants that define the database. An unknown value such as in $P(a,\omega)$, where a is a constant and ω a special symbol denoting unknown is, in fact, a logical statement $(\exists x)P(a,x)$. When the logical statement is transformed to clause form, a Skolem constant is introduced.

The statement $(\exists x)P(a,x)$ can be replaced by an equivalent statement, $P(a,a_1) \vee \ldots \vee P(a,a_n)$, where the a_i, $i = 1,\ldots,n$ are all of the finite number of constants in the database. Hence, a null value corresponds to an indefinite statement in the database.

Assume now that the only data in some database that contains constants a and b is $P(a,\omega)$, where ω is a Skolem constant. Since the database appears to be Horn, if one attempts to prove $P(a,b)$, one cannot do so and one is led, under the closed world assumption (CWA) to assume $\overline{P}(a,b)$. For the same reason one is led to assume $\overline{P}(a,a)$. But, then $P(a,\omega) \wedge \overline{P}(a,a) \wedge \overline{P}(a,b)$ is not consistent since ω is either a or b. However, using the generalized closed world assumption (GCWA), and recognizing that $P(a,\omega)$ is equivalent to $P(a,a) \vee P(a,b)$, we cannot conclude $\overline{P}(a,a)$ or $\overline{P}(a,b)$.

The GCWA is compatible, then, with null values in a database where the null value must be among the constants in the database. Null values in a formula such as $(\exists x)(P(a,x) \wedge Q(x,b))$ may be written as $P(a,\omega) \wedge Q(\omega,b)$. The Skolem constant ω must be the same in the two clauses which result from this formula, namely, $\{P(a,\omega), Q(\omega,b)\}$. However, given that null values arise from different formulae, the Skolem constants must be written as $R(\omega_1,b)$, where ω and ω_1 are distinct Skolem constants.

4. Summary

An extended definition has been given to the concept of a closed world assumption that applies to function-free clauses. Semantic and syntactic definitions have been provided for the GCWA that serves to generalize the CWA described by Reiter. It was shown that the syntactic and semantic definitions are equivalent. The generalization reduces to the CWA when one has Horn DBs. The concept of an open world assumption (OWA) makes no assumption about failure by negation and hence remains the same. We have also defined the concept of maximally complete, and have shown that DB $\cup$ $\overline{EDB}$ is maximally complete, where DB is the database and EDB the negated atoms defined by the GCWA. The GCWA was also shown to handle the concept of null values correctly.

What remains to be addressed is how the concept can be used in a practical sense.

Acknowledgements

I wish to express my appreciation to John Grant and Michael Hudak for stimulating discussions on this work. I also wish to express my appreciation to Jorges Vidart and the Universidad Simon Bolivar who provided support to me for a visit where collaborative work was performed with Phillipe Roussel. The concept of minimal models for non Horn databases as used in this paper is due to Phillipe Roussel. I also gratefully acknowledge support from the National Science Foundation under grant number SK-CRC6. Thanks are also due to Liz Bassett for her careful typing of this paper.

References

1. Clark, K.L. 1978, "Negation as Failure" In *Logic and Data Bases* (H. Gallaire and J. Minker, Eds.), Plenum Press, New York, N.Y., 1978, 293-322.

2. Davis, M. 1980, "The Mathematics of Non-Monotonic Reasoning", *Artificial Intelligence Journal*, 13, 2 (April 1980), 73-80.

3. Hewitt, C. 1972, "Description and Theoretical Analysis (Using Schemata) of PLANNER : A Language for Proving Theorems and Manipulating Models in a Robot", AI Memo No. 251, MIT Project MAC, Cambridge, Mass., April 1980.

4. Hill, R. 1974, "LUSH Resolution and its Completeness" <u>DCL Memo No. 78</u>, University of Edinburg School of Artificial Intelligence, August 1974.

5. Kowalski, R.A. and Kuehner, D. 1971, "Linear Resolution with Selection Function" <u>Artificial Intelligence Vol. 2</u>, 1971, 227-260.

6. Kowalski, R.A. 1978, "Logic for Data Description", In <u>Logic and Data Bases</u> (H. Gallaire and J. Minker, Eds.), Plenum Press, New York, N.Y., 1978, 77-103.

7. Loveland, D.W. 1968, "Mechanical Theorem Proving by Model Elimination", <u>JACM 15</u>, (April 1968) 236-251.

8. Loveland, D.W., Stickel, M.E. 1973, "A Hole in Goal Trees : Some Guidance from Resolution Theory", Reproduced in <u>IEEE Transactions on Computers, C-25</u>, April 1976, 335-341.

9. Loveland, D.W. 1978, <u>Automated Theorem Proving : A Logical Basis</u>, North Holland Publishing Co., New York, 1978.

10. McCarthy, J. 1980, "Circumscription - A Form of Non-Monotonic Reasoning", <u>Artificial Intelligence Journal, 13</u>, 2 (April 1980), 27-39.

11. McDermott, D. and J. Doyle 1980, "Non-Monotonic Logic I", <u>Artificial Intelligence Journal, 13</u>, 2 (Apirl 1980), 41-72.

12. McCarthy, J. 1980, "Addendum : Circumscription and other Non-Montonic Formalisms", <u>Artificial Intelligence Journal, 13</u>,2 (April 1980), 171-172.

13. Minker, J. and Zanon, G. 1979, "LUST Resolution : Resolution with Arbitrary Selection Function", <u>University of Maryland Technical Report TR-736</u>, February 1979.

14. Nicolas, J.M. and Syre, J.C. 1974, "Natural Question Answering and Automatic Deduction in the System Syntex", Proceedings IFIP Congress 1974, Stockholm, Sweden, August 1974.

15. Nicolas, J.M. and Gallaire, H. 1978, "Data Bases : Theory vs. Interpretation", In <u>Logic and Data Bases</u> (H. Gallaire and J. Minker, Eds.), Plenum Press, New York, N.Y. 1978, 33-54.

16. Reiter, R. 1971, "Two Results on Ordering for Resolution with Merging and Linear Format", <u>JACM 18</u>, (October 1971), 630-646.

17. Reiter, R. 1978a, "Deductive Question-Answering on Relational Data Bases", In <u>Logic and Data Bases</u> (H. Gallaire and J. Minker, Eds.), Plenum Press, New York, N.Y. 1978, 149-177.

18. Reiter, R. 1978b, "On Closed World Data Bases", In <u>Logic and Data Bases</u> (H. Gallaire and J. Minker, Eds.), Plenum Press, New York, N.Y., 1978, 55-76.

19. Reiter, R. 1980, "A Logic for Default Reasoning", <u>Artificial Intelligence Journal, 13</u>, 2 (April 1980), 81-132.

20. Roussel, P. 1975, <u>PROLOG - Manuel de Reference et de Utilisation</u>, Groupe d' Intelligence Artificielle, Universite d' Air Marseilles, Luminy, September 1975.

21. van Emden, M.H. 1977, "Computation and Deductive Information Retrieval", Department of Computer Science, University of Waterloo, Ont., Research Report CS-77-16, May 1977.

22. van Emden, M.H. and Kowalski, R.A. 1976, "The Semantics of Predicate Logic as a Programming Language" <u>JACM 23</u>, 4 (October 1976), 723-742.

23. Weyhrauch, R. W. 1980, "Prologomena to a Theory of Mechanized Formal Reasoning" <u>Artificial Intelligence Journal, 13</u>, 2 (April 1980), 133-170.

PROOF BY MATRIX REDUCTION AS PLAN + VALIDATION

Ricardo Caferra
Laboratoire IMAG - Université Scientifique et Médicale de Grenoble
BP 53 / 38041 GRENOBLE CEDEX - FRANCE

ABSTRACT

In the last years there has been a renewed of interest for methods based upon principles similar to those of matrix reduction. In this paper we first present matrix reduction and we characterize some of its features with respect to resolution. We then define an abstraction of connexion graphs for matrix reduction and, using this abstraction we introduce an implementation technique and a search of proofs which is separated into two parts : a plan search and a plan validation. The plan abstracts from the arguments of literals. Given a plan, a set of equations in the terms is determined. The existence of a solution of this set of equations validates the plan. The plan and the solution define a refutation of the initial set of clauses. When no solution exists, a subset of the equations is isolated which is responsible for the fact. After backtracking, future plans will never use this subset.
These ideas has been used for implementing the method of matrix reduction in Pascal on a microcomputer (LSI11). The separation into plan search and plan validation can be used (and is particularly adapted) for a parallel implementation.

1. INTRODUCTION

Since its publication in 1969, the method of matrix reduction [16], [17], has been viewed in two different ways : firstly it was considered uninteresting and was discarded in favour of the resolution method. The criticisms were mainly aimed at the first version of the method [15] which appeared before the resolution (for these criticisms see for example [24]). Secondly and more recently there has been a revival of interest for the method, or at least for methods using principles similar to that of the matrix reduction ([1], [2], [3], [5], [7]). To our knowledge, the only two works where the method is compared to that of resolution are [12] and [17]. In the latter the views are concise and general, in the former it is showed that very close ties exist ("isomorphisms") between matrix reduction, a form of linear resolution and the model elimination procedure [11], [13].
We consider that these data enable us to determine neither the relative value, with respect, i.e, to resolution, nor the possibilities of the method.

We have not been able to reveal limitations intrinsic to the method (the

problem of extra space which is needed may be solved by an adequate sharing technique, as will be seen) ; moreover it offers the possibility of using concepts such as those of matrix and paths (these concepts have been studied and developped by Bibel [2], [3] . This has lead us to clarify certain of its differences from the resolution and to define an implementation technique for it.
The implementation has been realized in PASCAL on a microcomputer (LSI11). It is interactive and has been designed to be easily adapted to a parallel environment which is not yet available to us. We shall first of all give the terminology and definition of notions used, then in the following order we shall speak of : matrix reduction in the ground case and some of its characteristics in relation to resolution ; matrix reduction in the general case ; the proposed implementation technique and finally we shall give a few details concerning the running of the program.

2. TERMINOLOGY AND DEFINITIONS

We have adopted the classical definitions of literal, complementary literals and unit clauses. We shall speak of *matrix* instead of set of clauses.

Def : Let M be a matrix $M = \{C_1, C_2, \ldots, C_m\}$ $c_i = \{L^i_1, \ldots L^i_{n_i}\}$, $1 \le i \le m$

a *path* P_k in M is a set of m literals :

$P_k = \{L^i_j / L^i_j \in C_i, 1 \le i \le m, 1 \le j \le n_i, \forall L, L' \in P_k$ if $L \in C_i$ and $L' \in C_j$ then $C_i \ne C_j\}$

The matrix contains therefore $NP(M) = n_1 \times n_2 \times \ldots \times n_m$ paths and its disjunctive normal form is obtained as follows :

$$dnf(M) = \bigvee_{k=1}^{NP(M)} \bigwedge_{L^i_j \in P_k} L^i_j$$

If V represents the set of all the variables and T the set of all terms in the language of predicate calculus then :

Def : A *substitution* is an application $\sigma : V \to T$ identity almost everywhere, that is to say, except in a finite set of points [8].

The extension of the definition of a substitution to endomorphisms of T is :

$$\sigma(f(t_1, \ldots, t_n)) = f(\sigma(t_1), \ldots, \sigma(t_n))$$

This, plus the fact that σ is a total function on V enables the *composition of two substitutions* σ and θ to be defined simply as another substitution γ (noted $\gamma = \sigma.\theta$) which is the composition of the functions σ and θ. That is one of the reasons we prefer this definition to that given, for example in [13], which is more common. As is usual the substitutions are represented as the set of pairs :

$$\sigma = \{t_1 \to x_1, \ldots, t_n \to x_n\} \text{ or } \sigma = \{<x_1, t_1>, \ldots, <x_n, t_n>\}$$

where $x_i \in V$ $t_i \in T$ $x_i \ne t_i$ $(1 \le i \le m)$

knowing that : $\sigma(x_j) = x_j$ $\forall x_j \ne x_i$ $(1 \le i \le n)$

The domain of substitutions is naturally extended to literals, clauses and set of clauses (in general called expressions).

Def : A substitution is *variable-pure* iff $t_i \in V$ $\forall i$ ($1 \leq i \leq n$).

Def : A substitution variable pure σ is a *renaming substitution* for an expression E iff for every variable x in E $\sigma(x)=y$ where y is a variable not in E.

Def : Given an equation in the terms :

$$e : t_1 = t_2$$

an *unifier* of t_1 and t_2 is any solution of the equation, that is to say any substitution σ such that :

$$\sigma(t_1) = \sigma(t_2)$$

Def : A *solution to a set of equations* is a solution to all the equations of the set.

Def : The set of two literals $\{P(t_1,\ldots t_n), \bar{P}(t'_1,\ldots,t'_n)\}$ is called a *possible contradiction*. The set of equations $S = \{t_1=t'_1,\ldots,t_n=t'_n\}$ is called the *corresponding substitution condition* for this possible contradiction.

If this set of equations has a solution, the given literals are effectively made contradictory when instantiated with the unifer substitution (the definition given in [16] is a little different).

For us an initial matrix never has pure literals. It is well known that if one is interested in satisfiability (unsatisfiability), one may always come back to such a matrix (this obviously corresponds to a linked conjunct [6]).

Def : Let M be a matrix, a *gp-abstraction of M* is a pair <G-M,S-C>, where :

. G-M is a not orientated labelled graph, whose nodes are literals of the clauses of M modified in the following way : any literal with the form $P(t_1,\ldots,t_n)$ is replaced by P and any literal with the form $\bar{P}(t_1,\ldots,t_n)$ is replaced by $\bar{P}$. The arcs link contradictory literals in the different clauses and they are labelled by integers (this restriction on labels is in no way necessary but it is practical to adopt it).

. S-C is an application $\mathbb{N} \to \varepsilon$

ε : set of all substitution conditions.

$S-C(n) = \{t_1=t'_1,\ldots,t_n=t'_n\}$

where n is the label of the arc linking the nodes wich generate the substitution condition $\{t_1=t'_1,\ldots,t_n=t'_n\}$

We shall speak only of substitution conditions corresponding to the literals possibly contradictory of clauses of M, knowing at the same time that they may be treated independently.

This idea corresponds to a link between two "ions" [9] ($P(-,-,\ldots,-)$ is an "ion" of a literal $P(t_1,t_2,\ldots,t_n)$).

G-M is an abstraction representing the infinity of matrices obtained by uniformly renaming all the literals of the clauses and changing their arities and arguments

(S-C changes for each case). The search of solutions to S-C(n) may be treated in different ways, in this sense this abstraction generalizes the following syntactic abstractions of Plaisted [14] :

1) The propositional abstraction

If $C = \{L_1, L_2, \ldots, L_k\}$ then $f(C) = C'$ $C' = \{L'_1, L'_2, \ldots, L'_k\}$

where L'_i ($1 \le i \le k$) is defined as :

P if L_i is of the form $P(t_1, \ldots, t_n)$

$\bar{P}$ if L_i is of the form $\bar{P}(t_1, \ldots, t_n)$

2) Renaming predicate and function symbols.
3) Permuting arguments.
4) Deleting arguments.

to which must be added :

5) Adding arguments.

The idea which leads us to this abstraction is similar to that expressed by Plaisted. It is based on the fact that the attempt to find a solution to a problem P may be divided into two parts : the search for a solution to a *simpler* problem P_1 whose solution is *necessary* for the solution of P (and which correspond moreover to an *infinity* of problems, each with the *same structure* as P) and another problem P_2 which tries to "particularize" (if possible) the solution proposed by P_1 for a class of problems to which P belongs.

Def : The solution to P_1 is called a *plan* and the solution to P_2, if there is one, a *validation of the plan* (if no solution to P_2 is found, we shall speak of a failure in the attemp of plan validation).

Under certain conditions (particulary in the case which interests us), this division naturally leads to the possibliity of seeking a solution P_1 in *parallel* with the attempts to validate the parts of P_1 which have already been found. The ideas of plan and plan validation are implicit in [22], [23] and is similar to those set out in [5] where the principal difference from the other methods is that a plan must be generated and the unifier is calculated as final step of the proof.

Note that a gp-abstraction is applied to a set of clauses, the abstract clauses are in general multisets of literals.

3. THE METHOD OF MATRIX REDUCTION. GROUND CASE

We shall present this method in a similar way to that of Loveland [12].
The method of matrix reduction is based on a property (a) and on an observation (b) :

(a) "A matrix M is unsatisfiable iff any path of M contains a pair of complementary literals".

(b) "In a matrix M a given pair of complementary literals may be formed of two literals simultaneously belonging to more than one M path, it will make contradictory all paths containing it. If such a pair is detected, the *number of tests* needed to check the contradiction of all the paths is reduced".

Def : A matrix with 2 unit clauses which are complementary literals is said to be a *simply unsatisfiable matrix*. Such a matrix is denoted "[]".

Def : Let M be a ground matrix (i.e. a matrix containing only ground clauses). The *matrix reduction rule* is an application :
$M : M^* \times C^* \times C^* \times L^* \times L^* \to (<M_1,M_2>)^* \cup []$, where
M^* : set of all matrices (we consider $\emptyset \in M^*$).
C^* : set of all clauses.
L^* : set of all literals.
$(<M_1,M_2>)^*$: set of all pairs of matrices.

Let C, C'∈M ; L∈C ; L'∈C' L and L' are complementary literals
There are 3 possibles values for the application M :
1) If C and C' are not unit clauses then :
   $M(M,C,C',L,L')=<(M-\{C\}-\{C'\})\cup\{\{L\}\}\cup\{C'-\{L'\}\}, (M-\{C\}-\{C'\})\cup\{\{L'\}\}\cup\{C-\{L\}\}>$
2) If C={L} or C'={L'} (but not both) then :
   2a) $M(M,\{L\},C',L,L') = <(M-\{C'\}) \cup \{C'-\{L'\}\},\emptyset>$
   2b) $M(M,C,\{L'\},L,L') = <\emptyset,(M-\{C\}) \cup \{C-\{L\}\}>$
3) If C={L} and C'={L'} then
   $M(M,\{L\},\{L'\},L,L') = []$

Prawitz [16], [17] presented the method in a visually more direct form :

$$M : \begin{vmatrix} P & \alpha \\ \bar{P} & \beta \\ \Gamma & \end{vmatrix} \quad M_L : \begin{vmatrix} P & \\ & \beta \\ \Gamma & \end{vmatrix} \quad M_R : \begin{vmatrix} & \alpha \\ \bar{P} & \\ \Gamma & \end{vmatrix}$$

where α, β denotes set of literals ; Γ denotes a set of clauses.
If $\alpha \neq \emptyset$ $\beta \neq \emptyset$ the pair $<M_L,M_R>$ is the result of applying matrix reduction M to M on the literals P and $\bar{P}$ (rule 1).
If $\alpha=\emptyset$ and $\beta \neq \emptyset$ M_L is the result of applying M to M (rule 2a).
If $\alpha \neq \emptyset$ and $\beta=\emptyset$ M_R is the result of applying M to M (rule 2b).
If $\alpha=\beta=\emptyset$ M is simply unsatisfiable (rule 3).
If the number of literals of α and β are (m-1) and (n-1) respectively and the number of paths in Γ is NP_Γ, then the number of paths to be tested without applying the principle of reduction is $m \times n \times NP_\Gamma$; the number of paths to be tested if the matrix reduction is applied is $NP_{M_L}+NP_{M_R}=(m+n-2) \times NP_\Gamma$. The reduction is notable, but *the method does not in fact test contradiction of paths* ; it applies recursively the matrix reduction until it reaches a simply unsatisfiable matrix, or a matrix to which a matrix reduction cannot be applied.

We shall denote $\{[\,]\}$ a set of simply unsatisfiable matrices.

A deduction is defined in the classical way :

Def : Let M be a matrix and $T = \{M_1, M_2, \ldots, M_p\}$ a set of matrices. A *gm-deduction* of T from M-noted $M \vdash_{gm} T$- is a finite sequence $B_1, B_2, \ldots, B_n$ of set of matrices such that :

(1) $T = B_n$

(2) $\forall i \ 1 \leq i \leq n$

$B_i = \{M\}$ or

$B_i = \{M_1, M_2, \ldots, M_k, M_{k+1}, \ldots, M_1\}$ and

$\exists j \ M_j \in B_{i-1} \ M(M_j) = <M_k, M_{k+1}>$

$B_i = (B_{i-1} - \{M_j\}) \cup \{M_k, M_{k+1}\}$

Def : A *gm-refutation* of M is a gm-deduction of $\{[\,]\}$ from M.

The matrix reduction method enables a refutation and a model for a set of clauses to be sought *simultaneously* :

Theorem : A matrix M is unsatisfiable iff $M \vdash_{gm} \{[\,]\}$. If i, j exists such that $M_j \in B_i$ and it is not possible to apply the matrix reduction rule to M_j, then M is satisfiable and evaluating to true all the literals of M_j we obtain a model for M.

Proof : It follows immediately from the equivalence :

(1) $M \equiv M_L \vee M_R$

where $\quad M : (L \vee \alpha) \wedge (\bar{L} \vee \beta) \wedge \Gamma$

$M_L : L \wedge \beta \wedge \Gamma$

$M_R : \alpha \wedge \bar{L} \wedge \Gamma$

This equivalence is very easy to prove by constructing a truth table. For M be a contradictory matrix, M_L and M_R must also be thus. With L and $\bar{L}$ present, M_L and M_R may not have a simultaneous true evaluation. On a matrix $M_j \in B_i$ in a gm-deduction a reduction may or may not be applied. If one may apply it one continues until "[]" is found, or until one falls in the second possibility. In this case, M_j contains at least one literal of each clause (by definition of matrix reduction rule), then by a true evaluation of each of the literals it contains, we have a model for M_j and by the equivalence (1) it is a model for M also. (Obviously, if the purity principle is applied we found an empty matrix ; the conclusion is that M is satisfiable but we have not construct a model for it). ∎

It is well known that it is difficult to compare proof methods bases on different inference rules. We shall try to point out some characteristics of the matrix reduction which we consider advantageous as regards the resolution (in this and the following section, we shall precede by "(*)" the points on which we compare the two methods).

(*) With the above theorem, it is of no use (or necessity) to define the equivalent of $R^n(M)$ for the resolution (i.e. the set of all pairs of matrices which may be obtained by applying the matrix reduction rule in every possible way). Every gm-deduction is thus equivalent to a *saturation search* (in the meaning of Loveland [13]).

One consequence of this is that no strategy applied to the matrix reduction may fall on an impasse, necessitating a backtracking (this is due to the fact that the matrix reduction rule retains more information than the resolution).

Example : We want to refute $M = \{PQ, \overline{P}R, \overline{Q}\overline{R}\}$ (circled literals are literals on which matrix reduction is applied).

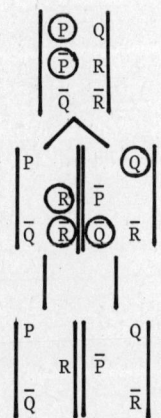

Refutation is not possible, M is satisfiable, $\{P, \overline{Q}, R\}$ and $\{Q, \overline{P}, \overline{R}\}$ are all the models of M.

(*) The matrix reduction rule does not generate tautology (this is obvious since the matrix reduction divides the clauses and the initial set of clauses contains no tautology).

(*) Search space : An indepth study of this problem would necessitate a separate work. Here we evaluate the variation in the number of possibles choices for applying the inference rule, before and after the application of matrix reduction and resolution to the initial matrix.

Let M be the initial matrix :

$$\begin{vmatrix} A & \alpha \\ \overline{A} & \beta \\ \Gamma & \end{vmatrix}$$

The number of possible applications of inference rule is :
$N = n_{\alpha\beta} + n_{\beta\Gamma} + n_{\alpha\Gamma} + n_{A\Gamma} + n_{\overline{A}\Gamma}^{-} + n_{\Gamma} + 1$
where n_{ij} : number of links between i and j
n_{Γ} : number of links between literals of clauses of Γ.

By resolving on A and $\overline{A}$ (and without taking into consideration any possibilities there are of applying the purity principle we shall speak on), the number of links becomes :

$$N_a = N_b - 1 + 2n_{\alpha\beta} + n_{\alpha\Gamma} + n_{\beta\Gamma}$$

By applying a matrix reduction on the same pair of complementary literals, the total number of links of M_L and M_R becomes :

$$N_a = N_b - 1 + n_\Gamma - n_{\alpha\beta}$$

We consider that for some kinds of problems the retention of more information and the *division* of the clauses carried out by the matrix reduction limits the growth of redundant deduction and thus "helps" the strategy which is used. The strategies used may be *simpler* (it is not necessary, for example, the no-tautology restriction). For example, for the set of clauses $S = \{PQ, \bar{P}, \bar{Q}R, \bar{R}S, \bar{R}\bar{S}T, P\bar{T}\}$ the *simplest refutations* by linear or SL-resolution needs 6 or 7 resolutions operations (Kowalski). All refutations by matrix reduction need 7 or 8 applications of this rule.

(*) Applications of purity principle : The division of clauses helps the application of purity principle, thus reducing the search space and simplifying the search for a solution. The resolution and the matrix reduction are applied to the initial matrix. We consider different possibilities of applying purity principle and we evaluate the number of links which remain after the *first* application of this principle.

1) A becomes pure :

   resolution : nblink $= n_{\bar{A}\Gamma} + 2n_{\beta\Gamma} + n_{\alpha\Gamma} + n_{\alpha\beta}$;

   matrix reduction : nblink $= n_{\beta\Gamma} + n_{\alpha\Gamma} + n_{\bar{A}\Gamma}$

2) $\bar{A}$ becomes pure :

   resolution : nblink $= n_{A\Gamma} + 2n_{\alpha\Gamma} + n_{\beta\Gamma} + n_{\alpha\beta}$;

   matrix reduction : nblink $= n_{\alpha\Gamma} + n_{A\Gamma} + n_{\beta\Gamma}$

3) A and $\bar{A}$ become pure :

   resolution : nblink $= n_{\alpha\Gamma} + n_{\beta\Gamma}$;

   matrix reduction : nblink $= n_{\alpha\Gamma} + n_{\beta\Gamma}$

4) Literals in clauses of Γ (say $\Delta \subseteq \Gamma$) become pure because

   i) their only links are with A :

   the purity principle does not apply in resolution ;

   in matrix reduction : nblink $= n_{\alpha\Gamma} + n_{\beta\Gamma} + n_{\bar{A}\Gamma} + (n_{A\Gamma} - n_{A\Delta})$

   ii) their only links are with $\bar{A}$:

   the purity principle does not apply in resolution

   in matrix reduction : nblink $= n_{\alpha\Gamma} + n_{\beta\Gamma} + n_{A\Gamma} + (n_{\bar{A}\Gamma} - n_{\bar{A}\Delta})$

5) The only links of Δ are with $\alpha(\beta)$

   the purity principle does not apply in resolution ;

   in matrix reduction :

   $$\text{nblink} = n_{A\Gamma} + n_{\beta\Gamma} + n_{\bar{A}\Gamma} + (n_{\alpha\Gamma} - n_{\alpha\Delta}) \text{ and}$$

   $$\text{nblink} = n_{A\Gamma} + (n_{\beta\Gamma} - n_{\beta\Delta}) + n_{\bar{A}\Gamma} + n_{\alpha\Gamma} \text{ respectively}$$

4. MATRIX REDUCTION METHOD. GENERAL CASE

Let M be a matrix of general clauses and $C \in M$, $C' \in M$, $L \in C$, $L' \in C'$, L and L' are possibly contradictory literals. If the substitution conditions corresponding to the pair L,L' have a solution α, the matrix reduction rule is as follows :

If C and C' are not unit clauses :

$M(M,C,C',L,L') = <(M-\{C\}-\{C'\}) \cup \{\{\alpha(L)\}\} \cup \{\alpha(C'-\{L'\})\}, (M-\{C\}-\{C'\}) \cup \{\{\alpha(L')\}\} \cup \{\alpha(C-\{L\})\}>$

It is defined in a similar way for the cases corresponding to 2a, 2b, 3, of the ground case.

For a matrice M to be unsatisfiable, the two matrices M_L and M_R obtained from it by matrix reduction must be unsatisfiable at the same time. If $\alpha_1(\alpha_2)$ is the substitution obtained as a solution to the substitution condition corresponding to a possible contradiction (on which we apply matrix reduction) in $M_L(M_R)$, then α_1 and α_2 must be *compatible* (this is analogous to the compatibility test in the interconnectivity graphs [24], [25]). In order to reduce the number of possible incompatibilities, the following equivalence may be used :

$\forall x \forall y \forall z \forall v ((P(x) \lor Q(y)) \land (\bar{P}(x) \land R(z)) \land S(v)) \equiv$

$\forall x \forall y \forall z \forall v \forall w ((P(x) \land R(z) \land S(v)) \lor (Q(y) \land \bar{P}(x) \land S(w)))$.

This enables the renaming in one of two matrices the variables of lines unaffected by the reduction applied to M.

We could have defined matrix reduction, as has been done in the resolution, by considering *sets of literals* instead of literals L and L', and afterwards define the above matrix reduction as the "binary matrix reduction" (one may prove that the two definitions are equivalent).

(*) It is interesting to note that the factorization notion is unnecessary in matrix reduction. This does away with difficulties and risks when it is used.

The example generally used to show necessity of factorization to the completeness of binary resolution $S = \{P(x) \lor P(y), \bar{P}(u) \lor \bar{P}(v)\}$ is treated very simply by the matrix reduction.

Def : A *m-deduction* (deduction in the general case) may be defined in the same way as the gm-deduction by adding the condition that the solution of substitution condition produced by applying the matrix reduction to a matrix in $B_i(\alpha)$ must be compatible with the solution corresponding to $B_1,\ldots,B_{i-1}$ (which may be called α_{i-1}) and $\alpha_i = \alpha \cdot \alpha_{i-1}$

Theorem : Let M be a matrix. M is unsatisfiable iff there exists a set of variants of clauses of M : M', such that $M' \vdash_m \{[\ ]\}$.

Proof : It follows directly from Herbrand's theorem and matrix reduction rule. ∎

We shall give an example of refutation using the matrix reduction. The graphic representation used and the way the proof is carried show the ideas used in the implementation.

Example [10] : Let M be :

$\bar{H}(u)$
$H(x)$ $\bar{P}(x)$
$H(z)$ $\bar{W}(y)$ $\bar{T}(z,y)$
$P(b)$ $W(b)$
$T(a,b)$

The gp-abstraction of M is :

S-C(1) = {u=x}
S-C(2) = {u=z}
S-C(3) = {x=b}
S-C(4) = {y=b}
S-C(5) = {z=a, y=b}

In the next refutation by matrix reduction purity principle is applied and we note :
— : selected link
≠ : deleted link by elimination of one node (application of matrix reduction or purity principle)

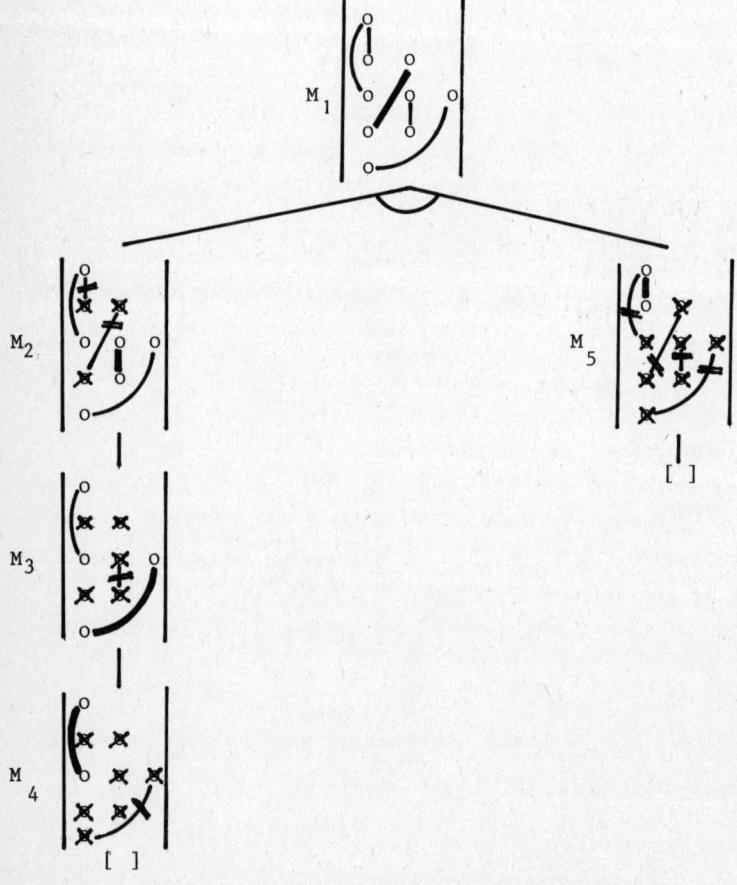

Plan : 3,4,5,2,1

Validation (renaming of S-C(1) on M5): The set $S = \{x=b, y=b, u_1=x, z=a, y=b, u=z\}$ has a solution.

We can note three principal differences from connexion graph [10] :
1) The links are not labelled by the m.g.u.
2) No m.g.u. is calculated until step validation.
3) Subgraphs of the initial graphs are tested.

5. IMPLEMENTATION TECHNIQUE

The technique used for the implementation is based on two principal points :
1) Structure sharing (for matrices and substitutions).
2) Separation of a refutation essay in a plan search and a plan validation.

A third point has to be underlined : a failure in the attempt of plan validation corresponds to a set S of equations in the terms that has not solution ; the program detects a subset of S (say S_1) that is responsible of this fact and that is "minimal" in the sense that all subsets of S_1 have a solution.

Pascal has been chosen because of its possibilities to use a varied data structure ; particularly the possibilities offered by Pascal sets are used as much as possible.

1) Structure sharing

   a) For matrices :
   . For a given set of clauses S, the gp-abstraction of S is obtained.
   . The matrices are represented as a set of integers (set of names of substitution conditions).

   b) For substitutions :
   . It is naturally done by the utilization of the chosen unification algorithm : the variables to be replaced by the same term t *are in the same equivalence class* that t (t is the representative element of the equivalence class).

2) Separation in plan + validation

   . From the gp-abstraction of S a *refutation plan* P is obtained :
   $$P : n_1, n_2, \ldots, n_p$$
   where $n_i \in \mathbb{N}$ ($1 \leq i \leq p$) are names of substitution conditions.

   . An attempt is made to *validate the plan* by seeking a solution to the set of equations :
   $$S = \{\eta_1(S-C(n_1)) \cup \eta_2(S-C(n_2)) \cup \ldots \cup \eta_p(S-C(n_p))\}$$
   where η_i ($1 \leq i \leq p$) is either a renaming substitution (if the substitution condition correspond to literals in lines which may be renamed) or the substitution identity everywhere.

A refutation is therefore a plan search *and* a plan validation.

. From among the various unifier algorithms which exist, we have chosen that of Huet [8]. Its principal characteristics are :

a) In a first step it constructs equivalence classes for the variables and terms. This characteristic is used so that clauses are not instantiated. The variable values are kept in equivalence classes.

b) In a second step it detects possible substitution component of the form $x \leftarrow f(\ldots,x,\ldots)$ (by looking for cycles in a graph).

This caracteristic may be used in an implementation of a language as PROLOG [10], [20] for an "optional occur check".
In the implementation the first step is carried out for *all* the substitution conditions corresponding to the plan and the cycles are searched when the first step has been successfully completed.

c) One difference from the original algorithm of Robinson [13], [18] is that the pairs of terms to be unified *may be chosen in any order* (the algorithm proposed in [19] also has these characteristics).

. In the case of a failure in the plan validation one detects a subset of the set of equations with no solution. This subset will never be chosen by other plans.

Example :
(The leftmost number is the order of the node in the tree traversal enumeration ; the second one is the name of applied reduction in this node and the pair between brackets, in matrix notation, is the place of each literal component of the reduction)

.
.
.

PROPOSED PLAN :

1 : 13 (52-63) : + R(X5,Z5) - R(X6,U6) ON MATRIX M1
2 : 12 (52-71) : + R(X5,Z5) - R(A,B)   ON MATRIX M2
3 : 9 (11-41) : + Q(A,F(C),F(B)) - Q(X4,Y4,Z4) ON MATRIX M3
4 : 3 (42-51) : + Q(Y4,X4,Z4) - Q(X5,Y5,Z5) ON MATRIX M4

REFUTATION TREE :

MATRIX	LEFT SON	RIGHT SON
M1	M2	M3
M2	∅	∅
M3	M4	∅
M4	∅	∅

DO YOU ACCEPT THIS PLAN (Y/N) ?
Y
COMPATIBLE REDUCTIONS :

```
1 :  13 (52-63) : + R(X5,Z5) - R(X6,U6)
2 :  12 (52-71) : + R(X5,Z5) - R(A,B)
3 :   9 (11-41) : + Q(A,F(C),F(B)) - R(X4,Y4,Z4)
```

INCOMPATIBLE WITH :

```
4 :   3 (42-51) : + Q(Y4,X4,Z4) - Q(X5,Y5,Z5)
```

INCOMPATIBLE SET OF REDUCTIONS : {3,9,12}.

OTHER IMPLEMENTATIONS DETAILS :

. Backtracking is carried out by direct access and not by following a chain of pointers. The backtracking is selective : the user (or strategy) may choose a restart point which is not necessarily the next choice in the order of nodes on the tree.

. In order for the plan not to choose certain literal pairs which may not be made complementary, certain non-applicable reductions are eliminated (i.e. reduction 5 in the example). Only the first step of unification algorithm is applied (that is why in the next example, 8 may be choosen and will be eliminated after it has caused an incompatibility).

. In some cases information from validation failures may be used to guide the introduction of variants of clauses.

. The set of support strategy may be used [20].

. Clearly it has been observed that the time required for a plan search is shorter than the time required for the two steps of plan validation ; therefore in a parallel implementation, several plans may be found while a validation of a proposed plan is essayed.

. As we work on a microcomputer, the problem of "garbage collection" is particulary important and an "ad hoc garbage collection" is carried out by the program.

6. A PROGRAM EXAMPLE

SET OF CLAUSES :

```
C 1 + P(G(X1,Y1),X1,Y1)
C 2 - P(X2,H(X2,Y2),Y2)
C 3 - P(X3,Y3,U3) + P(Y3,Z3,V3) - P(X3,V3,W3) + P(U3,Z3,W3)
C 4 - P(K(X4),X4,K(X4))
```

REDUCTIONS :

```
1 : 34-41 = + P(U3,Z3,W3) - P(K(X4),X4,K(X4))
2 : 34-21 = + P(U3,Z3,W3) - P(X2,H(X2,Y2),Y2)
3 : 32-41 = + P(Y3,Z3,V3) - P(K(X4),X4,K(X4))
```

```
4 : 32-21 = + P(Y3,Z3,V3) - P(X2,H(X2,Y2),Y2)
6 : 11-33 = + P(G(X1,Y1),X1,Y1) - P(X3,V3,W3)
7 : 11-31 = + P(G(X1,Y1),X1,Y1) - P(X3,Y3,U3)
8 : 11-21 = + P(G(X1,Y1),X1,Y1) - P(X2,H(X2,Y2),Y2)

PROPOSED PLAN :
1 : 8 (11-21) : + P(G(X1,Y1),X1,Y1) - P(X2,H(X2,Y2),Y2) ON MATRIX M1

REFUTATION TREE :
MATRIX      LEFT SON            RIGHT SON
  M1           Ø                    Ø

DO YOU ACCEPT THIS PLAN (Y/N) ?
Y
THERE IS A CYCLE (8)
DO YOU WANT FINISH OR MODIFY THE PLAN (F/M) ?
M
FROM WHICH REDUCTION (1,2,...) ?
1
PROPOSED PLAN :
1 : 6(11-33) : + P(G(X1,Y1),X1,Y1) - P(X3,V3,W3) ON MATRIX M1
2 : 7(11-31) : + P(G(X1,Y1),X1,Y1) - P(X3,Y3,U3) ON MATRIX M2
3 : 4(32-21) : + P(Y3,Z3,V3) - P(X2,H(X2,Y2),Y2) ON MATRIX M3
4 : 2(34-21) : + P(U3,Z3,W3) - P(X2,H(X2,Y2),Y2) ON MATRIX M4

REFUTATION TREE :
MATRIX      LEFT SON            RIGHT SON
  M1          M2                    Ø
  M2          M3                    Ø
  M3          M4                    Ø
  M4           Ø                    Ø

DO YOU ACCEPT THIS PLAN (Y/N) ?
Y

REFUTATION SUBSTITUTION :
X1 ← Y1
Y1 ← Y1
X2 ← Y1
Y2 ← Y1
X3 ← G(Y1,Y1)
Y3 ← Y1
U3 ← Y1
Z3 ← H(Y1,Y1)
```

V3 ← Y1
W3 ← Y1
X4 ← X4

Obviously, it is possible to print matrices of the refutation tree.
Another plan, and validation are possible :

PROPOSED PLAN :
1 : 6(11-33) : + P(G(X1,Y1),X1,Y1) - P(X3,V3,W3) ON MATRIX M1
2 : 7(11-31) : + P(G(X1,Y1),X1,Y1) - P(X3,Y3,U3) ON MATRIX M2
3 : 4(32-21) : + P(Y3,Z3,V3) - P(X2,H(X2,Y2),Y2) ON MATRIX M3
4 : 2(34-21) : + P(U3,Z3,W3) - P(X2,H(X2,Y2),Y2) ON MATRIX M4

REFUTATION TREE :

MATRIX	LEFT SON	RIGHT SON
M1	M2	∅
M2	M3	∅
M3	M4	∅
M4	∅	∅

DO YOU ACCEPT THIS PLAN (Y/N) ?
N
DO YOU WANT FINISH OR MODIFY THE PLAN (F/M) ?
M
FROM WHICH REDUCTION (1,2,...) ?
4
PROPOSED PLAN :
1 : 6(11-33) : + P(G(X1,Y1),X1,Y1) - P(X3,V3,W3) ON MATRIX M1
2 : 7(11-31) : + P(G(X1,Y1),X1,Y1) - P(X3,Y3,U3) ON MATRIX M2
3 : 4(32-21) : + P(Y3,Z3,V3) - P(X2,H(X2,Y2),Y2) ON MATRIX M3
4 : 1(34-41) : + P(U3,Z3,W3) - P(K(X4),X4,K(X4)) ON MATRIX M4

REFUTATION TREE :

MATRIX	LEFT SON	RIGHT SON
M1	M2	∅
M2	M3	∅
M3	M4	∅
M4	∅	∅

DO YOU ACCEPT THIS PLAN (Y/N) ?
Y
REFUTATION SUBSTITUTION :
X1 ← X1
Y1 ← K(H(X1,X1))

$X2 \leftarrow X1$
$Y2 \leftarrow X1$
$X3 \leftarrow G(X1,K(H(X1,X1)))$
$Y3 \leftarrow X1$
$U3 \leftarrow K(H(X1,X1))$
$Z3 \leftarrow H(X1,X1)$
$V3 \leftarrow X1$
$W3 \leftarrow K(H(X1,X1))$
$X4 \leftarrow H(X1,X1)$

REFERENCES

[1] ANDREWS P.B. : "Theorem proving via general mating",
JACM, Vol.28 N°2, April 1981 (193-214).

[2] BIBEL W. : "Tautology testing with a generalized matrix reduction method",
Theoretical Computer Science 8, 1979 (31-44).

[3] BIBEL W. : "On matrices with connections",
JACM, Vol.28 N°4, October 1981 (633-645).

[4] CHANG C. and LEE R. : "Symbolic logic and mechanical theorem proving",
Academic Press, 1973.

[5] CHANG C. and SLAGLE J.R. : "Using rewriting rules for connection graphs to prove theorems", Artificial Intelligence 12, 1979 (159-180).

[6] DAVIS M. : "Eliminating the irrelevant from mechanical proofs",
Symposia in Applied Mathematics Vol.15, 1963 (15-30).

[7] HENSCHEN L.J. : "Theorem proving by covering expressions",
JACM, Vol.26, N°3, July 1979 (385-400).

[8] HUET G. : "Résolution d'équations dans les langages d'ordre $1,2,\ldots,\omega$",
Thèse d'état, Université Paris VII, 1976.

[9] KLEENE S. : "Mathematical logic", John Wiley and Sons, 1967.

[10] KOWALSKI R. : "Logic for problem solving", North-Holland 1979.

[11] LOVELAND D.W. : "Mechanical theorem proving by model elimination",
JACM, Vol.15, N°2, April 1968 (236-251).

[12] LOVELAND D.W. : "A unified view of some linear Herbrand procedures",
JACM, Vol.19, N°2, April 1972 (366-384).

[13] LOVELAND D. W. : "Automated theorem proving : A Logical basis",
North-Holland, 1978.

[14] PLAISTED D.A. : "Theorem proving with abstraction", Artificial Intelligence 16, 1981 (47-108).

[15] PRAWITZ D. : "An improved proof procedure", Theoria 26, 1960 (102-139).

[16] PRAWITZ D. : "Advances and problems in mechanical proof procedures",
Machine Intelligence 4, Edinburgh Univ. Press 1969, (59-71).

[17] PRAWITZ D. : "A proof procedure with matrix reduction",
Symposium on Automatic Demonstration - Springer Verlag 1970, (207-214).

[18] ROBINSON J.A. : "A machine oriented logic based on the resolution principle", JACM, Vol.12, N°1, January 1965 (23-41).

[19] ROBINSON J.A. : "Logic : form and function, the mechanization of deductive reasoning", University Press Edinburgh, 1979.

[20] ROUSSELL Ph. : "PROLOG : Manuel de Référence et d'Utilisation", Groupe d'Intelligence Artificielle, Université d'Aix Marseille, Luminy, Sept. 1975.

[21] SAYA H. : "Complétude de la stratégie du support dans la méthode de réduction matricielle", R.R. IMAG N°93, Novembre 1977.

[22] SAYA H. and CAFERRA R. : "A structure sharing teohnique for matrices and substitutions in Prawitz's theorem proving method", R.R. IMAG N°101, Decem. 1977.

[23] SAYA H. et CAFERRA R. : "Représentation compacte de matrices et traitement de l'égalité formelle dans la méthode de Prawitz de démonstration automatique", Congrès AFCET 1978, tome 2 (371-381).

[24] SHOSTAK R.E. : "A graph-theoretic view of resolution theorem proving", TR-20-74 Center of Research in Computing Technology-Harvard University 1974.

[25] SICKEL S. : "A search technique for clause Interconnectivity Graphs", IEEE Transactions on Computers, Vol.C-25 N°8 August 1976 (823-835).

[26] SICKEL S. : "Variable range restrictions in resolution theorem proving", Machine Intelligence 8 - Ellis Horwood Ltd 1977 (73-85).

IMPROVEMENTS OF A TAUTOLOGY-TESTING ALGORITHM

K.M. Hörnig and W. Bibel[1)]

Technische Universität München

1.   In this paper we present an algorithm COMPLEMENTARY which tests whether or not formulas of propositional logic are tautologies. This algorithm is an improvement of (1.3) and (6.1) in /Bi 82/. We also discuss several possible enhancements first expressed in /Bi 82b/.

Algorithms for the tautology problem in propositional logic are basic tools for solving the corresponding problem for full first order logic. The algorithm discussed here is one of the main components of the <u>connection method</u> described in /Bi 82b/. More on the other components can be found in /Mü 81/.

As an aside, we mention that an investigation into the complexity of such algorithms becomes interesting in view of the still unsolved questions concerning $P \stackrel{?}{=} NP$.

In the following we shall give an intuitive description of the features which COMPLEMENTARY possesses. We also indicate, how one might proceed to prove soundness and completeness of COMPLEMENTARY. The exact proof of this is somewhat complicated and too long to be included. It may be found in /Hö 81/.

The algorithm itself is stated in an appendix.

An implementation of COMPLEMENTARY using UT-LISP has been carried out. It has proved to be a valuable help in the complicated analysis of COMPLEMENTARY needed to carry out the soundness proof.

2.   In /Bi 81/, the problem of checking, whether a given formula of propositional logic is a tautology, has been restated as the problem of checking, whether a path through a graph-like structure, called matrix, contains contradictory pairs of nodes labeled with literals. This point of view offers at least two advantages:
- thinking in categories of graphs seems to be easier for humans than thinking in terms of logical proof rules; in particular, we will make use of pictures throughout this paper, which convey much clearer the

---

[1)] This research has been supported by the Deutsche Forschungsgemeinschaft (DFG).

mathematical ideas than the cumbersome strict formalism;
- this representation suggests the development of new hardware suited to increase the efficiency of proof mechanisms.

We assume that the reader is familiar with the basic concepts of propositional logic.

A <u>literal</u> is a propositional variable or its negation. Literals will be denoted by $L_1, L_2, \ldots$. ^1L is the negation of the literal L.

In this paper we regard <u>matrices</u> as nested graphs built up using literals and interpreted as the following examples indicate.

Example 1:

$\begin{matrix} L_1 & L_3 & \\ L_2 & L_4 & L_5 \end{matrix}$ represents the formula

$$(L_1 \& L_2) \vee (L_3 \& L_4) \vee L_5$$

Example 2:

$\begin{matrix} L_1 & L_3 & L_4 & L_7 \\ & & L_5 & \\ L_2 & L_6 & & L_8 L_9 \end{matrix}$ represents the formula

$$(L_1 \& L_2) \vee ((L_3 \vee (L_4 \& L_5)) \& L_6)$$
$$\vee (L_7 \& (L_8 \vee L_9))$$

The columns of a matrix are called <u>clauses</u>.
It is a basic fact that any propositional formula can be represented by a matrix.

Examples of <u>paths</u> through matrices are given by:
$\{L_1, L_3, L_5\}$, $\{L_2, L_3, L_5\}$, $\{L_1, L_4, L_5\}$, and $\{L_2, L_4, L_5\}$ are the paths through the matrix of example 1 and $\{L_1, L_3, L_5, L_8, L_9\}$, $\{L_2, L_3, L_4, L_7\}$ are some of the paths through the matrix in example 2.

A path is <u>complementary</u> if it contains a literal L and its negation ^1L.
A matrix M is complementary if all paths through M are complementary.

The connection between path-search and proof-search is given by the following theorem.

<u>Theorem</u>: If F is a well-formed formula of propositional calculus and F is represented by the matrix M, then F is a tautology iff M is complementary.

This explains the I/O-specification of COMPLEMENTARY:
<u>input</u>: a matrix M

output: "complementary" if all paths through M are complementary
"non-complementary" otherwise.

3.   It is not hard to design a preliminary version of COMPLEMENTARY along the following lines.

Consider as an example the matrix M given by

$$\begin{array}{ccccc} {}^1L_1 & {}^1L_2 & {}^1L_1 & {}^1L_4 & {}^1L_5 \\ L_1 & L_2 & L_3 & & L_5 \\ L_4 & & & & \end{array}$$

We start by choosing a clause, say $\{L_1, L_4\}$ and choosing a literal in that clause, say L_1. Doing this, we have decided to study all paths passing through L_1 and put the rest of the clause ($\{L_4\}$) on a stack, where it remains as a subgoal. Now we try to complement as many paths as possible. This can be achieved by searching the rest of the matrix for ${}^1L_1$. In case of success, we have shown several paths through M to be complementary and turn to the next element in the clause $\{{}^1L_1, L_2\}$, namely L_2. This procedure is repeated as long as possible:

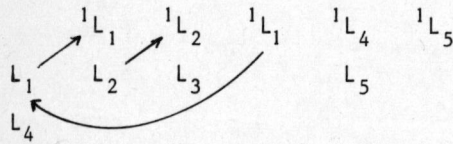

For L_3 we cannot find a complementary literal. At this point another possibility arises, namely to use the complementarity of L_1 and ${}^1L_1$:

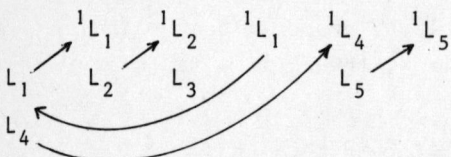

Since the fourth clause contains only ${}^1L_1$, we have in fact checked that every path through M beginning with L_1 is complementary. Therefore, we take up the subgoal L_4 and finish the procedure as follows:

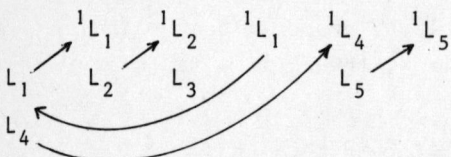

Now, we give a more abstract description of this process and introduce some important terms.

Let M_o be the input to COMPLEMENTARY.
At any step of the algorithm (after some initialization steps), we have:
- an actual clause C
- an actual matrix M
- ACT, an actual partial path in M_o s.t. the complementarity of its extensions through M_o has not yet been determined, called the <u>active</u> path
- WAIT, a stack which serves to store subgoals (and in the final version information about possible redundancies).

We picture states of the algorithm abstractly in the following way:

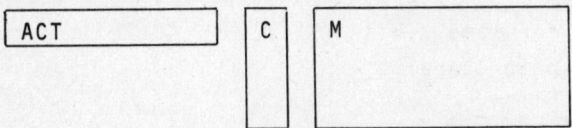

where an example of an interpretation of this picture is shown below

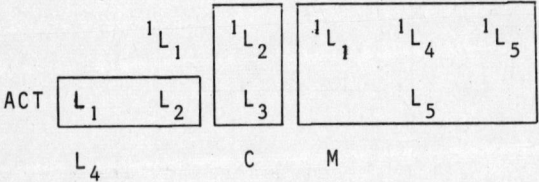

Such pictures are called snapshots.

The preliminary version of COMPLEMENTARY consists essentially of two loops, extension- and reduction-loop (see the figure in the appendix).

First, consider the extension-loop, where we have the following loop-invariants:
- all paths through M_o which have ACT as subpath have still to be checked
- all other paths through M_o have either been proved to be complementary or they are represented as subgoals on WAIT
- no literal in C is complementary to a literal in ACT.

We thus may choose a literal L from C and put ACT ← ACT ∪ {(L,I)}. I is a variable which serves as a counter and is increased with every pass through the bottom-loop.

If $C \smallsetminus \{L\} \neq \emptyset$, then we store (ACT', $C \smallsetminus \{L\}$, M) as a subgoal on WAIT (ACT' is the old value of ACT).

If possible, we continue in such a way that some more paths through M_o are proved to be complementary. This is the case if we can find a clause C' with $^1L \in C'$.

We put $C \leftarrow C'$, $M \leftarrow M \setminus \{C\}$, and $C \leftarrow C \setminus \{^1K\}$ for all literals K s.t. $(K,j) \in ACT$ for some j.

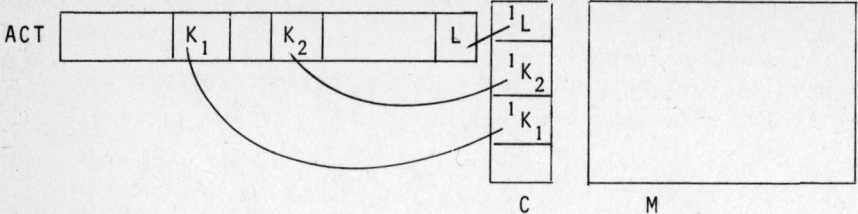

If we cannot find a clause $C' \in M$ with $^1L \in C'$, then there is a second possibility to reduce the number of unchecked paths, namely if there is a clause $C' \in M$ s.t. $^1K \in C'$ for some K with $(K,j) \in ACT$. Acting similarly as above, we get the following picture:

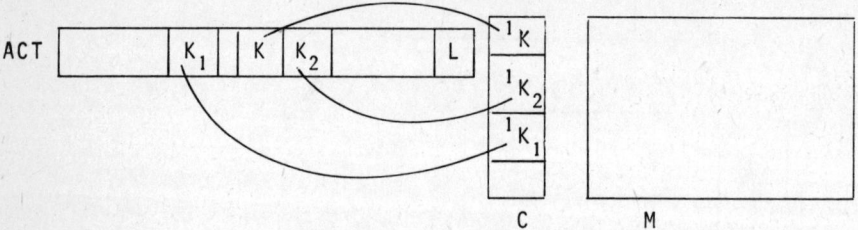

In both cases, after execution of the actions just described, the loop-invariants are fulfilled. Therefore, if $C \neq \emptyset$, we can execute another pass through the extension-loop. Otherwise we jump into the reduction-loop, since we have proved all paths complementary, which have ACT as sub-path.

During the execution of the extension-loop it may occur, that none of the two methods to reduce the number of unchecked paths is applicable. It is easy to see, that for the preliminary version of COMPLEMENTARY: M_o is complementary iff the algorithm applied to the actual value of M, i.e. the remaining matrix, (where the subgoals on WAIT are forgotten) yields "complementary".

This concludes the discussion of the extension-loop.

We now turn to the reduction loop. In the preliminary version of COMPLEMENTARY a pass through this loop consists of simply checking whether there is still a subgoal stored on WAIT and resuming the extension-loop with the subgoal as parameter. In the full version of COMPLEMENTARY several actions might be taken in the course of the reduction-loop which assure the correct removal of redundant subgoals (see section 4).

The preliminary version of COMPLEMENTARY can be applied to matrices in normal form (see example 1 above) only. To deal with matrices in non-normal form (example 2), some modifications are necessary, which can be seen from the listing of COMPLEMENTARY in the appendix.

In particular, it is no longer possible to drop all of WAIT, when applying COMPLEMENTARY recursively to a proper submatrix in the situation described just above. Instead, when we encounter a matrix in a clause, which is not a literal, then we store a <u>level-marker</u> (labeled "lm") on WAIT. In the situation above, it is only allowed to drop WAIT until we hit the next level-marker.

4.   This procedure can be considerably enhanced by adding further mechanisms which avoid several redundancies in the checking of the paths.

4.1   A first redundancy is seen in the following picture.

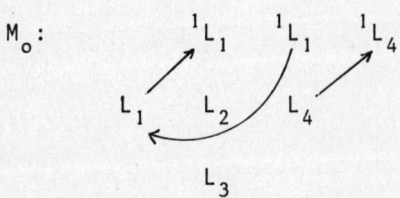

The preliminary version of COMPLEMENTARY would, after checking that all paths through M_o, having $\{L_1, {}^1L_1\}$ and $\{L_1, L_2\}$ as initial segments, are complementary, turn to the subgoal $\{L_1, L_3\}$. But this is redundant, since the same connections which prove that all paths extending $\{L_1, L_2\}$ are complementary, prove that all paths extending $\{L_1, L_3\}$ are complementary.

This redundancy can occur, when we have used the second possibility to extend ACT. Therefore we put some information about this on top of WAIT. This information is called a ditch-marker and labeled with "dm".

The following examples show, that we cannot discard all subgoals, when we encounter a ditch-marker on WAIT.

Example 3:

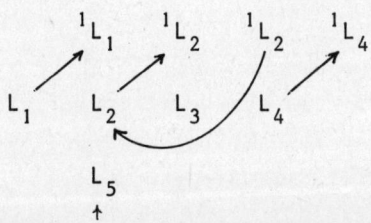

331

Example 4:

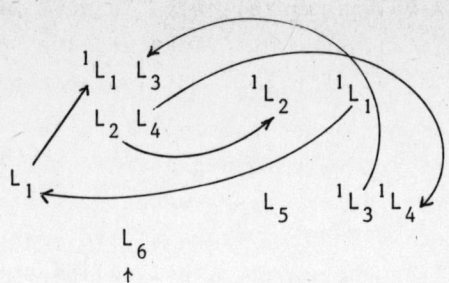

Both examples show that, intuitively speaking, the subgoal must be to the right of the left end of any connection with right end to the right of the subgoal.

In order to check this criterion in example 3, we keep track of the connections which have been established between elements of ACT and literals to the right of the actual clause. The variable SC contains the right coordinate of all elements of ACT which are the left end of such a connection. SC is used to disregard redundant subgoals, when we encounter a ditch-marker on WAIT.

It would be a very delicate, but not impossible task to generalize this idea to cope with the situation in example 4. We believe, however, that this would considerably increase the complexity of the statement of the algorithm and the proof of its soundness. We therefore have chosen a method which is cruder, but easier to realize:
it is allowed to remove subgoals until we hit the next level-marker. This precludes the removal of L_6 in example 4.

4.2 A second redundancy can occur in matrices in non-normal form. Consider the following example:

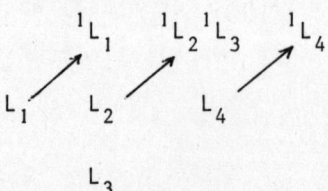

Again, the preliminary version would consider a redundant subgoal, namely paths extending $\{L_1,L_3\}$, even though it would suffice to notice, that connecting L_3 and $^1L_3$ would lead to the same procedure as had been performed when checking the paths extending $\{L_1,L_2\}$.

As in the preceding discussion, we note, that there are restrictions to the discard of subgoals in such cases. We demonstrate this by the following example, taken from /Bi 82a/.

Example 5:

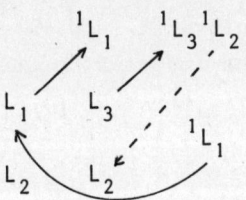

The subgoal involving L_2 in the second clause can be discarded, the subgoal involving L_2 in the first clause cannot.

It holds, that exactly those subgoals are redundant, which are to the right of the rightmost left end among all connections with a right end below or to the right of the origin of the dotted arrow (see /Bi 82a/).

This decision can be made during the course of the algorithm using the variables SC (see above) and PRSG, where all "potentially redundant subgoals" of this kind are stored.

5. It is rather obvious, that the preliminary version of COMPLEMENTARY has the correct I/O-behaviour. This is no longer clear, when we add the mechanisms introduced in 4.1 and 4.2. In fact, we shall see below in section 6 examples, where the addition of another mechanism interferes with them thus yielding an inconsistency.

We have, however, the following theorem, which is proved in /Hö 81/.

Theorem: If the input to COMPLEMENTARY is a complementary matrix, then COMPLEMENTARY yields "complementary". Otherwise, COMPLEMENTARY yields "non-complementary". In other words, COMPLEMENTARY satisfies the I/O-specification given at the end of section 2.

Here we present a few ideas used in this proof.

We give a number to each statement in COMPLEMENTARY. If x is a variable in the algorithm and s is the number of a statement, then x<s> is the value of x after execution of statement s (e.g. M<0> is the matrix, the complementarity of which is to be checked). A complete run of COMPLEMENTARY can be described by a map $n \mapsto s_n$, where s_n is the n-th statement carried out by the algorithm. It can be proved, that the algorithm terminates for all matrices. Therefore, there is some $n =: n_t$ s.t. s_n is the final step in the run of COMPLEMENTARY. Let $x(n) := x<s_n>$.

Since COMPLEMENTARY is sometimes applied recursively to proper submatrices and ACT thus reduced to ∅ again, ACT does not necessarily represent an initial segment of a path through the input matrix M_o. It turns out to be technically convenient to introduce a new variable NCPATH, which is enlarged like ACT, but not reset to ∅, when we apply COMPLEMENTARY recursively. NCPATH has the important property, that, if M_o is non-complementary, then $NCPATH(n_t)$ represents a non-complementary path through M_o. We call the partial path represented by NCPATH(n): $p_{NCPATH(n)}$.

Let $P(M)$, ($P^+(M)$, $P^-(M)$) be the set of all (non-complementary, complementary) paths through M.

A path $p \in P(M(0))$ is <u>represented after statement s_n</u>, if one of the following holds:

(R1) $\exists p'(p' \in P(M(n))$ & $p = p' \cup p_{NCPATH(n)})$

(R2) $\exists M_1 \in C(n), p_1 \in P(M_1), p_2 \in P(M(n)) : p = p_1 \cup p_2 \cup p_{NCPATH(n)}$

(R3)   there is a subgoal (C',ACT',NCPATH',M') on WAIT(n) s.t.
$\exists M_1 \in C', p_1 \in P(M_1), p_2 \in P(M') : p = p_1 \cup p_2 \cup p_{NCPATH'}$

Let $R(n) := \{p : p \in P(M(0))$ & p is represented after statement $s_n\}$.

R1,R2,R3 express formally, how those paths which have not been checked for complementarity in statements $s_o, \ldots, s_n$ are still represented for further examination.

We say that a statement is a <u>key-statement</u>, if it contains "goto" or "return". Let KS be the set of key-statements augmented by s_o.

The key to the proof of soundness of COMPLEMENTARY is the following lemma:

<u>Representation Lemma:</u>

If $P^+(M(0)) \neq \emptyset$, then for all $0 \leq n \leq n_t$: $s_n \in KS \rightarrow P^+(M(0)) \cap R(n) \neq \emptyset$.

It says roughly that, if M(0) is non-complementary, then we keep enough information about non-complementary paths until termination, so that we cannot terminate with "complementary".

Completeness of the procedure follows immediately from the fact, that COMPLEMENTARY always terminates, and the observation that, if COMPLEMENTARY stops with "non-complementary", then $NCPATH(n_t)$ codes a non-complementary path through M(0).

6.   In chapter 6 of /Bi 82b/. one finds several suggestions, how to remove further redundancies. We discuss here the possibilities to in-

clude two mechanisms, FACTORIZATION and CIRCUIT, into COMPLEMENTARY.

FACTORIZATION is the following rule:
In a situation like the following

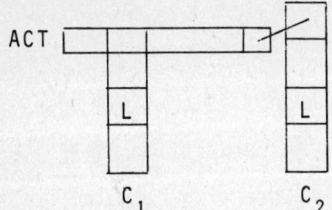

do <u>not</u> use the literal L in clause C_2 to continue the active path.

This rule is consistent with the preliminary version of COMPLEMENTARY, since, if we neglect a non-complementary path using this rule, we will take up another non-complementary path, when we deal with the subgoal L in clause C_1.

This argument breaks down for the full version of COMPLEMENTARY, as the following example shows.

Example 6:

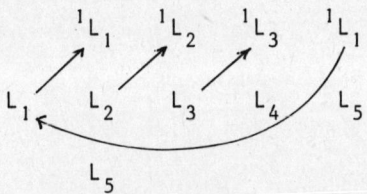

If we apply FACTORIZATION, then we neglect L_5 in the rightmost clause. But this subgoal is not taken up again later, since the mechanism described in section 4.1 makes COMPLEMENTARY drop the subgoal L_5 in the second clause. COMPLEMENTARY augmented with FACTORIZATION would therefore terminate with "complementary", which is clearly false.

Thus, in order to include FACTORIZATION into COMPLEMENTARY, it will be necessary to use more sophisticated bookkeeping, which prevents this kind of failure. We have no final solution to that problem.

CIRCUIT is the following rule:
In a situation like the following

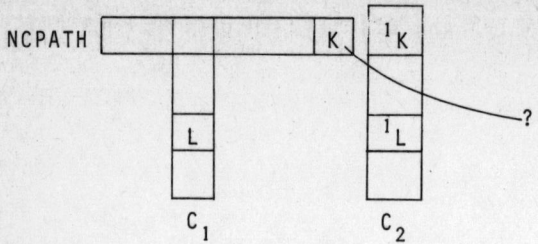

if L is not a member of NCPATH, then continue the search as if clause C_2 did not exist.

The name CIRCUIT is motivated by the following picture.

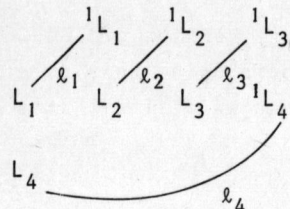

The connections $l_1,\ldots,l_4$ form a circuit and it can be shown (see /Bi 81/), that circuits can be avoided in proofs under certain conditions. CIRCUIT demands to leave out l_3. The following example shows, that CIRCUIT without FACTORIZATION is not even consistent with the preliminary version of COMPLEMENTARY.

Example 7:

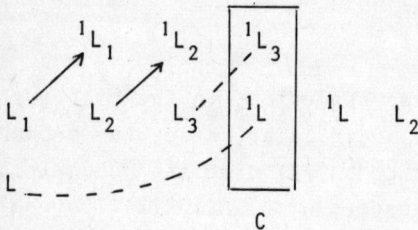

Clause C would be avoided and one would end up with the non-complementary path $\{L_1, L_2, L_3, L, L_2\}$ through $M \setminus \{C\}$.

With FACTORIZATION, however, one would stop at the fifth clause with "complementary".

A sketch of the proof, that indeed the combination of CIRCUIT and FACTORIZATION is consistent with the preliminary version of COMPLEMENTARY is easy to give using the following picture.

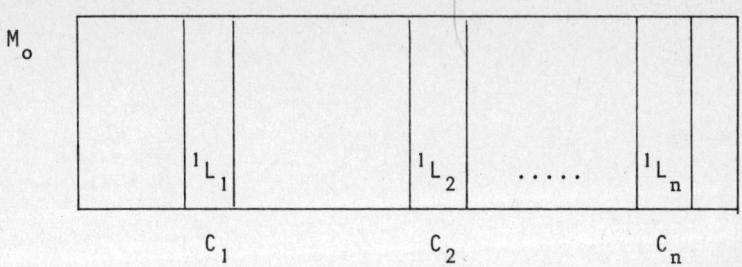

Given a non-complementary path p' through $M_o \setminus \{C_1,\ldots,C_n\}$, it can be extended to a non-complementary path p through M_o by letting $p \cap C_i = \{L_i\}$ for $1 \leq i \leq n$.

This works, since ^1L_i is not a member of p_{NCPATH}, when the clause C_i is discarded, and L_i is not added to NCPATH later by FACTORIZATION.

In view of the difficulties, which arise by the inclusion of FACTORIZATION into the full version of COMPLEMENTARY, we have not yet solved the problem of including CIRCUIT into this algorithm.

It will need considerable effort to design a correct extension of COMPLEMENTARY, which comprises the features above, and prove its correctness.

7. Although COMPLEMENTARY has become quite a complex deductive tool, it is built upon a rather simple idea, namely checking all the paths through a formula for connections. Its complexity derives from the ambitious attempt to eliminate as much redundancy as possible in this systematic process. This is not unlike the situation in resolution theory, where we also have a simple deductive rule, viz. resolution, which has provided only the basis for rather complex refinements.

In fact, there is even a close relationship between these two ideas, as has been shown in /Bi 82a/. But decisive differences remain. Since in contrast to resolution our basic idea applies to any formula (not only those in clausal form), it keeps the logical structure of the formula untouched during the proof process (rather than generating thousands of new clauses), and thus it can be easier grasped by mind. For these reasons it is certainly not accidental, that COMPLEMENTARY currently is the most advanced and least redundant deductive tool on the ground level, as has been demonstrated in /Bi 82a/ for key ATP-methods.

Appendix: The algorithm COMPLEMENTARY

We adopt the convention that all variables get value NIL in the beginning, except M which gets as value the matrix, the complementary of which is to be checked.

An element of WAIT has the form $B = (L\ I\ V)$, where L is the label of B, I the index of B, and V the value of B.

Abbreviations:

push (A,B)	push B on top of stack A
pop (A)	pop top element from stack A
M ∖ C	M∖{C}
^1K	complement of the literal K

COMPLEMENTARY:

0:   push (WAIT,("lm",0,∞));
1:   ACT ← NIL;
    I ← 0;
    SC ← NIL;
    PRSG ← NIL;
    M'' ← NIL;
2:   choose a clause C from M;
    M ← M ∖ C;
    for all literals K s.t. $^1K \in C$ and $(K,j) \in ACT$ for some j
    do ⌜C ← C ∖ ^1K;
        choose j s.t. $(K,j) \in ACT$;
        push (WAIT,("SC",I+1,{j}));⌝
    if C = NIL then goto 8;
3:   choose a matrix M' from C;
    C ← C ∖ M';
    if there exists $(l,M_o) \in PRSG$ s.t. $l \leq I$ and $M' \cup M_o$ is complementary
    then if C ≠ NIL then goto 3
                   else goto 8;
    if C ≠ NIL then push (WAIT, ("sg",I,(C,ACT,NCPATH,M)));
    if SC ≠ NIL then ⌜SC ← SC ∖ {j: j>I};
                  push (WAIT,("sc",I+1,SC));
                  SC ← NIL;⌝
    if PRSG ≠ NIL then ⌜PRSG ← PRSG ∖ $\{(l,M_o) : l>I\}$;
                     push (WAIT,("prsg",I+1,PRSG));
                     PRSG ← NIL;⌝

```
        if M' is not a literal then ⌈push (WAIT,("lm",I,∞));
                                    M ← M' ∪ M;
                                    goto 2;⌉
     I ← I+1;
     L ← M';
     ACT ← ACT∪{(L,I)};
     NCPATH ← NCPATH∪{(L,I)};
4:   if M = NIL then return ("non-complementary");
5:   if there is no clause C∈M s.t. ¹L occurs in C
     then if there is no clause C∈M s.t. ¹K occurs in C for some
                                                              (K,j)∈ACT
            then while the label of top of WAIT is not "lm"
                 do pop (WAIT);
                 goto 1;
            else ⌈choose C from M s.t. ¹K occurs in C for some (K,j)∈ACT;
                  push (WAIT,("dm",I,∞));⌉
     else choose C from M s.t. ¹L occurs in C;
     M ← M ∖ C;
6:   if ¹L occurs in C then choose a matrix M'∈C s.t. ¹L occurs in M'
                       else choose a matrix M'∈C s.t. ¹K occurs in M'
                                                       for some (K,j)∈ACT;
     C ← C ∖ M';
     for all literals K s.t. ¹K∈C or ¹K=M' and (K,j)∈ACT for some j
     do ⌈C ← C ∖ ¹K;
         choose j s.t. (K,j)∈ACT;
         push (WAIT,("sc",I,{j}));⌉
7:   if M' is a literal
     then if M'' ≠ NIL then ⌈push (WAIT,("prsg",I,{(0,M'')}));
                              M'' ← NIL;⌉
           if C ≠ NIL then goto 3
                      else goto 8;
     if C ≠ NIL then ⌈push (WAIT,("sg",I,(C,ACT,NCPATH,M)));
                       push (WAIT,("lm",I,∞));⌉
     if ¹L occurs in M'
     then choose a clause C from M' s.t. ¹L occurs in C
     else choose a clause C from M' s.t. ¹K occurs in C for some
                                                              (K,j)∈ACT;
     M'' ← M'∖C;
     M ← M''∪M;
     goto 6;
8:   if WAIT = NIL then return ("complementary");
```

9: if the label of top of WAIT is "sg"
    then if I = index of top of WAIT
         then ⌐(C-ACT,NCPATH,M) ← value of top of WAIT;
              pop (WAIT);
              if SC ≠ NIL
              then ⌐SC ← SC∖{j : j>I};
                   push (WAIT,("sc",I,SC));
                   SC ← NIL;⌐
              if PRSG ≠ NIL
              then ⌐PRSG ← PRSG∖{(1,M) : 1>I};
                   push (WAIT,("prsg",I,PRSG));
                   PRSG ← NIL;⌐
              goto 3;⌐
         else ⌐(C,ACT,NCPATH,M) ← value of top of WAIT;
              I ← index of top of WAIT;
              goto 3;⌐
10: if the label of top of WAIT is "sc"
    then ⌐SC ← SC ∪ value of top of WAIT;
         I ← index of top of WAIT;
         pop (WAIT);
         goto 8;⌐
11: if the label of top of WAIT is "dm"
    then ⌐SC ← SC∖{j : j >index of top of WAIT};
         while index of top of WAIT ≥ max of SC
               and label of top of WAIT ≠ "lm"
         do pop (WAIT);
         goto 8;⌐
12: if the label of top of WAIT is "lm"
    then ⌐I ← index of top of WAIT;
         pop (WAIT);
         goto 8;⌐
13: if the label of top of WAIT is "prsg"
    then ⌐SC ← SC∖{j : j > index of top of WAIT};
         m ← max (SC∪{0});
         for each (1,M_o) ∈ value of top of WAIT
         do if m>1 then substitute m for 1 in (1,M_o);
         PRSG ← PRSG ∪ value of top of WAIT;
         I ← index of top of WAIT;
         pop (WAIT);
         goto 8;⌐

Picture of the loops in COMPLEMENTARY:

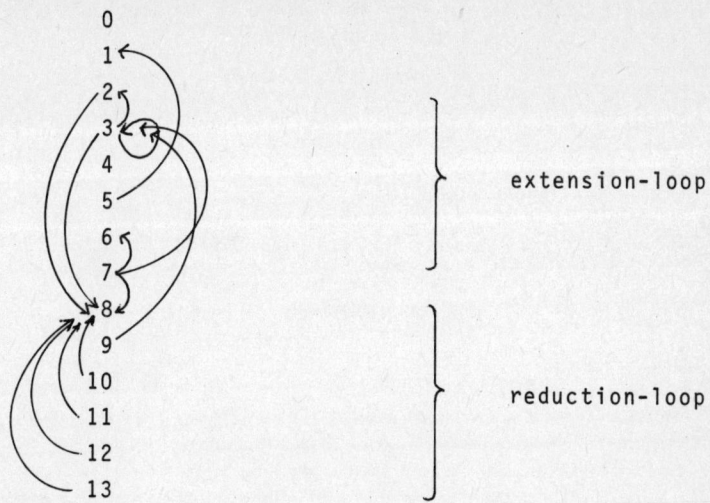

References:

/Bi 81/   W. Bibel
          On matrices with connections
          JACM Vol. 28 No. 4 (1981) 633-645

/Bi 82a/ W. Bibel
          A comparative study of several proof procedures
          Artif. Intelligence 18 (1982)

/Bi 82b/ W. Bibel
          Automated theorem proving
          Vieweg (to appear)

/Hö 81/   K.M. Hörnig
          Improvement and correctness proof
          of a tautology testing algorithm
          (in preparation)

/Mü 81/   A. Müller
          An implementation of a theorem prover based
          on the connection method
          Bericht ATP-12-XII-81, Projekt Beweisverfahren,
          Institut für Informatik, Technische Universität München (1981)

Author's address: Institut für Inforamtik
                    Technische Universität München
                    Postach 20 24 20
                    D-8000 München 2
                    Germany

Typescript: S. Tänzer

Representing Infinite Sequences of Resolvents in Recursive First-Order Horn Databases

Lawrence J. Henschen

Northwestern University
Evanston, Illinois 60201

Shamim A. Naqvi

Bell Laboratories
Murray Hill, New Jersey 07974

ABSTRACT

A First Order Database is defined as a function-free First-Order Theory in which the ground units serve as the Extensional Database and the proper non-logical axioms serve as the Intensional Database. This paper addresses the following problem: "Given a recursive non-logical axiom and a theorem to be proved which interacts with this axiom, can we describe a set of finite retrieval requests such that all and only the correct proofs to the theorem are found". Our solution uses resolution-proof techniques over connection graphs to derive a program of retrieval requests from the Extensional Database that gives all the answers to a query and has a well-defined termination condition.

1. Introduction

We define a relational database [9] as a function-free First-Order Horn Theory (FOHT) in which the set of ground units is called the Extensional Database (EDB) and the set of proper non-logical axioms is called the Intensional Database (IDB). The EDB corresponds to the notion of a relational database as a set of relations and where the relation names correspond to the predicate symbols of the FOHT. The problem of query answering is then analogous to that of finding all proofs of a theorem in the FOHT.

A problem with this formulation is that usually the EDB is a very large set of ground units and deductions involving it are bound to be time consuming. A solution proposed by Reiter [5] is to partition the FOHT into the EDB and the IDB components and to carry out deductions only over the IDB. Whenever a theorem cannot be proved over the IDB, deductive support is requested from the EDB. This deductive support is a set of requests for specific tuples (ground units) to be retrieved from the EDB. The EDB is assumed to be a highly optimized database query processor which can do these retrievals in an efficient manner.

A recursive intension is a sequence of clauses of the form

$\neg L11 \neg L12 \ldots L1n$
$\neg L21 \neg L22 \ldots L2n$
....
$\neg Lm1 \neg Lm2 \ldots Lmn$

such that the positive literal at the end of each clause unifies with the first literal of the next clause, and Lmn unifies with L11. In this paper we address the following problem. "Given a recursive intension in the IDB and a theorem to be proved which interacts with that intension, can we describe a set of finite retrieval requests to the EDB such that all and only the correct proofs to the theorem are found." The non-recursive case of this problem has been solved by Reiter [5]. As an example, consider the following set of clauses.

Example 1:
IDB: ¬P(x,y) ¬S(y,z) S(x,z)   ---(1)
       ¬P(x,y) S(x,y)   ---(2)
EDB: {P(a,b),P(c,d),...}
QUERY: S(?,d)

If we negate the query literal and attempt a refutation [6] with clause (1) given above, we get an infinite sequence of resolvents containing renamed instances of the S-literal. It should be noted that no answers can be produced if clause (2) were not in the IDB. In this paper we present a method of transforming such an infinite sequence of resolvents into a finite procedure (program). When invoked by a query, this procedure is executed by the EDB processor to retrieve all answers to the original query.

2. Preliminary Definitions

We shall represent the clauses of an IDB in the form of Connection Graphs (CG) [8]. A recursive intension occurs in a CG as a particular kind of cycle called a Potential Recursive Loop (PRL), or simply a loop.

Definition Let an edge between two nodes p1 and p2 in a CG be denoted by <p1,p2>. Then a Potential Recursive Loop (PRL) is a sequence of edges e1=<p1,p2>, e2=<p3,p4> ...... en=<p2n-1,p2n> such that the following conditions are satisfied:

1. The substitutions along the sequence of edges e1,e2,...,en-1 are mutually consistent.

2. For i≤k≤n-1 nodes p2k and p2k+1 belong to the same clause partition but p2k ≠ p2k+1.

3. p1 and p2n occur in the same clause partition but are not the same literal. Further, if one forms the resolvent of all the clauses on the pairs of literals <p2k-1,p2k> including two occurrences of the first clause, one for the occurrence of p1 and the other for the occurrence of p2n, then the two instances of the first clause are distinct.

Figure 1 shows an example PRL. The distinguishing feature of a recursive intension, from a theorem-proving point of view, is that we can continue resolving around the PRL generating longer and longer resolvents. The conditions in the definition of PRL above are those that guarantee this behavior.

In Figure 2 we show the general form of a PRL with a query node attached. In this figure we have shown only the query, the PRL itself, and the portion of the CG immediately connected to the PRL. The clauses Di" ¬Ci may in general be connected to other parts of the CG. Further, the expressions ¬Ci and ¬Ei are meant to represent the remainder of the clauses and need not be just a single literal. Finally, there need not be a clause Di" ¬Ci for every literal ¬Di. We now make the following definitions with reference to Figure 2.

Definition The clause to which the query attaches is called the start clause. In Figure 2 this is the clause D0 ¬D1 ¬E1.

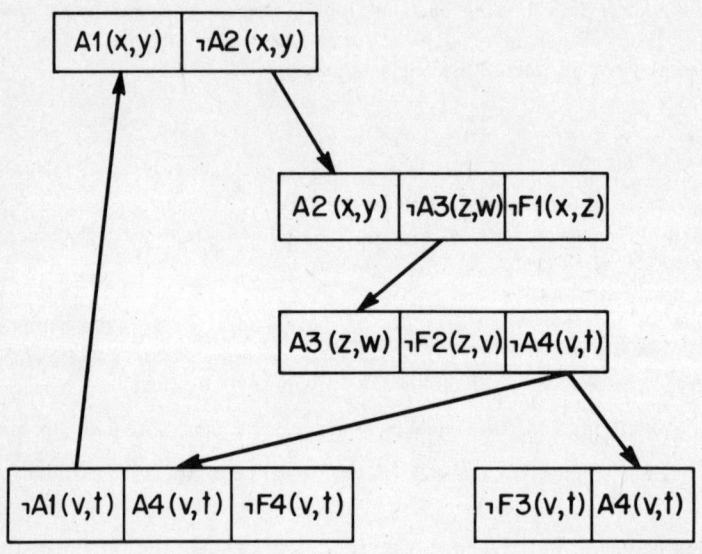

FIGURE 1: EXAMPLE PRL

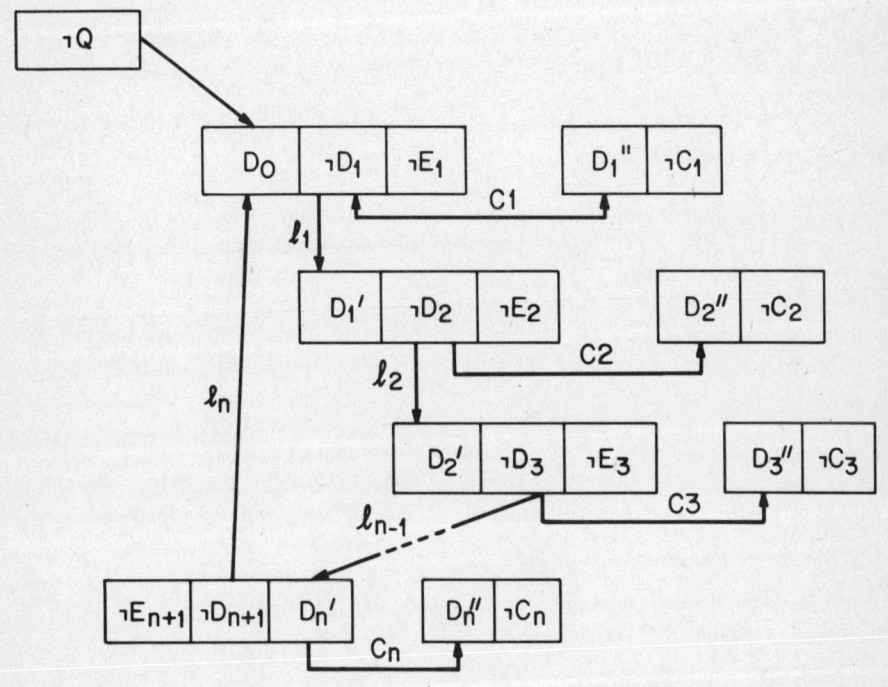

FIGURE 2: GENERAL FORM OF A PRL

Definition The literal D0 in the start clause to which the query attaches is called the start literal.

Definition The literal within the PRL that attaches to the start literal is called the end literal. In Figure 2, this is the literal $\neg D_{n+1}$.

Definition The clause $\neg E_{n+1}\ \neg D_{n+1}\ D_n'$ containing the end literal is called the end clause.

Definition The literals $\neg D_i$ and D_i' as shown in Figure 2 are called cycle literals.

Definition The closing edge of the PRL is the edge connecting the end literal to the start literal. This is marked as ln in Figure 2.

Definition An edge leading from a cycle literal out of the PRL is called an exit edge. Thus $c_1, c_2, \ldots, c_n$ are cycle edges in Figure 2.

Definition The augmented loop residue is the resolvent obtained by resolving the start clause with the clause $D_1'\ \neg D_2\ \neg E_2$, then resolving the result with the clause $D_2'\ \neg D_3\ \neg E_3$, etc., up to and including the end clause.

Definition An augmented exit expression is any resolvent obtained by resolving the start clause with $D_1''\ \neg C_1$, or by resolving the start clause with $D_1'\ \neg D_2\ \neg E_2$, and then resolving with $D_2''\ \neg C_2$, or ....

Note we do not include resolution with the query in the above two definitions. In Figure 2, the edges connecting the literals $\neg D_i$ to D_i'' are exit edges. Assuming that all the variables in the clauses of Figure 2 have been separated, the augmented loop residue is the clause

$$(D0\ \neg E1\ \neg E2\ ..\ \neg E_{n+1}\ \neg D_{n+1})\sigma$$

where σ is the simultaneous **mgu** of all the pairs (D_i, D_i'). Note that the end literal is in the augmented loop residue. Similarly any clause of the form

$$(D0\ \neg E1\ \ldots\ \neg E_j\ \neg C_j)\sigma j$$

is an exit expression where σj is the simultaneous **mgu** of the pairs (D_i, D_i') for $i=1,\ldots,j-1$ and (D_j, D_j'').

Definition The variables of the start literal, $D0\sigma$, in the augmented loop residue that unify with the constants in the query are called determined variables. If x is a determined variable and $L(....x..y..)$ is a literal of the augmented loop residue, other than the start or end literal, then y is also a determined variable.

Definition The set of literals of the augmented loop residue that contain determined variables, excluding the start and end literals, is called the Determined Part.

Definition The set of literals of the augmented loop residue, excluding the start and end literals, which do not contain determined variables is called the Induced Part.

We note that the definitions of Determined and Induced Parts are independent of any exits. By the commutativity and associativity of the logical OR, we may write the augmented loop residue as S ¬A ¬F ¬B, where S is the instance of the start literal, ¬F is the instance of the end literal, ¬A is the disjunction of the literals in the Induced Part, and ¬B is the disjunction of the literals in the Determined Part.

Definition A PRL is said to be stable if the determined variables in the end literal occur in the same positions as the determined variables in the start literal.

3. Temporary Simplifying Assumptions

In this section we make some simplifying assumptions in order to make the initial presentation clear. In the next section we shall eliminate the more important assumptions. To see how the remaining assumptions are relaxed, we refer the reader to [3]. In most cases, relaxing the assumptions is straight-forward or even trivial. Only one case, mutual recursion, is complicated, but this case, while important in general, occurs relatively infrequently in database applications. The assumptions are as follows:

A1 Assume that there is only one exit edge from a PRL.

A2 Assume that except for the start and end literals the loop residue consists entirely of literals occurring in the EDB.

A3 Assume that there are no isolated literals, i.e., a set of literals which has no variables in common with its complement set. This, in effect, says that the loop residue represents a well-formed join expression [9].

A4 Assume that there is only one PRL in the CG.

A5 Assume that there are no constants in the substitutions that unify the loop or the closing edge. Further, assume that the simultaneous **mgu** of the cycle literals $((D_i, D_i')$ in Figure 2) do not equate any variables in the start literal. These two assumptions imply that the instance S of the start literal in the augmented loop residue (and the instance in any exit expression, as well) is just the start literal itself up to alphabetic variance.

A6 Assume that the PRL is stable.

Reiter [5] has shown how queries composed of more than one predicate can be broken down into unit queries; the final result is obtained by combining the results from these unit queries through set operations. Thus, in this paper we shall restrict ourselves to unit queries only.

4. Notation

In the remainder of this paper we use the upper case letters X,Y,Z,W to denote vectors of variables, lower case letters x,y,z,... to indicate individual variables, lower case letters a,b,c,... from the start of the alphabet to denote individual constants, and the upper case letter C to denote a vector of constants. The form of the incoming query will be assumed as ¬Q(?,C).

Recall that the augmented loop residue can be partitioned into an Induced Part (containing non-determined variables) and a Determined Part (containing determined variables) and the start and end literals. In order to analyze the general pattern of resolvents from the PRL we need to be more specific about the positions of these variables. We, therefore, write an augmented loop residue as

$$Q(W',Z') \neg A(Y,W') \neg Q(W'',Z'') \neg B(X,Z')$$

where A and B represent sets of literals. Further, we write the augmented exit expression as

$$Q(W',Z') \neg E(U,V,W',Z').$$

If S is the start literal, $\neg Q(W',Z')$, and F is the end literal, $\neg Q(W'',Z'')$, of the PRL, then the variables have the following meaning:

- $Z = \{z \mid z \text{ is a determined variable}\}$
- $W = \{w \mid w \text{ is a non-determined variable}\}$
- $Z' = \{z \mid z \in Z \text{ and } z \text{ occurs in S}\}$
- $W' = \{w \mid w \in W \text{ and } w \text{ occurs in S}\}$
- $Z'' = \{z \mid z \in Z \text{ and } z \text{ occurs in F}\}$
- $W'' = \{w \mid w \in W \text{ and } w \text{ occurs in F}\}$
- $X = Z - Z'$ and $Y = W - W'$.
- U and V are those variables from the partially unified loop (i.e., loop unified up to the exit) that corresponds to X and Y.

5. Pattern of Data Retrievals for a PRL

Recall that we are interested in retrieving the same set of answers that a resolution program would retrieve for the same query (i.e. the set of answers logically implied by the database). Such answers can only be derived in resolution by resolving all the way around a PRL zero or more times and then out an exit because any other resolvent will contain a cycle literal.

In forming resolvents below, the scheme for separation of variables is as follows. We label the variables of the loop with subscript 1; when preparing to resolve along the closing edge, we increment the variables of the resolvent and leave the variables in the loop alone. Thus, as will be seen below, the two literals that need to be unified at the closing edge will always be $Q(W1',Z1')$ and $\neg Q(W2'',Z2'')$, the latter being the renamed version of $\neg Q(W1'',Z1'')$. For this unification, we chose to replace the variables $W1'$ and $Z1'$, by the variables $W2''$ and $Z2''$ respectively. We have defined the augmented loop residue and exit expression as resolvents and therefore have already taken into account the separation of variables within the PRL.

We begin our analysis by considering the resolvents generated. In order to see explicitly how the query interacts with the expressions derived below, we ignore the resolution of the query with the start clause for the moment. Thus, we are interested in the augmented loop residue and corresponding exit expression. The resolvent obtained by going directly from the start clause to the exit is

$$Q(W1',Z1',) E(U,V,W1',Z1') \quad \text{---(1)}$$

which is just the exit expression with the variables subscripted. For the resolvent obtained by going around the loop once and then out the exit, we first resolve up to the closing edge. This gives

$$Q(W1',Z1',) \neg A(Y1,W1') \neg Q(W1'',Z1'',) \neg B(X1,Z1',) \text{ ---}(2)$$

namely, the augmented loop residue with variables subscripted. In preparing to resolve along the closing edge, we number the variables in (2) by incrementing the subscripts to obtain

$$Q(W2',Z2') \neg A(Y2,W2') \neg Q(W2'',Z2'') \neg B(X2,Z2') \text{ ---}(3)$$

We now use (3) to continue resolving up to and out the exit giving

$$Q(W2',Z2') \neg A(Y2,W2') \neg E(U1,V1,W2'',Z2'') \neg B(X2,Z2') \text{ ---}(4)$$

Note that in (4) it is the literal $Q(W2',Z2')$, the descendant of the original start literal, that resolves with the query. Also, the variables $W2''$ and $Z2''$ in the E-literal form the links in the above join expression between the Induced and Determined Parts. Similarly, to go around the loop twice, we would use (3) to resolve along the closing edge, and then resolve all the way to the closing edge again, to get

$$Q(W2',Z2') \neg A(Y2,W2') \neg A(Y1,W2'') \neg Q(W1'',Z1'')$$
$$\neg B(X1,Z2'') \neg B(X2,Z2') \text{ ---}(5)$$

In order to continue resolving across the closing edge the second time, we increment the subscripts of (5) and resolve with (1). This gives

$$Q(W3',Z3') \neg A(Y3,W3') \neg A(Y2,W3'') \neg E(U1,V1,W2'',Z2'')$$
$$\neg B(X2,Z3'') \neg B(X3,Z3') \text{ ---}(6)$$

In these derivations, formula (1) corresponds to resolving directly from the start clause to the exit, formula (4) corresponds to resolving from the start clause all the way around the PRL once and then to the exit, formula (6) corresponds to traversing the PRL twice and then out the exit. These correspond to join expressions that may be used to retrieve answers. Formulas (2), (3) and (5) are presented only to clarify the connection of variables with subscript i to those with subscript i+1.

We now consider how the query interacts with expressions (1), (4) and (6). Recall the form of the query is $\neg Q(?,C)$ and in each case, resolves with the $Q(Wi',Zi')$ literal where $i = 1, 2, \ldots$. Thus, in each case the vector Zi' receives the constants from the query, and the vector Wi' corresponds to the answers being sought. The three resulting formulas are

$$\neg E(U1,V1,W1',Z1') \text{ ....}(7)$$
$$\neg A(Y2,W2') \neg E(U1,V1,W2'',Z2'') \neg B(X2,Z2') \text{ ....}(8)$$
$$\neg A(Y3,W3') \neg A(Y2,W3'') \neg E(U1,V1,W2'',Z2'') \neg B(X2,Z3'') \neg B(X3,Z3') \text{ ....}(9)$$

with the above understanding that in each case Zi' gets assigned the constants C and Wi' are the answers.

We now make two crucial observations.

Observation 1: First, the pattern of retrieval expressions is clearly evident in formulas (7),

(8) and (9). In each of these expressions, the leftmost part is of the form $\neg A(Yi,Wi')$ and the rightmost part is of the form $\neg B(Xi,Zi')$. In going from one of these formulas to the next, these two parts are replaced by $\neg A(Yi+1,Wi+1')$ $\neg A(Yi,Wi+1'')$ and $\neg B(Xi,Zi+1'')$ $\neg B(X1,Zi+1')$ respectively.

Observation 2: Second, the evaluation of the B-part of formula $i+1$ includes the evaluation of the B-part of formula i. For example, in (8) we must evaluate $\neg B(X2,Z2')$ with the values C for $Z2'$. In (9), we evaluate $\neg B(X3,Z3')$ with values C for $Z3'$, the same data retrievals as for (8) albeit with differently numbered variables. Similarly $\neg B(X2,Z3'')$ $\neg B(X3,Z3')$ of (9) represents the same data retrievals as $\neg B(X3,Z4'')$ $\neg B(X1,Z4')$ of the next resolvent in the series of resolvents.

As the last step towards deriving the desired program format, we consider again how the redundancy in evaluation of B-parts above can be eliminated. We begin the data retrieval process by first evaluating (7) with values C for $Z1'$. We may immediately print out a first group of answers, namely those values retrieved for $W1'$. Let us also immediately evaluate

$$\neg B(X2,Z2') \text{ ---(10)}$$

We may now proceed to evaluate the formula

$$\neg A(Y2,W2') \neg E(U1,V1,W2'',Z2'') \text{ ---(11)}$$

using the values of $Z2''$ obtained from (10). As before, such an evaluation corresponds exactly to evaluating (8) with values C substituted for $Z2'$. Of course, there may have been several tuples retrieved for $Z2''$ and (11) is evaluated for each such tuple. Further, the values retrieved for $W2'$ in each case may be printed as additional answers. Now we take each of these tuples retrieved for $Z2''$ in (10), substitute for $Z2'$ and evaluate (10) a second time. We will obtain a new, second set of tuples corresponding to the vector $Z2''$. This sequence of evaluations of (10) corresponds to the same data retrieval as the formula $\neg B(X2,Z3'')$ $\neg B(X3,Z3')$ with $Z3'$ replaced by C, that is, to the data retrievals for the B- part of formula (9). The above process continues for successive traversals of the PRL.

We may now describe the general program format for performing the above sequence of retrievals. There is an initialization phase which includes the evaluation of (7) and (10) and the printing of an initial set of answers. We also initialize a formula variable, F, to be the join expression (11) which will be the first in a sequence of such expressions to be evaluated by the program. The main part of the program is a loop whose purpose is to evaluate the current form of F and print all the answers so obtained, and then to evaluate (10) in preparation for the next iteration. Also, at the end of each iteration the formula F is extended. These comments lead to the general program format shown in Appendix 1a. The algorithm for deriving programs for this simplified case is shown in Appendix 1c.

6. Relaxing the Simplifying Assumptions

In this section we discuss the relaxation of some of the simplifying assumptions made in Section 4. For a complete treatment, the reader is referred to [3].

6.1 The Single PRL Assumption

Clearly, PRLs with no clauses in common can be compiled separately. Intersecting PRLs which have different start clauses lead to mutually recursive programs and will not be discussed here. We therefore consider PRLs with a common start clause. An example is shown in Figure 3.

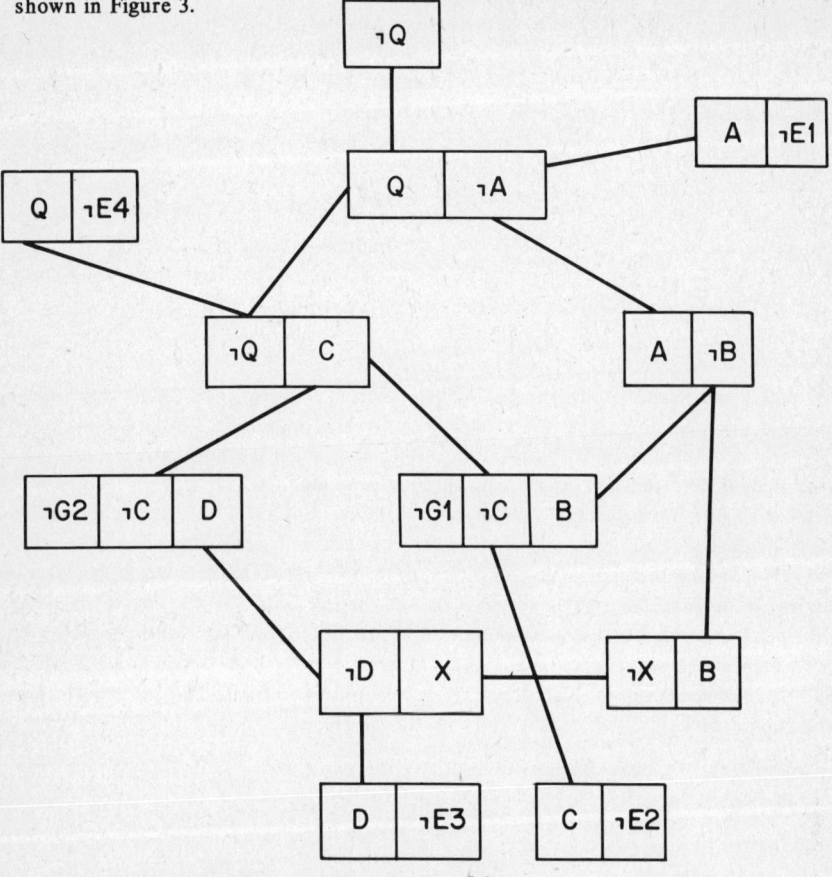

FIGURE 3: MULTIPLE PRLs

Here we have excluded the individual variables so as to concentrate on the form of the PRLs. There are two paths leading from the start clause and back to the start clause again. In this case the two paths join together in a common end clause, but in general this need not occur. Further, as shown there may be several exits, some common and some independent, for each of the two PRLs. It should be noted that an exit edge from a portion of the graph after two paths have rejoined, leads to two distinct exit expressions because of the different clauses formed by resolving along the separate paths to that point in the graph. In Figure 3, the exit to the clause Q ¬E4 yields two different resolvents depending on which path was used. Finally, each path will have its own Determined and Induced Parts. For the present example let us call these B1, B2 and A1, A2 respectively. Thus, we get two augmented loop residues --

$$Q \neg A1 \neg Q \neg B1$$
$$Q \neg A2 \neg Q \neg B2$$

In resolution one can resolve around these PRLs independently. That is, one could begin with the start clause, resolve around path 1, say, to the end, and then continue resolving following either of the two paths independently for successive traversals. This leads to expressions of the form

$$S \neg Ai1 \neg Ai2 \ldots \neg Ain \neg F \neg Bin \ldots \neg Bi2 \neg Bi1$$

where each ij is either 1 or 2. Similarly we get exit expressions of the form

$$S \neg Ai1 \neg Ai2 \ldots \neg Ain \neg E \neg Bin \ldots \neg Bi2 \neg Bi1$$

where E is some arbitrary exit.

We attach to each tuple placed into the queue the sequence of subscripts corresponding to the sequence of paths (i.e. sequence of B-parts) used to derive the tuple. This can be achieved in our program format by having an EVAL(Bi) for $i=1,2$; each tuple currently in the queue is used with each EVAL and the new tuples will inherit the sequences from the old tuples, of course, extended by one more subscript. Note that two tuples that differ only in the subscript sequence must be considered different by the ENQUE function. Regarding the expanding set of formulas corresponding to the F variable in the derived programs, we may either store the expanding set of these formulas or regenerate them as needed. Note also that in the scheme for EVAL(Bi) above, tuples with like subscripts can be easily grouped together in the queue or might even be distributed in a tree whose branching is based on the subscript sequence. This makes it a simple matter to guarantee that tuples from the queue are used with the correct answer expression.

6.2 The Constants in the PRL Unifier Assumption

We assumed that there were no constants in the substitution that unifies the loop or in the substitution along the closing edge. Further, that the simultaneous **mgu** of the pairs (Di,Di′), in Figure 2, do not equate any variables in the start literal. If two variables in the start literal are equated by the PRL unifier, we merely include a condition that the two corresponding attribute values be equal. We consider the remaining cases that arise:

1. A determined variable z is replaced by a constant in the PRL or the exit unifier.

    Suppose the PRL unifier provides a value, say b. Figure 4a shows such a case. Here we have purposely separated the variables and not unified the loop or the exit. In this case, unifying the loop forces z to become b. Clearly, the main loop of the derived program should not be completed if the value supplied by the query is different than b. Similar remarks apply for the values provided at the closing edge (in this case by the variable 'v'). Note, however, that any exit expressions which occur in the PRL before the incompatible substitution may still yield values corresponding to the resolvent obtained by going from the start clause directly to that exit. In Figure 4a, if

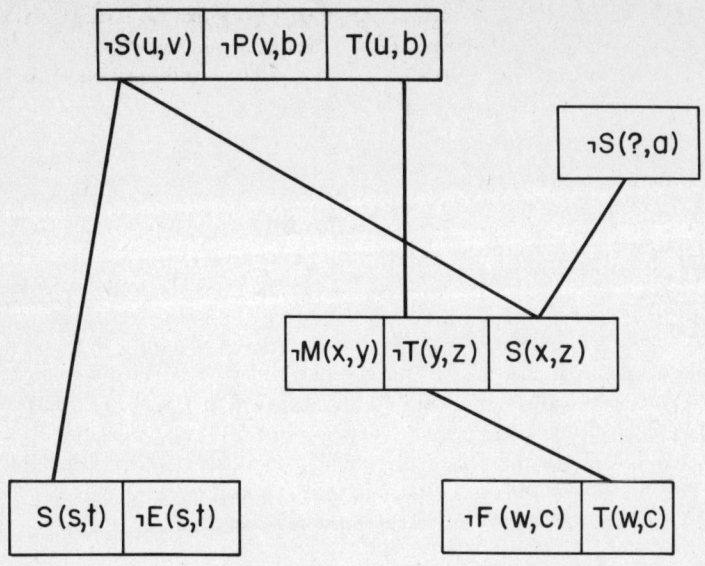

FIGURE 4a: CONSTANTS IN THE PRL AND EXIT UNIFIERS

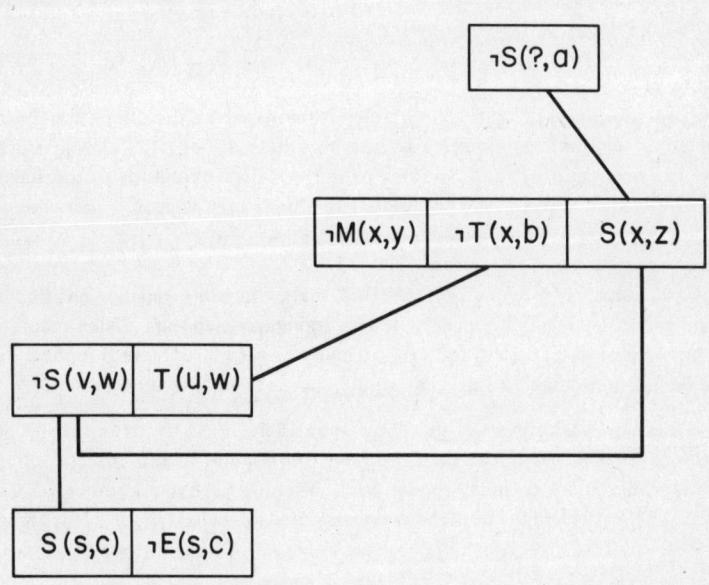

FIGURE 4b: CONSTANTS IN THE PRL AND EXIT UNIFIERS

z gets a value c then answers can be obtained by evaluating ¬M(x,y) ¬F(y,z). Therefore, in the general program format, at the point corresponding to the assignment of the variable, we insert a process that removes incompatible tuples from the queue. For Figure 4a, the outer loop looks as follows:

>   initialization
>   loop for first exit expression
>   eliminate incompatible tuples from Q
>   loop for second exit expression
>   loop for B-part
>   extend F
>   etc.

The second possibility is that a determined variable may be assigned a value by the exit unifier. In Figure 4a, assume that T(u,b) is changed to T(u,w). The determined variable z is assigned the value c by the exit to ¬F(w,c) T(w,c). No tuple from the queue should be used with the corresponding F-expression whose value for z is different than c. The solution is to test for such a conflict by inserting a conditional statement before each EVAL(F) statement in the program form.

2. A constant occurs in the unifier for the closing edge.

If the variable being replaced by the constant occurs in the start literal, Q, then it must be a determined variable, and as such is free to receive any value subject to the restrictions listed above and in the following. If the variable occurs in the end literal, ¬Q, then this case requires a consistency check on the tuples retrieved from the EVAL(B) statement.

6.3 The Stability Assumption

The stability assumption implies that the variables instantiated by the query are the same as those instantiated by determined variables by the substitution along the closing edge of the PRL. In effect, we are requiring that the form of the ¬Q literal presented to the start clause, at any time, is the same. If this is not the case, two things may happen. First, the question marks may disappear as the following example shows.

Example 5.3.1. Consider Figure 5. The question mark shifts its relative position at each traversal around the PRL and finally, on the fourth traversal disappears. Using resolution, we can generate the form of database requests for as many traversals as there is a question mark. Beyond this point, further database retrievals would appear to be meaningless.

Second, if the question mark does not disappear then there must be some cycling behavior exhibited by the it. That is, given the fixed number of variables in the Q literal of the start clause, the pattern of variables must repeat itself after some fixed number of traversals around the PRL. Thus, we might find the following kind of behavior for some integers 'm' and 'n'.

1. From the initial query we traverse the PRL m times.

2. After n subsequent traversals, we obtain the same pattern of variables and values, as after the first m traversals.

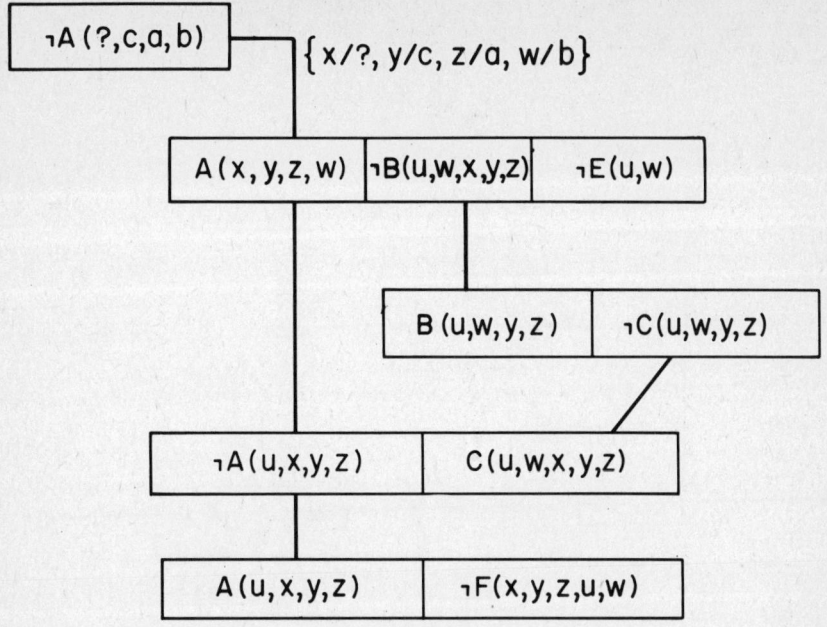

FIGURE 5

Example 5.3.2 Consider Figure 6. Using the form 'variable.i' to represent the separated variables for the ith traversal around the PRL of Figure 6, we have the following residues and the corresponding query forms (where 'v' stands for a non-determined variable and 'd' for for a determined variable).

First traversal:
    Form of incoming query: A(v,v,d,d,d)
    Loop residue: ¬G(x.1,t.1) ¬A(t.1,a,y.1,w.1,c) ¬E(b,w.1)

Second traversal:
    Form of incoming query: A(v,d,v,d,d)
    Loop residue: ¬G(x.1,t.1) ¬G(t.1,t.2) ¬A(t.2,y.1,a,w.2,c) ¬E(w.1,w.2) ¬E(b,w.1)

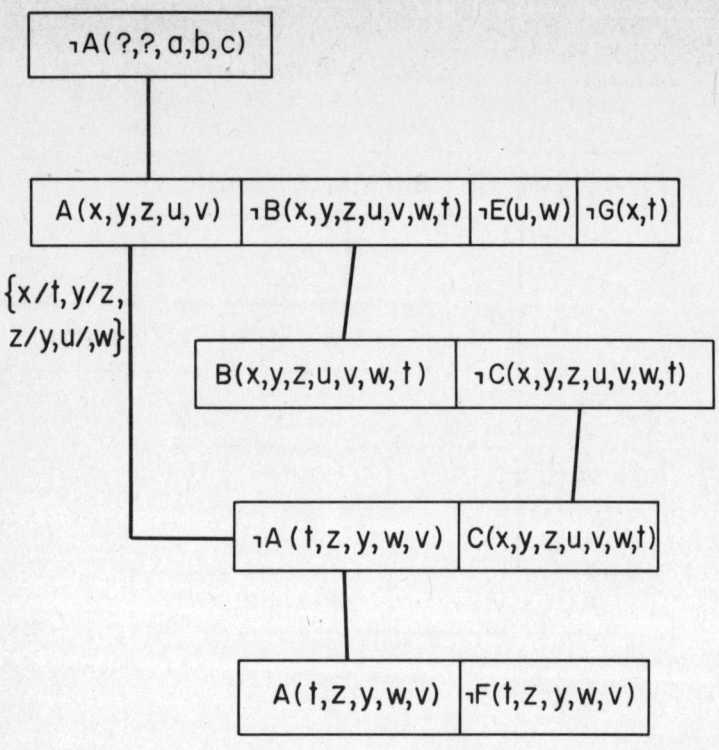

FIGURE 6: UNSTABLE PRL

Third traversal:
    Form of incoming query: A(v,v,d,d,d)
    Loop residue: ¬G(x.1,t.1) ¬G(t.1,t.2) ¬G(t.2,t.3) ¬A(t.3,a,y.1,w.3,c)
        ¬E(w.2,w.3) ¬E(w.1,w.2) ¬E(b,w.1)

Fourth traversal
    Form of incoming query: A(v,d,v,d,d)
    Loop residue: ¬G(x.1,t.1) ¬G(t.1,t.2) ¬G(t.2,t.3) ¬G(t.3,t.4) ¬A(t.4,y.1,a,w.4,c)
        ¬E(w.3,w.4) ¬E(w.2,w.3) ¬E(w.1,w.2) ¬E(b,w.1)

At this point, it is obvious that the form of the query will oscillate between A(v,v,d,d,d) and A(v,d,v,d,d). Next, the answer expressions that are obtained for each traversal are as follows.

answer 0: ¬G(x.1,t.1) ¬F(t.1,a,y.1,w.1,c) ¬E(b,w.1)

answer 1: ¬G(x.1,t.1) ¬G(t.1,t.2) ¬F(t.2,y.1,a,w.2,c) ¬E(w.1,w.2) ¬E(b,w.1)

answer 2: ¬G(x.1,t.1) ¬G(t.1,t.2) ¬G(t.2,t.3)
¬F(t.3,a,y.1,w.3,c) ¬E(w.2,w.3) ¬E(w.1,w.2) ¬E(b,w.1)

answer 3: ¬G(x.1,t.1) ¬G(t.1,t.2) ¬G(t.2,t.3) ¬G(t.3,t.4) ¬F(t.4,y.1,a,w.4,c)
¬E(w.3,w.4) ¬E(w.2,w.3) ¬E(w.1,w.2) ¬E(b,w.1)

We see, in this case, that $m=0$ and $n=2$. We treat the first m traversals as straight line code. Then, the following n traversals form the body of the outer loop of the database program. Within this outer body, there will be n inner loops, each one corresponding to a single PRL traversal and having its own ENQUE, DEQUE, and PRINT statements. The program derived for the CG of Figure 6 is shown in Appendix 1b.

7. Conclusion

Correctness and Termination of the generated procedures is discussed in [3]. We have presented a technique for transforming the infinite set of resolvents generated in a recursive intension into a finite procedure. The method as presented can be used for first-order theories containing function signs. Of course, in this case the finiteness of the search process cannot be guaranteed.

8. Acknowledgements

The authors thank Dan Fishman and the referees for valuable comments.

9. Appendix 1

a) General Format of Derived Programs

/* ENQUE(Q,x) puts those values of vector x on a queue, Q, which are not already on Q; DEQUE(Q) returns the front element of the queue Q*/

```
Z1'=c
EVAL(¬E(U1,V1,W1',Z1'))
print(w1')
EVAL(¬B(X2,Z2'))
ENQUE(Q,Z2")
F=¬A(Y2,W2') V¬E(U1,V1,W2",Z2")
i:=2
while (Q ≠ empty) do
   Q1:=Q
     while (Q1 ≠ empty) do /* first Inner Loop */
        Z2":= DEQUE(Q1)
        eval(F)
        print (Wi')
     end
   Q':=Q
```

```
        Q:=empty
        while (Q' ≠ empty) do  /* second inner loop */
              Z2':=DEQUE(Q')
              EVAL(B(X2,Z2'))
              ENQUE(Q, Z2")
        end
    REPLACE ¬A(Yi,Wi') by ¬A(Yi+1,Wi+1') V¬A(Yi,Wi") in F.
    i:=i+1
end
```

1b) Program Derived for CG of Figure 6

```
z.1:=a; u.1:=b; v.1:=c;
F:= ¬G(t.1,t.0) ¬F(t.0,z.1,y.1,w.1,v.1) ¬E(u.1,w.1)
i:=2
ENQUE (Q,(z.1,u.1,v.1))
while (Q ≠ empty) do
   Q1:=Q
   Q2:=empty
   while (Q1 ≠ empty) do
      (z.2,u.2,v.2) :=DEQUE(Q1)
      eval(F)
      print(t.i,y.2)
      ENQUE(Q2,(z.2,w.2,v.2))
   end
i:=i+1
F:=F V¬G(t.i,t.i-1)
Q:=empty
while (Q2 ≠ empty) do
     (y.2,u.2,v.2) :=DEQUE(Q2)
     eval(F)
     print (t.i,z.i)
     ENQUE (Q,(y.2,w.2,v.2))
   end
i:=i+1
F:= F V¬G(t.i,t.i-1)
end
```

/* Each inner loop processes all the tuples from the previous outer loop, i.e., from the previous PRL traversal. It will also enque tuples for the next inner loop.*/

1c) Algorithm for Deriving Database Programs

Input: A PRL on literal R and a query literal Q.

Output: A database retrieval program for resolving query Q.

Method:

1. Attach the query node to the CG and derive the loop residue and the answer expressions;

2. Identify the determined and nondetermined variables of the PRL.

3. Derive the appropriate program by using the skeleton shown in Appendix 1a.

10. References

[1] A. Aho and J. Ullman, "The Theory of Parsing, Translation, and Compiling, Vol. 1: Parsing," Prentice-Hall, Englewood Cliffs, N.J., 1972.

[2] C. Chang, "On Evaluation of Queries Containing Derived Relations in a Relational Data Base," Workshop on Formal Basis for Databases, Toulouse, France, 1979.

[3] L. Henshen, S. Naqvi, "On Compiling Queries in Recursive First-Order Databases", submitted to JACM.

[4] R. Reiter, "Deductive Question Answering on Relational Data Bases," in Logic and Data Bases (ed. H. Gallaire and J. Minker), Plenum Press, 1978.

[5] R. Reiter, "Equality and Domain Closure for First-Order Data Bases," Journal of the ACM, Vol. 27, No. 2, 1980.

[6] J. Robinson, "A Machine-Oriented Logic Based on the Resolution Principle," Journal of the ACM, Vol. 12, No. 1, 1965.

[7] S. Shapiro, and D. McKay, "Inference with Recursive Rules," Proc. NCAI, Stanford University, 1980.

[8] S. Sickel, "A Search Technique for Clause Interconnectivity Graphs," IEEE transactions on computers, Vol. C-25, No. 8, 1976.

[9] J. Ullman, "Principles of Data Base Systems," Computer Science Press, Potomac, Md., 1980.

THE POWER OF THE CHURCH-ROSSER PROPERTY
FOR STRING REWRITING SYSTEMS[†]

Ronald V. Book

Department of Mathematics
University of California at Santa Barbara
Santa Barbara, Ca. 93106, U.S.A.

INTRODUCTION

The Church-Rosser property has been shown to be extremely powerful when possessed by term-rewriting systems, tree-manipulating systems, and other types of replacement systems [9,10,13,18,21]. It is used in the study of matching and unification problems [20]. When applied to string-replacement systems such as Thue systems, the first use of this property was in the study of formal language theory where the specification of formal languages as finite unions of congruence classes of Thue systems with the Church-Rosser property was investigated by Nivat and his co-workers [6,7,16,17]. (See [1] for a survey of much of the work in this area.) Recent work on Thue systems with the Church-Rosser property extends the earlier work on specification of formal languages, studies the complexity of word problems and variations on word problems, and investigates the structure of the monoids presented by these systems [2-5,12,19].

In this paper it is shown that there is a decision procedure for a specific class of true sentences about congruences that are Church-Rosser. The class of sentences is defined syntactically and has prenex form $\exists^p \forall^q$ or $\forall^p \exists^q$, linear terms, positive matrices, and other restrictions. The decision procedure depends on the Thue system being finite, monadic, and Church-Rosser. The procedure decides questions that are PSPACE-complete and can be carried out in polynomial space.

A Thue system presents a monoid. Thus, examples of applications of the main result are stated in terms of the monoid so presented. If T is a finite Thue system that is monadic and Church-Rosser, then the following questions about the monoid presented by T can be decided by applying this procedure:

[†]This research was supported in part by the National Science Foundation under grants MCS80-11979 and MCS81-16327.

(a) Given a finite set, is the submonoid generated by this set a group?

(b) Given a finite set, is the submonoid generated by this set a right ideal (resp., a left ideal, a two-sided ideal)?

(c) Given two elements x and y, which (if any) of Green's relations hold for x and y?

SECTION 1. PRELIMINARIES

It is assumed that the reader is familiar with the basic results in the theories of automata, computability, and formal language theory as covered in a text such as [8]. In this section notation is established and the basic definitions and properties of Thue systems and congruences are described.

If Σ is a finite alphabet, then Σ^* denotes the free monoid with identity e generated by Σ. If w is a string, then the <u>length</u> of w is denoted by $|w|$: $|e| = 0$; $|a| = 1$ if $a \in \Sigma$; $|wa| = |w| + 1$ if $w \in \Sigma^*$, $a \in \Sigma$.

A <u>Thue system</u> T on a finite alphabet Σ is a subset of $\Sigma^* \times \Sigma^*$. The <u>Thue congruence generated by</u> T is the reflexive transitive closure $\overset{*}{\longleftrightarrow}_T$ of the relation $\longleftrightarrow_T$ defined as follows: for any u, v such that $(u,v) \in T$ or $(v,u) \in T$ and any $x, y \in \Sigma^*$, $xuy \longleftrightarrow_T xvy$. If $x \longleftrightarrow_T y$ and $|x| > |y|$, then write $x \rightarrow_T y$, and write $\overset{*}{\rightarrow}_T$ for the reflexive transitive closure of the relation denoted by $\rightarrow_T$.

Let T be a Thue system on the alphabet Σ. If $x, y \in \Sigma^*$ and $x \overset{*}{\longleftrightarrow}_T y$, then x and y are <u>congruent</u> (<u>mod</u> T). If $x \in \Sigma^*$, then the <u>congruence class of</u> x (<u>mod</u> T) is $[x] = \{y \in \Sigma^* \mid x \overset{*}{\longleftrightarrow}_T y\}$. This notation is extended to sets $A \subseteq \Sigma^*$ as $[A] = \{y \in \Sigma^* \mid \text{there exists } x \in A \text{ such that } x \overset{*}{\longleftrightarrow}_T y\}$, so that $[A] = \cup\{[x] \mid x \in A\}$. If $x, y \in \Sigma^*$ and $x \overset{*}{\rightarrow}_T y$, then x is an <u>ancestor</u> of y and y is a <u>descendant</u> of x; the set of all ancestors of x is denoted $\langle x \rangle$ and the set of all descendants of x is denoted $\Delta^*(x)$. This notation is also extended to sets $A \subseteq \Sigma^*$ as $\langle A \rangle = \{y \in \Sigma^* \mid \text{there exists } x \in A \text{ such that } y \overset{*}{\rightarrow}_T x\}$ and $\Delta^*(A) = \{y \in \Sigma^* \mid \text{there exists } x \in A \text{ such that } x \overset{*}{\rightarrow}_T y\}$, so that $\langle A \rangle = \cup\{\langle x \rangle \mid x \in A\}$ and $\Delta^*(A) = \cup\{\Delta^*(x) \mid x \in A\}$.

Let T be a Thue system on the alphabet Σ. If $x \in \Sigma^*$ and there is no y such that $x \rightarrow y$, then x is <u>irreducible</u> (<u>mod</u> T). Let IRR(T) denote the set of all strings that are irreducible (mod T).

If T is a Thue system on the alphabet Σ, then the set of congruence classes of the congruence $\overset{*}{\longleftrightarrow}_T$ forms a monoid $\mathcal{M}_T$. This monoid is the quotient of Σ^* by $\overset{*}{\longleftrightarrow}_T$ and so has [e] as identity.

The multiplication in $\mathcal{M}_T$ is given by $[x] \circ [y] = [xy]$. The Thue system T is a <u>presentation</u> of $\mathcal{M}_T$.

For background on semigroups and monoids, see [14].

When there is no ambiguity regarding the Thue system T under consideration, the subscript T will be omitted.

SECTION 2. CHURCH-ROSSER SYSTEMS

In this section we describe some of the properties of Thue systems with the Church-Rosser property.

Let T be a Thue system on the alphabet Σ. The system T is said to be <u>Church-Rosser</u> (or to <u>have the Church-Rosser property</u>) if for all $x, y \in \Sigma^*$, if $x \stackrel{*}{\longleftrightarrow} y$, then there exists $z \in \Sigma^*$ such that $x \stackrel{*}{\to} z$ and $y \stackrel{*}{\to} z$.

If a Thue system is Church-Rosser, then every congruence class has a unique irreducible string [7,9]. For any finite Thue system there is a linear-time algorithm that will compute on input x an irreducible string y congruent to x [2]. Thus, if a finite Thue system is Church-Rosser, then its word problem (i.e., "given x, y, is x congruent to y?") is not only decidable but also decidable in linear time [2].

Notice that if T is Church-Rosser, then for any set $A \subseteq \Sigma^*$, the set $[A] = \{y \mid \text{for some } x \in A, x \stackrel{*}{\longleftrightarrow} y\}$ is exactly $\langle \Delta^*(A) \cap \text{IRR}(T) \rangle = [\Delta^*(A) \cap \text{IRR}(T)]$ since $x \stackrel{*}{\longleftrightarrow} y$ if and only if there is a (unique) irreducible z such that $x \stackrel{*}{\to} z$ and $y \stackrel{*}{\to} z$.

It is decidable whether a finite Thue system is Church-Rosser [17]; in fact, this question is tractable.

<u>Proposition 1</u> [5]. There is a polynomial-time algorithm to decide the following question: "given a finite Thue system T, is T Church-Rosser?"

In this paper we focus on congruences generated by certain restricted Church-Rosser systems.

A Thue system T on an alphabet Σ is <u>monadic</u> if $(u,v) \in T$ implies $|u| > |v|$ and $v \in \Sigma \cup \{e\}$.

One result on monadic Thue systems is of particular interest here.

<u>Proposition 2</u> [4]. Let T be a finite Thue system on the alphabet Σ. Suppose that T is monadic. For every regular set $R \subseteq \Sigma^*$, the set $\Delta^*(R)$ of descendants of R is a regular set and a finite-state acceptor (regular expression) for $\Delta^*(R)$ can be effectively constructed from T and a finite-state acceptor (regular expression) for R.

Berstel [1] has pointed out that if T is a finite Thue system, IRR(T) is a regular set. Combining this fact with Proposition 2, we have the following.

Proposition 3. Let T be a finite Thue system on the alphabet Σ. Suppose that T is monadic. Then for every regular set $R \subseteq \Sigma^*$, the set of irreducible descendants of R, $\Delta^*(R) \cap IRR(T)$, is a regular set.

SECTION 3. THE MAIN RESULT

In this section the class of sentences to be studied is defined syntactically and then an interpretation is given in terms of an arbitrary Thue congruence. The main result states that there is a decision procedure for the class of true sentences when the congruence is generated by a finite monadic Church-Rosser system.

3.1 The Class of Sentences

Let Σ be a finite alphabet.

(i) Variables. Let V_E and V_U be two disjoint countable sets of symbols such that $V \cap \Sigma = \phi$ where $V = V_U \cup V_E$. A symbol $v(U,i)$ in V_U is a universal variable and a symbol $v(E,i)$ in V_E is an existential variable. For each variable $v(U,i)$ ($v(E,i)$), there is a regular set $D(U,i) \subseteq \Sigma^*$ (resp., $D(E,i) \subseteq \Sigma^*$), the domain of the variable, that is effectively specified by specifying a regular expression.

(ii) Terms. A constant term is a string in Σ^*. A universal term is a nonempty string in $(\Sigma \cup V_U)^*$. An existential term is a nonempty string in $(\Sigma \cup V_E)^*$.

(iii) Formulas. A constant formula is an ordered pair of constant terms. A mixed formula is an ordered pair of terms, one of which is existential and the other is universal. A universal formula is an ordered pair of terms such that either both are universal or one is universal and the other is constant. An existential formula is an ordered pair of terms such that either both are existential or one is existential and the other is constant. A formula is linear if no variable occurs twice in it. A formula is written (t_1, t_2).

(iv) Clauses. A formula is a clause. If C_1 and C_2 are clauses such that no existential variable occurs in both C_1 and C_2, then $(C_1 \wedge C_2)$ is a clause. If C_1 and C_2 are clauses such that no universal variable occurs in both C_1 and C_2, then $(C_1 \vee C_2)$ is a clause.

(v) Sentences. If C is a clause with existential variables $v(E,i_1), \ldots, v(E,i_p)$ and universal variables $v(U,j_1), \ldots, v(U,j_q)$, then $\forall v(U,j_1) \ldots \forall v(U,j_q) \exists v(E,i_1) \ldots \exists v(E,i_p) C$ and

$\exists v(E,i_1) \ldots \exists v(E,i_p) \forall v(U,j_1) \ldots \forall v(U,j_q)C$ are sentences. Let SEN(Σ) be the set of all sentences and let LINSEN(Σ) be the set of all sentences containing only linear formulas.

3.2 The Interpretation for Thue System T

For any given Thue system T on the alphabet Σ, define the following interpretation:

(i) Each variable takes values in its domain.

(ii) Let (t_1, t_2) be a formula. Let $val(t_1)$ ($val(t_2)$) be a string in Σ^* resulting from substituting for each variable occurring in t_1 (resp., t_2) a string in its domain. Then (t_1, t_2) is interpreted as the set of all statements $val(t_1) \xleftrightarrow{*}_T val(t_2)$.

(iii) The symbol $\wedge$ is interpreted as conjunction and the symbol $\vee$ is interpreted as disjunction.

Under this interpretation each sentence in SEN(Σ) is either true or false as a statement about the congruence $\xleftrightarrow{*}$ and hence about the monoid $\mathcal{M}_T$ presented by T (recall: $\mathcal{M}_T$ is $\Sigma^*/\xleftrightarrow{*}_T$). Let SEN(T,$\Sigma$) denote the set of all statements about $\xleftrightarrow{*}$ obtained from SEN(Σ) under this interpretation, and let LINSEN(T,Σ) denote the set of all statements about $\xleftrightarrow{*}$ obtained from LINSEN(Σ) under this interpretation

3.3 The Result

<u>Theorem 1</u>. Let T be a Thue system on the alphabet Σ. If T is finite, monadic, and Church-Rosser, then there is a decision procedure for the set of all <u>true</u> sentences in LINSEN(T,Σ), that is, there is an algorithm that on input a sentence in LINSEN(Σ) will eventually halt and correctly answer the question of whether that sentence is true or false under the interpretation for T described above.

SECTION 4. APPLICATIONS

Before sketching the proof of Theorem 1, we briefly describe some applications.

Let T be a Thue system on the alphabet Σ and let $\mathcal{M}_T = \Sigma^*/\xleftrightarrow{*}$ be the monoid presented by T.

A set $A \subseteq \Sigma^*$ is <u>independent in</u> $\mathcal{M}_T$ if $x \in A$ implies that x is not congruent to any string in $(A - \{x\})^*$.

For a finite set A we can construct a sentence S representing the property that A is not independent in $\mathcal{M}_T$. Let $A = \{x_1, \ldots, x_n\}$. For each $i = 1, \ldots, n$, let v_i be an existential variable with domain the regular set $(A - \{x_i\})^*$. Then A is not independent in $\mathcal{M}_T$ if and only if $\exists v_1 \ldots \exists v_n ((x_1 \xleftrightarrow{*} v_1) \vee \ldots \vee (x_n \xleftrightarrow{*} v_n))$ is true if and only if the sentence $\exists v_1 \ldots \exists v_n ((x_1, v_1) \vee \ldots \vee (x_n, v_n))$

is true under the interpretation for $\xleftrightarrow{*}$. Let S be the sentence
$\exists v_1 \ldots \exists v_n ((x_1,v_1) \vee \ldots \vee (x_n,v_n))$.

Theorem 2. Let T be a Thue system on the alphabet Σ. Suppose that T is finite, monadic, and Church-Rosser. Then the following question is decidable: for finite $A \subseteq \Sigma^*$, is A independent in $\mathcal{M}_T$?

For any $A \subseteq \Sigma^*$, let $\mathcal{M}(A)$ be the smallest submonoid of $\mathcal{M}_T$ containing $\{[x] \mid x \in A\}$, i.e., $\mathcal{M}(A) = \{[e]\} \cup \{[x_1] \circ \ldots \circ [x_n] \mid n \geq 1$, each $x_i \in A\}$ where $\circ$ is the multiplication of $\mathcal{M}_T$. Thus, $\mathcal{M}(A) = \{[w] \mid w \in A^*\}$.

Theorem 3. Let T be a Thue system on an alphabet Σ. Suppose that T is finite, monadic, and Church-Rosser. Then the following question is decidable: for a finite set $A \subseteq \Sigma^*$, is the submonoid $\mathcal{M}(A)$ generated by A a subgroup of $\mathcal{M}_T$?

If T is a Thue system on a finite alphabet Σ, then $\mathcal{M}_T$ is a group if and only if $\mathcal{M}(\Sigma) = \mathcal{M}_T$ is a subgroup of $\mathcal{M}_T$.

Corollary. Let T be a Thue system on a finite alphabet Σ. Suppose that T is finite, monadic, and Church-Rosser. Then the following question is decidable: is $\mathcal{M}_T$ a group?

The questions posed in Theorem 3 and its Corollary can be answered in polynomial time by using a different technique [3].

A <u>right</u> (<u>left</u>) <u>ideal</u> of a monoid $\mathcal{M}$ is a nonempty set $\mathcal{I} \subseteq \mathcal{M}$ such that for every $x \in \mathcal{I}$ and $y \in \mathcal{M}$, $xy \in \mathcal{I}$ (resp., $yx \in \mathcal{I}$). A <u>two-sided</u> <u>ideal</u> is a subset that is both a left and right ideal.

Theorem 4. Let T be a Thue system on an alphabet Σ. Suppose that T is finite, monadic, and Church-Rosser. Then the following questions is decidable: for a regular set $A \subseteq \Sigma^*$, is the submonoid $\mathcal{M}(A)$ generated by A a right (left, two-sided) ideal of $\mathcal{M}_T$?

In a monoid $\mathcal{M}$, <u>Green's relations</u> are defined as follows:
(a) $x \mathcal{R} y$ if and only if $\{xz \mid z \in \mathcal{M}\} = \{yz \mid z \in \mathcal{M}\}$;
(b) $x \mathcal{L} y$ if and only if $\{zx \mid z \in \mathcal{M}\} = \{zy \mid z \in \mathcal{M}\}$;
(c) $x \mathcal{J} y$ if and only if $\{z_1 x z_2 \mid z_1, z_2 \in \mathcal{M}\} = \{z_1 y z_2 \mid z_1, z_2 \in \mathcal{M}\}$;
(d) $x \mathcal{D} y$ if and only if $x \mathcal{R} y$ or $x \mathcal{L} y$;
(e) $x \mathcal{H} y$ if and only if $x \mathcal{R} y$ and $x \mathcal{L} y$.

Theorem 5. Let T be a Thue system on an alphabet Σ. Suppose that T is finite, monadic, and Church-Rosser. Then Green's relations for $\mathcal{M}_T$ are decidable, i.e., it is decidable given x and y whether $x \mathcal{R} y$ or $x \mathcal{L} y$ or $x \mathcal{J} y$ or $x \mathcal{D} y$ or $x \mathcal{H} y$ holds.

SECTION 5. PROVING THE MAIN RESULT

This section is devoted to a sketch of the proof of Theorem 1.

Consider sentences in LINSEN(Σ). No term has both existential and universal variables and every formula is linear. If $(C_1 \wedge C_2)$ is a clause, then no existential variable occurs in both C_1 and C_2, and if $(C_1 \vee C_2)$ is a clause, then no universal variable occurs in both C_1 and C_2; thus, one can distribute the quantifiers over $\wedge$ and $\vee$ so that any sentence in LINSEN(Σ) is equivalent to a positive (i.e., using only conjunctions and disjunctions) combination of sentences in LINSEN(Σ) each having only a single formula. This means that it is sufficient to restrict attention to those sentences in LINSEN(Σ) with only a single formula.

Consider the formula (t_1, t_2). With t_1 (t_2), associate a regular set R_1 (resp., R_2) as follows:
 (i) if t_1 is a constant term, then $t_1 \in \Sigma^*$ so let $R_1 = \{t_1\}$;
 (ii) if t_1 is not a constant term, then t_1 is a concatenation of variable symbols and constants, that is, $t_1 \in (V \cup \Sigma)^*$; let R_1 be the regular set obtained from t_1 by substituting for each variable symbol v in t_1 the regular set that is the domain of v, so that R_1 is the concatenation of regular sets (domains) and sets containing only a single string (constants) in the order specified by t_1.

In order to determine whether the sentence containing (t_1, t_2) is true for the congruence generated by T, it is necessary to obtain information about the sets $[R_1] = \{y \mid \text{for some } x \in R_1, x \xleftrightarrow{*} y\}$ and $[R_2] = \{y \mid \text{for some } x \in R_2, x \xleftrightarrow{*} y\}$. Since T is finite, monadic, and Church-Rosser, it is sufficient to consider the sets of irreducible representatives of R_1 and R_2. Since T is Church-Rosser, the set of irreducible representatives of R_1 (R_2) is the set of irreducible descendants of R_1 (resp., R_2).

Let $IR_1 = \Delta^*(R_1) \cap IRR(T)$ and let $IR_2 = \Delta^*(R_2) \cap IRR(T)$. By Proposition 3, IR_1 and IR_2 are regular sets of irreducible strings. Further, from T and regular expressions for R_1 and R_2, one can construct regular expressions for IR_1 and IR_2.

Consider the formula (t_1, t_2). Based on whether (t_1, t_2) is universal, existential, constant, or mixed, one determines the appropriate relation between IR_1 and IR_2. If (t_1,t_2) is universal, then the sentence is true if and only if $IR_1 = IR_2$ and both IR_1 and also IR_2 are singleton sets. If (t_1,t_2) is existential, then the sentence is true if and only if $IR_1 \cap IR_2 = \phi$. If (t_1,t_2) is constant, then the sentence is true if and only if $IR_1 = IR_2$ (in this case, IR_1 and IR_2 are automatically singleton sets). If (t_1,t_2) is mixed, then either the prefix is $\exists^p \forall^q$, t_1 is existential, and t_2 is universal, in which case the sentence is true if and only if IR_2 is a singleton set and $IR_2 \subseteq IR_1$, or the prefix is $\forall^p \exists^q$, t_1 is existential, and t_2 is universal, in which case the sentence is true if and only if $IR_1 \subseteq IR_2$. All of these conditions on IR_1 and IR_2 are decidable by using standard techniques.

SECTION 6. REMARKS

The trivial Thue system with no rules is monadic and Church-Rosser. In this case, this procedure reduces to the equivalence of regular expressions, a PSPACE-complete problem [15].

Proposition 2 can be carried out in polynomial space and the various decisions about regular sets can be answered using polynomial space. Thus this procedure can be carried out in polynomial space.

The procedure also applies to infinite regular monadic Thue systems T that are Church-Rosser since in this case, if $R \subseteq \Sigma^*$ is regular, then so is $\Delta^*(R) \cap IRR(T)$ [19].

The condition that the Thue system is Church-Rosser cannot be omitted since it is known that the question "does a finite monadic Thue system present a group?" is undecidable.

REFERENCES

1. J. Berstel, Congruences plus que parfaites et langages algébrique, Seminaire d'Informatique Théorique (1976-77), Institut de Programmation, 123-147.

2. R. Book, Confluent and other types of Thue systems, J. Assoc. Comput. Mach. 29 (1982), 171-182.

3. R. Book, When is a monoid a group? The Church-Rosser case is tractable, Theoret. Comput. Sci. 18 (1982), to appear.

4. R. Book, M. Jantzen, and C. Wrathall, Monadic Thue systems, Theoret. Comput. Sci. 19 (1982), to appear.

5. R. Book and C. Ó'Dúnlaing, Testing for the Church-Rosser property, Theoret. Comput. Sci. 16 (1981), 223-229.

6. Y. Cochet, Sur l'algébricité des classes de certaines congruences definies sur le monoide libre, Thèse 3e cycle, Rennes, 1971.

7. Y. Cochet and M. Nivat, Une generalization des ensembles de Dyck, *Israel J. Math.* 9 (1971), 389-395.

8. J. Hopcroft and J. Ullman, *Formal Languages and Their Relation to Automata*, Addison-Wesley, 1969.

9. G. Huet, Confluent reductions: abstract properties and applications to term-rewriting systems, *J. Assoc. Comput. Mach.* 27 (1980), 797-821.

10. G. Huet and D. Oppen, Equations and rewrite rules: a survey, in R. Book (ed.), *Formal Language Theory: Perspectives and Open Problems*, Academic Press, 1980, 349-405.

11. H. Hunt, D. Rosenkrantz, and T. Szymanski, On the equivalence, containment, and covering problems for the regular and context-free languages, *J. Comput. Syst. Sci.* 12 (1976), 222-268.

12. M. Jantzen, On a special monoid with a single defining relation, *Theoret. Comput. Sci* 16 (1981), 61-73.

13. D. Knuth and P. Bendix, Simple word problems in universal algebras, in J. Leech (ed.), *Computational Problems in Abstract Algebra*, Pergamon Press, 1970, 263-297.

14. G. Lallement, *Semigroups and Combinatorial Applications*, Wiley-Interscience, 1979.

15. A. Meyer and L. Stockmeyer, The equivalence problem for regular expressions with squaring requires exponential space, *Proc. 13th IEEE Symp. Switching and Automata Theory* (1972), 125-129.

16. M. Nivat, On some families of languages related to the Dyck languages, *Proc. 2nd ACM Symp. Theory of Computing* (1970), 221-225.

17. M. Nivat (avec M. Benois), Congruences parfaites et quasi-parfaites, *Seminaire Dubreil*, 25e Année (1971-72), 7-01-09.

18. M. O'Donnell, *Computing in Systems Described by Equations*, Lecture Notes in Computer Science 58 (1977), Springer-Verlag.

19. C. Ó'Dúnlaing, Finite and Infinite Regular Thue Systems, Ph.D. dissertation, University of California at Santa Barbara, 1981.

20. P. Raulefs, J. Siekmann, P. Szabo, and E. Unvericht, A short survey of the state of the art in matching and unification problems, *SIGSAM Bulletin* 13 (1979), 14-20.

UNIVERSAL UNIFICATION AND A
CLASSIFICATION OF EQUATIONAL THEORIES

J. Siekmann, P. Szabó
University of Karlsruhe
Institut für Informatik I
Postfach 6380
D-7500 Karlsruhe 1
West Germany

> *"However, to generalize, one needs experience ..."*
>
> Grätzer,
> Universal Algebra, 1968

0. MOTIVATION

Unification of two terms under a theory T amounts to solving an equation in that theory. However, the mathematical investigation of equation solving in certain theories is a subject as old as mathematics itself and, right from the beginning, very much at the heart of it: It dates back to Babylonian mathematics (about 2000 B.C.).

Universal unification carries this activity on in a more abstract setting: just as universal algebra abstracts from certain properties that pertain to specific algebras and investigates issues that are common to all of them, universal unification addresses problems, which are typical for equation solving as such.

Just as traditional equation solving drew its impetus from its numerous applications (the - for those times - complicated division of legacies in Babylonian times and the application in physics in more modern times), unification theory derives its impetus from its numerous applications in computer science, artificial intelligence and, last but not least, in the field of computational deduction.

Central to unification theory are the notion of a *set of most general unifiers* µUΣ (traditionally: the set of base vectors spanning the solution space) and the *hierarchy of unification problems* based on µUΣ:

  (i) a theory T is *unitary* if µUΣ always exists and has at most one element;
 (ii) a theory T is *finitary* if µUΣ always exists and is finite;
(iii) a theory T is *infinitary* if µUΣ always exists and there exists a pair of terms such that µUΣ is infinite for this pair;
 (iv) a theory T is of *type zero* otherwise
      (see below for an exact definition of this hierarchy).

In this paper we characterize the borderline between (i) and (ii) for equational theories. Major open research problems concern the boundary between (ii) and (iii) as well as further existence proofs for µUΣ.

In the first part we briefly survey unification theory as it stands today. The second part gives a classification of equational theories from the view point of unification and the third part contains the main result of this paper: a characterization of unitary theories.

1. INTRODUCTION TO UNIFICATION THEORY

Unification theory is concerned with problems of the following kind: Let f and g be function symbols, a and b constants and let x and y be variables and consider two *first order terms* built from these symbols; for example:
$$t_1 = f(x,g(a,b))$$
$$t_2 = f(g(y,b),x).$$
The first question which arises is whether or not there exist terms which can be substituted for the variables x and y such that the two terms thus obtained from t_1 and t_2 become equal: in the example $g(a,b)$ and a are two such terms. We shall write
$$\sigma_1 = \{x \leftarrow g(a,b), y \leftarrow a\}$$
for such a unifying substitution: σ_1 is a *unifier* of t_1 and t_2 since $\sigma_1 t_1 = \sigma_1 t_2$.

In addition to the *decision problem* there is also the problem of finding a *unification algorithm* which generates the unifiers for a given pair t_1 and t_2.

Consider a variation of the above problem, which arises when we assume that f is commutative:
(C) $\qquad\qquad f(x,y) = f(y,x).$
Now σ_1 is still a unifying substitution and moreover $\sigma_2 = \{y \leftarrow a\}$ is also a unifier for t_1 and t_2, since
$$\sigma_2 t_1 = f(x,g(a,b)) =_C f(g(a,b),x) = \sigma_2 t_2.$$
But σ_2 is *more general* than σ_1, since σ_1 is an instance of σ_2 obtained as the *composition* $\lambda \circ \sigma_2$ with $\lambda = \{x \leftarrow g(a,b)\}$; hence a unification algorithm only needs to compute σ_2.

There are pairs of terms which have more than one most general unifier (i.e. they are not an instance of any other unifier) under commutativity, but they always have at most *finitely many*. This is in contrast to the first situation (of free terms), where every pair of terms has at most *one most* general unifying substitution.

The problem becomes entirely different when we assume that the function denoted by f is associative:
(A) $\qquad\qquad f(x,f(y,z)) = f(f(x,y),z).$
In that case σ_1 is still a unifying substitution, but
$$\sigma_3 = \{x \leftarrow f(g(a,b), g(a,b)), y \leftarrow a\}$$
is also a unifier:

$\sigma_3 t_1 = f(f(g(a,b), g(a,b)), g(a,b)) =_A f(g(a,b), f(g(a,b), g(a,b))) = \sigma_3 t_2$

But $\sigma_4 = \{x \leftarrow f(g(a,b), f(g(a,b), g(a,b))), y \leftarrow a\}$ is again a unifying substitution and it is not difficult to see that there are *infinitely many* unifiers, all of which are most general.

Finally, if we assume that both axioms (A) and (C) hold for f then the situation changes yet again and for any pair of terms there are at most *finitely many* most general unifiers *under (A) and (C)*.

Many special unification algorithms for common theories have been developed in the past, since they have important applications in computer science [RSS79]. We shall now briefly review a formalism to express problems of the above kind.

1.1 *Unification from an Algebraic Point of View*

As usual let $\mathbb{N}$ be the set of natural numbers. A set of 'symbols with arity' is a mapping $\Omega: M \to \mathbb{N}$, where M is some set. For $f \in M$ Ωf is the *arity* of f. The *domain of* Ω is used to denote certain n-ary operations and is sometimes called a *signature*. $(f,n) \in \Omega$ is abbreviated to $f \in \Omega$.

A *Universal Algebra A* is a pair (A, Ω), where A is the *carrier* and $f \in \Omega$ denotes a mapping

$$f: A^n \to A, \text{ where } \Omega f = n \text{ (and if } a_1, \ldots, a_n \in A$$

then we write $f_A(a_1, \ldots, a_n)$ for the *realization* of the denoted mapping). Note that if $\Omega f = 0$ then f is a distinguished constant of the algebra A. $COD(\Omega)$, the *codomain* of Ω, is its *type*.

If A and B are algebras, $\varphi: A \to B$ is a *homomorphism* if $\varphi f_A(a_1, \ldots, a_n) = f_B(\varphi a_1, \ldots, \varphi a_n)$; a bijective homomorphism is called an *isomorphism*, in symbols $\simeq$.

For a subset $A_o \subseteq A$, $\varphi_o = \varphi|_{A_o}$ is the *restriction* of φ to A_o. An equivalence relation ρ is a *congruence* relation iff $a_1 \rho b_1, \ldots, a_n \rho b_n$ implies $f_A(a_1, \ldots, a_n) \rho f_A(b_1, \ldots, b_n)$.

$A/_\rho = (A/_\rho, \Omega)$ is the *quotient algebra modulo* ρ. $[a]_\rho$ is the congruence class generated by $a \in A$.

For a class of algebras $\mathbf{K}_o$ of fixed type, the algebra $A = (A, \Omega)$ is *free*

in K_o on the set X, in symbols $A_{K_o}(X)$, iff
  (i) $(A,\Omega) \in K_o$
  (ii) $X \subseteq A$
  (iii) if $B \in K_o$ and $\varphi_o: X \to B$ is any mapping, then there exists a unique
       homomorphism $\varphi: A \to B$ with $\varphi_o = \varphi|_X$.

If K is the class of *all* algebras of the fixed type, then $A_K(X)$ is *the*
(since it exists and is unique up to isomorphism) *absolutely free algebra
on X*. The elements of $A_K(X)$ are called *terms* and are given a concrete
representation W_Ω^X by:
  (i) $x \in X$ is in W_Ω^X
  (ii) if $t_1, t_2, \ldots, t_n$ are terms and $\Omega f = n$, $n \geq 0$, then $f(t_1, \ldots, t_n)$ is in
       W_Ω^X.

We assume that Ω consists of the disjoint sets Φ and Γ such that
          $f \in \Phi$ iff $\Omega f \geq 1$ and
          $f \in \Gamma$ iff $\Omega f = 0$.

Φ is called the set of function symbols, Γ the set of constants and X
the set of variables.

We define operations
$$\hat{f}: (W_\Omega^X)^n \to W_\Omega^X \quad \text{for } n = \Omega f$$
by $\hat{f}(t_1, \ldots, t_n) = f(t_1, \ldots, t_n)$. Let $\hat{\Omega}$ be the set of these (term
building) operations. Let $\emptyset$ denote the empty set.

$F_\Omega^X = (W_\Omega^X, \hat{\Omega})$ is isomorphic to $A_K(X)$ and hence is called the absolutely
free *term algebra* on X. $F_\Omega^\emptyset$ is the *initial* term algebra (or Herbrand
universe). We shall write F_{Ω_o} for $F_\Omega^\emptyset$. Our interest in F_{Ω_o} is motivated
by the fact that for every algebra $A = (A,\Omega)$ there exists a unique
homomorphism
$$h_A: F_{\Omega_o} \to A.$$
But then instead of investigating A, we can restrict our attention to a
quotient of F_{Ω_o} modulo the congruence induced by h_A.

In order to have variables at our disposal in the initial algebra we
define $\Omega_X = \Omega \cup X$, that is we treat variables as special constants. Since
$$F_\Omega^X \simeq F_{\Omega_X}^\emptyset$$
we simply write F_Ω if $X \neq \emptyset$ and $X \subset \Omega$ and
F_{Ω_o} if $X = \emptyset$. Because terms are objects in F_Ω we shall write $t \in F_\Omega$ instead
of $t \in W_\Omega^X$.

An *equation* is a pair of terms $s, t \in F_\Omega$, in symbols $s = t$. The equation
$s = t$ is *valid* in the algebra A (of the same type), in symbols
$$A \models s = t \quad \text{iff}$$

for every homomorphism $\varphi: F_\Omega \to A$

$$\varphi s = \varphi t \text{ in } A.$$

Let $\bar{\sigma}: X \to F_\Omega$ be a mapping which is equal to the identity mapping almost everywhere. A *substitution* $\sigma: F_\Omega \to F_\Omega$ is the homomorphic extension of $\bar{\sigma}$ and is represented as a finite set of pairs

$$\sigma = \{x_1 \leftarrow t_1, \ldots, x_n \leftarrow t_n\}.$$

Σ is the set of substitutions on F_Ω. The identity mapping on F_Ω, i.e. the *empty substitution*, is denoted by ε. If t is a term and σ a substitution, define

$V: F_\Omega^n \to 2^X$ by $V(t)$ = set of variables in t and $V(t_1, \ldots, t_n) = \bigcup_{i \leq n} V(t_i)$.

$|t| \in \mathbb{N}$ denotes the length of t (i.e. the number of symbols in t)

$\text{DOM}(\sigma) = \{x \in X: \sigma x \neq x\}$

$\text{COD}(\sigma) = \{\sigma x: x \in \text{DOM}(\sigma)\}$

$\text{XCOD}(\sigma) = V(\text{COD}(\sigma))$

$\Sigma_o \subset \Sigma$ is the set of *ground* substitutions, i.e. $\sigma \in \Sigma_o$ iff $\text{COD}(\sigma) \subset F_{\Omega_o}$.

An equation $s = t$ is *unifiable* (is *solvable*) in A iff there exists a homomorphism $\xi: F_\Omega \to A$ such that $\xi s = \xi t$ is valid in A. A set of equations T induces a congruence $=_T$ in F_Ω and $F_\Omega/_{=_T}$ is the quotient algebra modulo $=_T$.

A *unification problem for* T, denoted as

$$\langle s = t \rangle_T,$$

is given by the equation $s = t$, $s, t \in F_\Omega$. <u>The problem is to decide whether or not $s = t$ is unifiable in $F_\Omega/_{=_T}$.</u>

We denote the constituent parts of the initial algebra $F_\Omega/_{=_T}$ as

$$(F_\Omega, T)\text{-}algebra.$$

1.2 Unification from a Logical Point of View

1.2.1 EQUATIONAL LOGIC

The *well formed formulas* of our logic are equations defined as pairs in $W_\Omega^X \times W_\Omega^X$ and denoted as $s = t$.

A substitution σ is a finite set of pairs in $W_\Omega^X \times W_\Omega^X$ (i.e. classical work confuses the issue a little by identifying the representation with the mapping that is being represented). The application of $\sigma = \{x_1 \leftarrow t_1, \ldots, x_n \leftarrow t_n\}$ to a term t, σt is obtained by simultaneously replacing each x_i in t by t_i.

Let T be a set of equations. The equation $p = q$ *is derivable from* T, $T \vdash p = q$, if $p = q \in T$ or $p = q$ is obtained from T by a finite sequence of the following operations:
  (i) $t = t$ is an axiom
  (ii) if $s = t$ then $t = s$
  (iii) if $r = s$ and $s = t$ then $r = t$
  (iv) if $s_i = t_i$, $1 \leq i \leq n$ then $f(s_1,\ldots,s_n) = f(t_1,\ldots,t_n)$, $n = \Omega f$
  (v)   if $s = t$ then $\sigma s = \sigma t$ where $\sigma \in \Sigma$.

For a set of equations T, $T \models s = t$ iff $s = t$ is valid in all *models* of T.

Theorem (Birkhoff): $T \models s = t$ iff $T \vdash s = t$

We shall abbreviate $T \models s = t$ (and hence $T \vdash s = t$) by $s =_T t$. An equation $s = t$ is *T-unifiable*, iff there exists a substitution σ such that $\sigma s =_T \sigma t$.

Although this is the traditional view of unification, its apparent simplicity is deceptive: we did not define what we mean by a 'model'. In order to do so we should require the notion of an interpretation of our well formed formulas, which is a 'homomorphism' from W_Ω^X to certain types of algebras, thus bringing us back to section 1.1.

Since neither $\models$ nor $\vdash$ are particularily convenient for a computational treatment of $=_T$, a third method is presented below.

1.2.2 COMPUTATIONAL LOGIC

For simplicity of notation we assume we have a box of symbols, GENSYM, at our disposal, out of which we can take an unlimited number of "new" symbols. More formally: for F_Ω, let $\Omega = \Phi \cup \Gamma \cup X$, where $X = X_o \cup$ GENSYM with $\Omega x = 0$, $x \in X$.

We shall adopt the computational proviso that whenever GENSYM is referenced by $v \in$ GENSYM it is subsequently 'updated' by GENSYM' = GENSYM $- \{v\}$ and $X_o' = X_o \cup \{v\}$ and $\Omega' = \Phi \cup \Gamma \cup X'$, where $X' = X_o' \cup$ GENSYM'. Since $F_\Omega \simeq F_{\Omega'}$, we shall not always keep track of the '-s and just write F_Ω.

A *renaming substitution* $\rho \in \Sigma_X \subset \Sigma$ is defined by

(i) $COD(\rho) \subset X$
(ii) $\forall x,y \in X$: if $x \neq y$ then $\rho x \neq \rho y$.

For $s,t \in F_\Omega$: $s \sim_\rho t$ if $\exists \rho \in \Sigma_X$ such that $\rho s = \rho t$. If $\rho s = t$ then t is called an *X-variant* of s, if in addition $COD(\rho) \subset GENSYM$ then t is called a *new* X-variant of s.

In order to formalize the accessing of a subterm in a term, let $\mathbb{N}^*$ be the set of sequences of positive integers, Λ the empty sequence in $\mathbb{N}^*$ and let · be the concatenation operation on sequences. Members of $\mathbb{N}^*$ are called *positions*, and denoted by $\pi \in \mathbb{N}^*$. They are used as follows: for any $t \in F_\Omega$ let $\Pi(t) \subset \mathbb{N}^*$, the set of *positions in* t, be:
(i) if $\Omega t = 0$ then $\Pi(t) = \{\Lambda\}$
(ii) if $t = f(t_1,\ldots,t_n)$ then $\Pi(t) = \{\Lambda\} \cup \{i \cdot \pi : 1 \leq i \leq n, \pi \in \Pi(t_i)\}$.

For example: $f(g(a,y),b) = \{\Lambda,1,2,1 \cdot 1,1 \cdot 2\}$.
The *subterm of* $t = f(t_1,\ldots,t_n)$ *at* π, $t|\pi$, is defined as:

(i) $t|\pi = t$ for $\pi = \Lambda$ or $\pi \notin \Pi(t)$
(ii) $t|i \cdot \pi' = t_i|\pi'$ for $\pi = i \cdot \pi'$.

For example: $f(g(a,y),b)|1.2 = y$.

A *subterm replacement* of t by s at π, $\hat{\rho}t$, with $\hat{\rho} = [\pi \leftarrow s]$ is defined as:
(i) $\hat{\rho}t = s$ if $\pi = \Lambda$
(ii) $\hat{\rho}t = f(t_1,\ldots,\hat{\delta}t_i,\ldots,t_n)$ if $t = f(t_1,\ldots,t_n)$ and
         $\pi = i \cdot \pi'$ and $\hat{\delta} = [\pi' \leftarrow s]$
(iii) $\hat{\rho}t = t$ if $\pi \notin \Pi(t)$.

We denote replacements by $\hat{\sigma},\hat{\rho},\hat{\delta}$, etc. and substitutions by σ,ρ,δ etc.

A relation $\to \subseteq F_\Omega \times F_\Omega$ is *Noetherian* (terminating) if there are no infinite sequences: $s_1 \to s_2 \to s_3 \to \ldots$. As usual $\xrightarrow{+}$ is the transitive and $\xrightarrow{*}$ the reflexive and transitive closure of $\to$. A relation $\to$ is *confluent* if for every $r,s,t \in F_\Omega$ such that $r \xrightarrow{*} s$ and $r \xrightarrow{*} t$ there exists a $u \in F_\Omega$ such that $s \xrightarrow{*} u$ and $t \xrightarrow{*} u$. A confluent Noetherian relation is *canonical*.

We define two important relations $\to_R$ and $\twoheadrightarrow_R$ on $F_\Omega \times F_\Omega$ as follows:

A *rewrite system* $R = \{l_1 \Rightarrow r_1,\ldots,l_n \Rightarrow r_n\}$ is any set of pairs $l_i, r_i \in F_\Omega$, such that $V(r_i) \subseteq V(l_i)$, $1 \leq i \leq n$.

For two terms s and t we say s is *rewritten* to t, $s \rightarrow_R t$, if there exists $\pi \in \Pi(s)$, $\sigma \in \Sigma$ and $l_i \rightarrow r_i \in R$ such that $s|\pi = \sigma \tilde{l}_i$ and $t = \hat{\sigma} s$, where $\hat{\sigma} = [\pi \leftarrow \sigma \tilde{r}_i]$ and $\tilde{l}_i, \tilde{r}_i$ are new X-variants of l_i, r_i. Occasionally we keep track of the information by writing $s \xrightarrow[[\pi,i,\sigma]]{} t$, $s \xrightarrow[[\pi,i]]{} t$, $s \xrightarrow[[\pi]]{} t$ etc.

For two terms s and t we say s is *paramodulated* to t, $s \rightarrowtail_R t$, if there exist $\pi \in \Pi(s)$, $l_i \rightarrow r_i \in R$, $\sigma \in \Sigma$ such that $\sigma(s|\pi) = \sigma \tilde{l}_i$ and σ is most general (see 1.3 below), $\tilde{l}_i$ is a new X-variant of l_i and $\sigma s \xrightarrow[[\pi,i]]{} t$.

For example for $R = \{g(x,0) \rightarrow 0\}$ we have
$s = f(g(a,y),y) \rightarrowtail_R f(0,0) = t$
with $\pi = 1$ and $\sigma = \{x \leftarrow a, y \leftarrow 0\}$.
But note $s \not\rightarrow_R t$, since we are not allowed to substitute into s.

The notation and definitions of term rewriting systems are consistent with [HT80]; the importance of term rewriting systems (demodulation) for theorem proving was first noticed in [WR67]. Suppose for an equational theory T there is a rewrite system R_T such that for $s,t \in F_\Omega$:

$s =_T t$ iff $\exists p \in F_\Omega$ such that $s \xrightarrow[R_T]{*} p$ and $t \xrightarrow[R_T]{*} p$.

In that case we say T is *embedded into* R_T and write

$$T \hookrightarrow R_T .$$

For an equational theory T there are techniques to obtain a system R_T such that $T \hookrightarrow R_T$; moreover for many theories of practical interest it is possible to obtain a rewrite system R_T such that $\xrightarrow[R_T]{}$ is canonical [KB70], [HT80], [PS81], [HU80.1]. Canonical relations $\xrightarrow{}$ are an important basis for *computations* in *equational logics*, since they define a unique *normal form* $\|t\|$ for any $t \in F_\Omega$, given by $t \xrightarrow{*} \|t\|$ and $\not\exists s \in F_\Omega$ such that $\|t\| \rightarrow s$. Hence
(i) $s =_T t$ iff $\|s\| = \|t\|$.

In case R_T is Noetherian (i.e. R defines the Noetherian relation $\rightarrow_{R_T}$), we also say it is a *reduction system*.

As a consequence of Birkhoffs theorem we have
(ii) $\exists \sigma \in \Sigma : \sigma s =_T \sigma t$ iff
$\exists \bar{p}, p \in F_\Omega$ and $\exists \delta \in \Sigma : h(s,t) \xrightarrow[R_T]{*} h(\bar{p},p)$ with $\delta \bar{p} =_\emptyset \delta p$
where h is an 'auxiliary' function symbol not in Ω, i.e. because h is '*new*', h is not 'affected' by R_T.

In [FA79], [LB79], [HU80], [SS81.5], [H82] this theorem is used as a basis for a *universal unification* algorithm: just as a universal Turing machine takes as its input a specific argument and the description of a special Turing machine, a universal unification algorithm has an input pair consisting of a special unification problem and an equational theory T.

1.3 Universal Unification

An equational theory T is *decidable* iff $s =_T t$ is decidable for any $s, t \in F_\Omega$. Let $\mathsf{T}_=$ denote the family of decidable finitely based equational theories.

A *T-unification problem* $<s = t>_T$ consists of a pair of terms $s, t \in F_\Omega$ and a theory $T \in \mathsf{T}_=$.

A substitution $\sigma \in \Sigma$ is a *T-unifier* for $<s = t>_T$ iff $\sigma s =_T \sigma t$. The subset of Σ which unifies $<s = t>_T$ is $U\Sigma_T(s,t)$, the *set of unifiers* (for s and t) *under T*. It is easy to see that $U\Sigma_T$ is recursively enumerable (r.e.) for any s and t: Since F_Ω is r.e. so is Σ; Now for any $\delta \in \Sigma$, check if $\delta s =_T \delta t$ (which is decidable since $T \in \mathsf{T}_=$) then $\delta \in U\Sigma_T(s,t)$ otherwise $\delta \notin U\Sigma_T(s,t)$.

We shall omit the subscript T and (s,t) if they are clear from the context. The composition of substitutions is defined by the usual composition of mappings: $(\sigma \circ \tau)t = \sigma(\tau t)$. If $W \subseteq X$, then T-equality is extended to substitutions by

$$\sigma =_T \tau \ [W] \quad \text{iff} \quad \forall x \in W \quad \sigma x =_T \tau x,$$

σ *and* τ *are T-equal in W*. We say σ is an *instance* of τ and τ is *more general* than σ, in symbols

$$\sigma \leq_T \tau \ [W] \quad \text{iff} \quad \exists \lambda \in \Sigma:$$

$$\sigma =_T \lambda \circ \tau [W] \text{ for some } W \subset X.$$

If $\sigma \leq_T \tau[W]$ and $\tau \leq_T \sigma[W]$ then $\sigma \approx_T \tau[W]$, σ and τ are *T-equivalent in W*.

For $\Sigma_1, \Sigma_2 \subseteq \Sigma$ we define $\Sigma_1 \circ \Sigma_2 = \{\sigma_1 \circ \sigma_2 : \sigma_1 \in \Sigma_1, \sigma_2 \in \Sigma_2\}$.
$\Sigma_1 \subseteq_T \Sigma_2 \ [W]$ iff $\forall \sigma_1 \in \Sigma_1 \ \exists \sigma_2 \in \Sigma_2$ s. th. $\sigma_1 =_T \sigma_2 \ [W]$,
$\Sigma_1 =_T \Sigma_2 \ [W]$ iff $\Sigma_1 \subseteq_T \Sigma_2 \ [W]$ and $\Sigma_2 \subseteq_T \Sigma_1 \ [W]$.

Universal unification is concerned with two fundamental problems:

PROBLEM ONE (*Decidability Problem*)

> For a given equational theory $T \in \mathsf{T}_=$, *is it decidable for any s and t whether s and t are unifiable?*

That is, we are interested in classes of theories such that "s and t are unifiable under T" is decidable for every T in that class.

A unifier σ for $\langle s = t \rangle_T$ is called *a most general unifier* (mgu) if for any unifier $\delta \in U\Sigma_T(s,t)$: $\delta \leq_T \sigma [\![W]\!]$, where $V(s,t) = W$. Since in general a single most general unifier does not exist for $\langle s = t \rangle_T$, we define $\mu U\Sigma_T(s,t)$, the *set of most general unifiers*, as:

(i) $\mu U\Sigma \subseteq U\Sigma$ (correctness)

(ii) For every δ with $\delta s =_T \delta t$ there
exists $\sigma \in \mu U\Sigma$ and $\lambda \in \Sigma$ such that
$$\delta =_T \lambda \circ \sigma \ [\![W]\!]$$ (completeness)

(iii) $\sigma \neq_T \tau \ [\![W]\!]$: $\sigma, \tau \in \mu U\Sigma$ (minimality).

From condition (ii) it follows in particular that $U\Sigma =_T \Sigma \circ U\Sigma \ [\![W]\!]$, i.e. $U\Sigma$ is a *left ideal* in the semigroup $(\Sigma, \circ)$ and $U\Sigma$ is *generated by* $\mu U\Sigma$.

For practical applications these conditions are sometimes too general and there are additional technical requirements on $DOM(\sigma)$, $COD(\sigma)$ and $XCOD(\sigma)$ for $\sigma \in \mu U\Sigma$, which we shall state when the need arises.

$\mu U\Sigma_T$'s do not always exist; when they do then they are unique up to the equivalence $\approx_T$, see [HT76], lemma 2.16. For that reason it is sufficient to generate one $\mu U\Sigma_T$. In the following we always take $\mu U\Sigma_T$ as the representative of the equivalence class $[\mu U\Sigma_T]_\approx$.

PROBLEM TWO (*Existence Problem*):

For a given equational theory $T \in \mathcal{T}_=$, *does* $\mu U\Sigma_T(s,t)$ *always exist for every* $s, t \in F_\Omega$.

PROBLEM THREE (*Enumeration Problem*):

For a given equational theory $T \in \mathcal{T}_=$, *is*
$$\mu U\Sigma_T(s,t)$$
recursively enumerable for any $s, t \in F_\Omega$?

That is, we are interested in algorithms which generate all mgu's for a given problem $\langle s = t \rangle_T$. TABLE 1 summarizes the major results that have been obtained for special theories, which consist of combinations of the following equations:

A	(associativity)	$f(f(x,y),z) = f(x,f(y,z))$
C	(commutativity)	$f(x,y) = f(y,x)$
D	(distributivity)	$f(x,g(y,z)) = g(f(x,y),f(x,z))$
		$f(g(x,y),z) = g(f(x,z),f(y,z))$
H,E	(homomorphism, endomorphism)	$\varphi(x \circ y) = \varphi(x) \cdot \varphi(y)$
I	(idempotence)	$f(x,x) = x$

Theory T	Type of T	Unification decidable	$\mu U\Sigma_T$ recursive	$\mathcal{A}_T$	References
∅	1	Yes	Yes	Yes	[HE30][RO65][RO71][KB70][G67][PR60] [BA73][HT76][MM79][PW78]
A	∞	Yes	Yes	Yes	[HM67][PL72][SI75][LS75][MA77]
C	ω	Yes	Yes	Yes	[SI82]
I	ω	Yes	Yes	Yes	[RS78][SS82.2][SZ82]
A+C	ω	Yes	Yes	Yes	[ST75][LS76][HU79]
A+I	?	Yes	Yes	No	[SS82.3][SZ82][SS82.1]
C+I	ω	Yes	Yes	Yes	[RS78]
A+C+I	ω	Yes	Yes	Yes	[LS76]
D	∞	?	Yes	Yes	[SZ82]
D+A	∞	No	Yes	Yes	[S78][SZ82]
D+C	∞	?	Yes	Yes	[SZ82]
D+A+C	∞	No	Yes	Yes	[SZ82]
D+A+I	?	Yes	Yes	No	[SZ82]
H,E	1	Yes	Yes	Yes	[VO78]
H+A	∞	Yes	Yes	Yes	[VO78]
H+A+C	ω	Yes	Yes	Yes	[VO78]
E+A+C	∞	?	?	No	[VO78]
QG	ω	Yes	Yes	Yes	[HU80]
AG	ω	Yes	Yes	Yes	[LA79]
H10	?	No	?	No	[MA70][DA73]
FPA	ω	Yes	Yes	Yes	[LA80]
Sot, T = ∅	?	No	-	-	[GO81]
Hot, T = ∅	0	No	-	-	[HT73][HT75][HT76][BA78]

TABLE 1

Abbreviations:

FPA: Finitely Presented Algebras
QG: Quasi-Groups
AG: Abelian-Groups
H10: Hilbert's 10th Problem
Sot: Second order terms
Hot: Higher order terms (i.e. ≥ 3rd order)

The column under $\mathcal{A}_T$ indicates whether or not a type conformal algorithm has been presented in the literature. The 'type of a theory' and 'type conformal' are defined below.

Except for Hilbert's tenth problem, we have not included the classical work on equation solving in 'concrete' structures such as rings and fields, which is well known.

The relationship of universal unification to these classical results is similar to that of universal algebra [GR79] to classical algebra.

The central notion $\mu U\Sigma_T$ induces a hierarchy of classes of equational theories (*the unification-relevant theories*):

(i) A theory T is *unitary* if $\forall s,t\ \mu U\Sigma_T(s,t)$ exists and has at most one element. The class of such theories is U_1 (*type one*).

(ii) A theory T is *finitary* if it is not unitary and if $\forall s,t\ \mu U\Sigma_T(s,t)$ exists and is finite.
The class of such theories is U_ω (*type ω*).

(iii) A theory T is *infinitary* if $\forall s,t\ \mu U\Sigma_T(s,t)$ exists and there exists $<p = q>_T$ such that $\mu U\Sigma_T(p,q)$ is infinite.
The class of such theories is U_∞ (*type ∞*).

(iv) A theory T is of *type zero* if it is not in one of the above classes. The class of these theories is U_o.

There are several examples for unitary, finitary and infinitary theories in the above table. An example of a type zero theory due to F. Fages [FA81] is:
$$T = \{f(1,x) = x;\ g(f(x,y)) = g(y)\}$$
since $\mu U\Sigma_T$ does not exist for the problem $<g(x) = g(a)>_T$. Define a partial order $\leq_T$ on terms by: $s \leq_T t$ iff $\exists \delta \in \Sigma$ satisfying $s =_T \delta t$. A *matching problem* $<s \geq t>_T$ consists of a pair of terms and a theory $T \in T_=$. A substitution $\nu \in \Sigma$ is a T-*matcher* (or one-way unifier) if $\nu s =_T t$. If s and t are matchable we shall write $s \leq_T t$, resp. $s \geq_T t$.

The notion $\mu M\Sigma_T$ induces a hierarchy (*the matching relevant theories*) similar to the hierarchy based on $\mu U\Sigma_T$: a theory T is *unitary matching* if $\mu M\Sigma_T$ always exists and has at most one element. The class of such theories is M_1. Analogeously we define M_ω, M_∞ and the class M_o. Note by the same example as above: $M_o \neq \emptyset$.

A *unification algorithm* $\mathcal{A}_T$ (a *matching algorithm* $\mathcal{M}_T$) for a theory T is an algorithm which takes two terms s and t as input and generates a set $\Psi_T \subseteq U\Sigma_T$ ($\subseteq M\Sigma_T$) for $<s = t>_T$ (for $<s \geq t>_T$). A *minimal* algorithm $\mu \mathcal{A}$ ($\mu \mathcal{M}_T$) is an algorithm which generates a $\mu U\Sigma_T$ ($\mu M\Sigma_T$).

For many practical applications this requirement is not strong enough, since it does not imply that the algorithm terminates for theories $T \in U_1 \cup U_\omega$. On the other hand, for $T \in U_\omega$ it is sometimes too rigid, since an algorithm which generates a finite superset of $\mu U\Sigma_T$ may be far more efficient than the algorithm $\mu \mathcal{A}_T$ and for that reason preferable. For that reason we define:

An algorithm $\mathcal{A}_T$ is *type conformal* iff:

(i) $\mathcal{A}_T$ generates a set Ψ_T with $U\Sigma_T \supseteq \Psi_T \supseteq \mu U\Sigma_T$ for some $\mu U\Sigma_T$.

(ii) $\mathcal{A}_T$ terminates and Ψ_T is finite if $T \in U_1 \cup U_\omega$ and

(iii) if $T \in U_\infty$ then $\Psi_T = \mu U\Sigma_T$ for some $\mu U\Sigma_T$.

Similarly: algorithm $\mathcal{M}_T$ is type conformal iff (i) - (iii) hold with U replaced by M.

Experience shows that unification algorithms for different theories are usually based on entirely different methods. For theoretical reasons as well as for heuristic purposes it would however be interesting to have a universal unification algorithm for a whole class of theories, however inefficient it might be. A *universal unification algorithm* (*a universal matching algorithm*) *for a class of theories* $\mathbf{T} \subset \mathbf{T}_=$ is an algorithm which takes as input a pair of terms (s,t) *and a theory* $T \in \mathbf{T}$ and generates a complete set of unifiers (matchers) for $<s = t>_T$ (for $<s \geq t>_T$).

Since $U\Sigma_T$ is trivially r.e. for any $T \in \mathbf{T}$, there is the important requirement that a universal unification algorithm is either minimal or at least type conformal. See [FA79], [LB79], [HU80], [SS81.5], [H82], [SZ82].

2. A CLASSIFICATION OF EQUATIONAL THEORIES

In this section some important subclasses of equational theories are defined, which turned out to be of practical interest as well as being useful as "basic building blocks" for other equational classes. We shall first present the definitions and then show a few theorems in order to demonstrate the descriptive value of these theories.

As in section 1.3 let $T_=$ be the class of *equational theories*, which are finitely based and have a decidable word problem. At present the most important subclass is

$$T_\Rightarrow := \{T \in T_= : \text{there exists a term rewriting system R s.th. } T \hookrightarrow R\}$$

A theory T is *regular* iff for every $l = r \in T$ $V(l) = V(r)$; we shall write T^* if T is a class of regular theories. As an immediate result we have: $T_=^* \subset T_\Rightarrow$.

An important requirement with respect to unification theory is that the matching problem is decidable for T; let $T_\leq$ denote this class. The class $A := T_\Rightarrow \cap T_\leq$ is the class of *admissible* theories. Defining $T_\downarrow \subset T_\Rightarrow$ as the subclass with a *confluent* rewriting system and $R \subset T_\Rightarrow$ as the subclass with a *Noetherian* rewriting system and abbreviating $R_\downarrow = RT_\downarrow$ (throughout this section we use the denotational proviso that juxtaposition abbreviates intersection of classes) as the name for the *canonical theories in a generalized sense* (i.e. any canonicalization is allowed). Defining $C \subset R_\downarrow$ as the class having a (standard) canonicalization and let $M_{\omega 1} = M_\omega \cup M_1$ we have the classes $ACM_{\omega 1}$ and $AC^*M_{\omega 1}$, which turned out to be important for universal unification algorithms: in [SS81], [H82] it is shown that $\mu U\Sigma_T$ is recursively enumerable for a subclass of $AC^*M_{\omega 1}$. Calling this subclass $AC^*M_\downarrow$ we have

2. *Theorem 1:* $$U_o AC^*M_\downarrow = \emptyset$$

i.e. $\mu U\Sigma_T$ *exists* for any $T \in AC^*M_\downarrow$. An example by F. Fages [FA81] shows: $U_o AC^*M_{\omega 1} \neq \emptyset$. The class of Ω-*free* theories, F_Ω turned out to be important for its descriptive worth:

$$F_\Omega := \{T \in T_= : \text{if } f(t_1,\ldots,t_n) =_T f(s_1,\ldots,s_n) \text{ then } t_i =_T s_i; 1 \leq i \leq n\}$$

The following four lemmata characterize F_Ω with respect to the basic hierarchy:

2. *Lemma 1:* $$U_1 F_\Omega \neq \emptyset \text{ and } F_\Omega \neq U_1$$
i.e. there exists an Ω-free unitary theory, but F_Ω is different from U_1

2. *Lemma 2:* $\quad U_\omega F_\Omega \neq \emptyset$ and $F_\Omega \neq U_\omega$
i.e. there exists an Ω-free finitary theory, but F_Ω is different from U_ω

2. *Lemma 3:* $\quad U_\infty F_\Omega \neq \emptyset$ and $F_\Omega \neq U_\infty$
i.e. there exists an Ω-free infinitary theory, but F_Ω is different from U_∞

But:
Problem: $\quad U_0 F_\Omega = \emptyset$?
i.e. does $\mu U \Sigma$ exists for every Ω-free theory?

In other words F_Ω is somehow 'diagonal' to the basic hierarchy of equational classes. But we have the surprising result:

2. *Theorem 2:* $\quad M_1 = F_\Omega$
i.e. F_Ω constitutes exactly the class of unitary matching theories.

Continuing the description of the building blocks of unification theory let $\quad U = U_1 \cup U_\omega \cup U_\infty$ be the class of *unificationrelevant* theories and defining $\quad N := AU$ as the class of *normal* theories we have by 2. theorem 1: $AC^*M_\downarrow = NC^*M_\downarrow$.

As an important concept we define the notion of a *local subclass* λT of a class T:

For a theory T let $I(T) := \{\sigma l, \sigma r : l = r \in T, \sigma \in \Sigma\}$ (i.e. all the instances of l and r) and let

$G(T) := \{\hat{\sigma} l, \hat{\sigma} r : l = r \in T, \hat{\sigma} = [\pi \leftarrow x], \pi \in \Pi(l) \text{ or } \pi \in \Pi(r), x \in X\}$

(i.e. the finite set of all generalizations of l and r). We assume terms equal under renaming to be discarded, i.e. $G(T)/_\approx$. Now define the *characteristic set* of an equational theory T as:

$$\chi(T) := I(T) \cup G(T).$$

Let $\mathscr{E}(T)$ be some first order property of T. If the property $\mathscr{E}$ is only considered with respect to a subset Θ of F_Ω, $\Theta \subset F_\Omega$, we shall write $\mathscr{E}(T)|_\Theta$.

Definition: A theory T is χ-*reducible* iff
$$\mathscr{E}(T)|_{\chi(T)} \text{ implies } \mathscr{E}(T)$$
Let $T_\mathscr{E}$ be the class of theories having property $\mathscr{E}$ then the χ-*subclass* $\lambda T_\mathscr{E} \subseteq T_\mathscr{E}$ is the set:
$$\lambda T_\mathscr{E} := \{T \in T_\mathscr{E} : T \text{ is } \chi\text{-reducible}\}$$
In [SS81.5] it is shown that

$$AC^*M_\infty = \chi AC^*M_\infty$$

and hence we have $AC^*M_{\omega 1} = \chi AC^*M_{\omega 1}$. This already simplifies the test for $T \in AC^*M_\infty$, since we only have to show that it holds for matching problems on $\chi(T)$; i.e. $<s \geq t>_T$ $s,t \in \chi(T)$.

Since $\chi(T)$ is infinite in general the problem remains of how to show that T is $\mathcal{E}$, which may sometimes be possible using the compactness theorem.

For certain classes of theories (see § 3 below) it is however possible to reduce this problem entirely to a *finite test set* $TEST(T) \subset \chi(T)$ such that

$$\mathcal{E}(T)|_{\Theta_1} \text{ implies } \mathcal{E}(T)|_{\Theta_2}$$

where $\Theta_1 = TEST(T)$ and $\Theta_2 = \chi(T)$.

Let $T_\mathcal{E}$ be the class of theories having property $\mathcal{E}$ then the *local subclass with finite test set* is the set:

$$\lambda T_\mathcal{E} := \{T \in \chi T_\mathcal{E} : a \textit{ finite } TEST(T) \subset \chi(T) \text{ exists}\}$$

In particular we say a theory T is *locally unitary* λU_1, *locally finitary* λU_ω, *locally infinitary* λU_∞.

Per definition we have

$$\lambda U_i \subseteq U_i, \text{ where } i \in \{1,\omega,\infty\},$$

but of course the main interest is in the inclusion relation in the other direction. Our main result in the section below shows such a result for unitary theories.

Another example is the Knuth-Bendix theorem: $\lambda C = C$ [KB70].

Let $M = M_1 \cup M_\omega \cup M_\infty$ be the class of *matching relevant* theories (i.e. $\mu M \Sigma_T$ always exists). Then we have the following existence theorem:

2. *Theorem 3:* $$T^*_= \subseteq M$$
   i.e. the most general matching set always exists for regular theories.

2. *Theorem 4:* $$F_\Omega \subset T^*_=$$
   i.e. every Ω-free theory is regular.

Finally we define the *permutative theories* P as those that have a finite equivalence class:

$$\forall T \in P : \forall t \in F_\Omega \; [t]_{=_T} \text{ is finite.}$$

For this class we have

2. *Lemma 4:* $\quad P = P^*$
   the regular permutative theories are exactly the permutative theories.

2. *Theorem 4:* $\quad P \subset U$
   i.e. $\mu U\Sigma_T$ always exists for a theory $T \in P$.

Corollary: $\quad P \subset M$

2. *Theorem 5:* $\quad P \subset A$
   i.e. the permutative theories are admissible and hence since $N = AU$ by definition:

Corollary: $\quad P = NP$
   i.e. the permutative theories are normal theories.

[SZ82] contains many more examples (theorems), which amplify our intention: the above theories are some of the building blocks of a 'meccano kit' out of which new problems and theories may be constructed at will.

Proofs as well as additional results concerning the classification of equational theories may be found in [SS82.4].

SUMMARY

$T_=$: equational theories, finitely based, decidable
T^* : regular theory
$T_\to$: T has a rewriting system
$T_\ne$: matching problem decidable
A : admissible theories $A = T_\to T_\ne$
$T_\downarrow$: confluent theories
R : Noetherian theories
$R_\downarrow$: canonical theories
C : $C \subset R_\downarrow$: standard canonicalization
$U_1, U_\omega, U_\infty, U_0$: unitary, finitary, infinitary, type zero theories
$M_1, M_\omega, M_\infty, M_0$: unitary matching, etc.
U : $U = U_1 \cup U_\omega \cup U_\infty$, $\mu U\Sigma_T$ exists, unification relevant
M : $M = M_1 \cup M_\omega \cup M_\infty$, $\mu U\Sigma_T$ exists, matching relevant
F_Ω : Ω-free theories
$\lambda T, \chi T$: local theories, χ-reducible theories
P : permutative theories (finite equivalence class)
N : normal theories: $N = AU$

3. A CHARACTERIZATION OF UNITARY THEORIES

In 1975 P. Hayes conjectured that Robinson's unification algorithm for free terms may well be the only case with *at most one* most general unifier, which is false: For example let $T_a := \{a = a\}$ for any constant a, then $T_a \in U_1$.

But the problem turned out to be more complex than anticipated at the time: for example let $T_{aa} := \{f(a,a) = a\}$ for any constant a, then
$$T_{aa} \in U_1.$$

The following two theorems finally settle the boundary between finitary and unitary theories, but the formal development for their proofs is too voluminous for inclusion in this paper.

3. Theorem 1: $\qquad \lambda U_1 \subset U_1 \subset F_\Omega$

3. Theorem 2: $\qquad \lambda U_1 = U_1$

The finite test set TEST(T) for $T \in U_1$ is obtained as
$$\text{TEST}(T) := \{t \in I(T): |t| \leq \max(T)\} \cup G(T)$$
where $\max(T) := \max\{|l|, |r|: \; l = r \in T\}$.

To illustrate the use of the above theorems let us consider the empty theory T_ε, i.e. the Robinson-unification problem for free terms. In order to show $T_\varepsilon \in U_1$, in the stone age of unification theory one had to invent a special algorithm and then prove its completeness and correctness [RO65], [KB70].

A more elegant method is contained in [HT76]: factoring F_Ω by $\approx$, it is possible to show that $F_\Omega/_\approx$ forms a complete lattice under $\leq_T$. Hence if two terms are unifiable there exists a common instance and hence there exists a l.u.b., which is the most general such instance: thus follows $T_\varepsilon \in U_1$.

However using the above theorems, this result is immediate: Since the absolutely free algebra of terms is in particular Ω-free: $T_\varepsilon \in F_\Omega$. Now since $\chi(T_\varepsilon)$ is empty every TEST set is empty. Hence there does not exist a pair in TEST with more than one mgu, thus follows $T_\varepsilon \in U_1$.

6. REFERENCES

[BA73] L.D. Baxter, An Efficient Unification Algorithm, Univ. of Waterloo, Techn. Report CS-73-23, 1973

[BA78] L.D. Baxter, The Undecidability of the Third Order Dyadic Unification Problem, Information and Control, vol 38, no 2, 1978

[DA73] M. Davis, Hilbert's tenth Problem is unsolvable, Amer. Math. Monthly, vol 80, 1973

[FA79] M. Fay, First Order Unification in an Equational Theory, Proc. 4th Workshop on Autom. Deduction, Texas, 1979

[G67] J.R. Guard, F.C. Oglesby, J.H. Benneth, L.G. Settle, Semi-Automated Mathematics, JACM 1969, vol 18, no 1

[GO81] D. Goldfarb, The Undecidability of the Second Order Unification Problem, J. of Theor. Comp. Sci. 13, 1981

[GO66] W.E. Gould, A Matching Procedure for ω-order Logic (thesis), Air Force Cambridge Research Labs, 1966

[GR79] G. Grätzer, Universal Algebra, Springer Verlag, 1979

[HE30] J. Herbrand, Recherches sour la theorie de la demonstration, Travaux de la Soc. des Sciences et des Lettres de Varsovie, no 33, 128, 1930

[HM67] J. Hmelevskij, The Solution of certain Systems of Word Equations, Dok. Akad. Nauk SSSR (1964), (1966), (1967), (Soviet Math. Dokl.)

[HT73] G. Huet, The Undecidability of Unification in Third Order Logic, Information and Control 22, 1973

[HT76] G. Huet, Résolution d'équations dans des langages d'ordere $1,2,\ldots,\omega$, Thèse d'Etat, Univ. de Paris, VII, 1976

[HT75] G. Huet, A Unification Algorithms for Typed Lambda Calculus, J. Theoretic. Comp. Sci., 1, 1975

[HT80] G. Huet, D.C. Oppen, Equations and Rewrite Rules, In "Formal Languages: Perspectives and Open Problems", Ed. R. Book, Academic Press, 1980

[HT80] G. Huet, Confluent Reductions: Abstract Properties and Applications to Term Rewriting Systems, JACM, vol 27, no 4, 1980

[HU80] J.M. Hullot, Canonical Forms and Unification, Proc. of 5th Workshop on Automated Deduction, Springer Lecture Notes, 1980

[HU80.1] J.M. Hullot, A Catalogue of Canonical Term Rewriting Systems, Research Rep. CSL-113, SRI-International, 1980

[H82] A. Herold, Universal Unification and extended regular ACFM Theories, Univ. Karlsruhe, 1982

[KB70] D. Knuth, P. Bendix, Simple Word Problems in Universal Algebras, in: Comp. Problems in Abstract Algebra. J. Leech (ed), Pergamon Press, 1970

[KO75] R. Kowalski, A Proof Procedure based on Connection Graphs, JACM, vol 22, no 4, 1975

[LA79]   D.S. Lankford, A Unification Algorithm for Abelian Group Theory, Rep. MTP-1, Louisiana Techn. Univ., 1979

[LB79]   D.S. Lankford, M. Ballantyne, The Refutation Completeness of Blocked Permutative Narrowing and Resolution, 4th Workshop on Autom. Deduction, Texas, 1979

[LS75]   M. Livesey, J. Siekmann, Termination and Decidability Results for String-unification, Univ. of Essex, Memo CSM-12, 1975

[LS76]   M. Livesey, J. Siekmann, Unification of Sets and Multisets, Univ. Karlsruhe, Techn. Report, 1976

[LO80]   D. Loveland, Automated Theorem Proving, North Holland, 1980

[MA77]   G.S. Makanin, The Problem of Solvability of Equations in a Free Semigroup, Soviet Akad. Nauk SSSR, Tom 233, no 2, 1977

[MM79]   A. Martelli, U. Montaneri, An Efficient Unification Algorithm, University of Pisa, Techn. Report, 1979

[MA70]   Y. Matiyasevich, Diophantine Representation of Rec. Enumerable Predicates, Proc. of the Scand. Logic Symp., North Holland, 1978

[NI80]   N. Nilsson, Principles of Artificial Intelligence, Tioga Publ. Comp., California, 1980

[PW78]   M. Paterson, M. Wegman, Linear Unification, J. of Comp. and Syst. Science, 1978, 16

[PR60]   D. Prawitz, An Improved Proof Procedure, Theoria 26, 1960

[PS81]   G. Peterson, M. Stickel, Complete Sets of Reductions for Equational Theories with Complete Unification Algorithms, JACM, vol 28, no 2, 1981

[PL72]   G. Plotkin, Building in Equational Theories, Machine Intelligence, vol 7, 1972

[RS78]   P. Raulefs, J. Siekmann, Unification of Idempotent Functions, Universität Karlsruhe, 1978

[RSS79] P. Raulefs, J. Siekmann, P. Szabó, E. Unvericht, A short Survey on the State of the Art in Matching and Unification Problems, SIGSAM Bulletin, 13, 1979

[RO65]   J.A. Robinson, A Machine Oriented Logic based on the Resolution Principle, JACM 12, 1965

[RO71]   J.A. Robinson, Computational Logic: The Unification Computation, Machine Intelligence, vol 6, 1971

[SI75]   J. Siekmann, Stringunification, University of Essex, Memo CSM-7

[SL74]   J. Slagle, ATP for Theories with Simplifiers, Commutativity and Associativity, JACM 21, 1974

[ST75]   M. Stickel, A complete Unification Algorithm for assoc. commutative functions, Proc. of 4th IJCAI, Tbilisi, USSR, 1975

[SI82]   J. Siekmann, Unification of Commutative Terms (submitted), Universität Karlsruhe, 1982

[SS82.2]   P. Szabó, J. Siekmann, A Minimal Unification Algorithm for Idempotent Functions, Universität Karlsruhe, 1982 (forthcoming)

[SS82.3]   P. Szabó, J. Siekmann, Unification in Idempotent Semigroups (forthcoming), Universität Karlsruhe, 1982

[SS82.4]   P. Szabó, J. Siekmann, Universal Unification and a Classification of Equational Theories, Universität Karlsruhe, 1982 (full paper submitted)

[SS81.5]   P. Szabó, J. Siekmann, Universal Unification and Regular ACFM Theories, Proc. of IJCAI-81 (full paper submitted)

[S78]   P. Szabó, The Undecidability of the D_A-Unification Problem, Universität Karlsruhe, 1978

[SZ82]   P. Szabó, Theory of First Order Unification (in German, thesis), Universität Karlsruhe, 1982

[VA75]   J. van Vaalen, An Extension of Unification to Substitutions with an Application to Automatic Theorem Proving, IJCAI-4, Proc. of 1975

[VO78]   E. Vogel Unifikationsalgorithmen für Morphismen, Universität Karlsruhe (in German, Diploma thesis), 1978

[WR73]   L. Wos, G. Robinson, Maximal Models and Refutation Completeness: Semi-decision Procedures in Automatic Theorem Proving, in: Word problems (W.W. Boone, F.B. Cannonito, R.C. Lyndon, eds), North Holland, 1973

[WR67]   L. Wos, G.A. Robinson, D. Carson, L. Shalla, The Concept of Demodulation in Theorem Proving, JACM, vol 14, no 4.

[SS82.1]   P. Szabó, J. Siekmann, A Noetherian and Confluent Rewrite System for Idempotent Semigroups, Semigroup Forum 1982

[LA80]   D. Lankford, A new complete FPA-Unification algorithm, Jan. 1980, MTP-8

[FA81]   F. Fages, private communication (to appear as INRIA report, France)